1982

PROJECTIVE PLANES

A SERIES OF BOOKS IN MATHEMATICS

EDITORS: R. A. Rosenbaum
G. Philip Johnson

PROJECTIVE PLANES

Frederick W. Stevenson
University of Arizona

W. H. Freeman and Company
San Francisco

Printed in the United States of America
International Standard Book Number: 0-7167-0443-9
Library of Congress Catalog Card Number: 72-156824

AMS 1970 Subject Classifications: 50D20, 50D25, 50D30, 50D35.

1 2 3 4 5 6 7 8 9

CONTENTS

PREFACE

This book attempts to examine in depth the consequences of the simple axiom system that describes the mathematical structure known as the projective plane. Numerous books have been written on the projective plane. Pickert's *Projektive Ebenen* and, more recently, Dembowski's *Finite Geometries*, provide as complete a coverage of a single mathematical subject as can be found in the literature. However, these books are not textbooks. They are invaluable for the advanced graduate student but are well beyond the reach of the average undergraduate. Textbooks on projective geometry tend to cover a wide range of topics. This is appropriate because projective geometry provides a natural springboard for the study of non-Euclidean geometries and even linear algebra. Nevertheless, the result is that the projective plane is not studied for its own sake.

It is generally conceded that an undergraduate majoring in mathematics should study in depth at least one area of the subject. Often this area is algebra — perhaps group theory or ring theory. The projective plane is also a suitable area for such a concentrated study. Like group theory and ring theory, it has a simple axiomatic foundation. It also introduces the student to several basic concepts, the most important being the transformation group. The projective plane shares a remarkable relationship with algebraic structures of two binary operations, such as the field, division ring, semifield, nearfield, and quasifield. Furthermore, its study exposes the student to several other areas of mathematics, such as combinatorial analysis, linear algebra, and number theory. Finally, there are many unsolved problems in this area; some are probably workable, others are classics that may never be solved.

The book is divided into three parts. Part 1 introduces the student to examples and techniques; Part 2 examines the classical theorems of Desargues and Pappus; Part 3 is devoted to the study of non-Desarguesian planes. The first two parts can be taught comfortably in a one-semester, upper-division course for undergraduates. The whole book could be covered in a one-semester course at the graduate level. Perhaps Part 3 and selected portions of Parts 1 and 2 would be appropriate for a seminar. Naturally, some sections are more important than others; for example, Section 2.2, 2.3 (the part dealing with affine difference sets only), 3.3, 4.6, 5.3, and 6.3 can be omitted without significantly affecting the student's understanding of the other sections. The exercises generally are not of the drill variety; more-difficult ones are indicated by an asterisk. A careful selection of 100 or less of the more than 350 exercises would be sufficient for one semester.

I wish to thank my wife, Cheryl, for her invaluable assistance in the formulation, preparation, and completion of this book.

October 1970   *Frederick W. Stevenson*

PROJECTIVE PLANES

PART 1

FUNDAMENTALS

In this part the basic concepts are introduced
and the methods necessary for an analysis
of the planes of Parts 2 and 3
are developed.

CHAPTER 1

BASIC CONCEPTS

1.1 INTRODUCTION

The field of geometry, perhaps more than any other area of study, has provided the key ideas in the growth of mathematics as an abstract discipline. Euclid's *Elements*, a treatise written about 300 B.C., represents the first attempt to organize a field of knowledge deductively. His five postulates for plane geometry served as the outstanding model of an axiom system for two thousand years. The celebrated fifth postulate has perhaps been responsible for more controversy than any other single mathematical axiom; in large part this postulate gave birth to the idea that an axiom system need not be grounded in conceptually familiar ideas.

Euclid's fifth postulate is best known in the form known as Playfair's Axiom: *For any line L and point p not on L there exists one and only one line M parallel to L and passing through p*. After centuries of frustration, attempts to prove the fifth postulate from the four more "obvious" postulates were abandoned and other postulates were substituted in its place. For example, in the early nineteenth century,

Hungary's Bolyai Janos (1802–1860) and Russia's Nikolai Lobachevski (1793–1856) developed geometries that allowed more than one line parallel to a given line to pass through a given point; Germany's Bernhard Riemann (1826–1866) developed a geometry with no parallel lines. Today many different non-Euclidean geometries are being studied and developed; the projective plane is one such geometry.

The projective plane is an example of a non-Euclidean geometry having no parallel lines. Its origins are grounded both in the Euclidean plane and in the familiar plane of common sense. No clear distinction has been found between these two planes, but if one were found it would probably rest on the validity of the fifth postulate in the common-sense world. Such familiar examples of parallel lines as railroad tracks appear to converge as they recede from the observer. If we were to assume that they eventually meet at some point we could build a new geometry in the following way. Choose any point p in the Euclidean plane and consider all the lines through p. To each of these lines add an extra point called an ideal point or point at infinity and stipulate that each different line through p contain a different ideal point. For any line L not on p, attach the ideal point that belongs to the unique line through p parallel to L. Thus all the lines in the Euclidean plane have been assigned an ideal point and families of parallel lines share the same ideal point. Finally, to ensure that every two points determine a line, create a new line called the ideal line or line at infinity, made up of all the ideal points. The resulting plane is a projective plane called the real projective plane or the extended Euclidean plane.

The use of the word "projective" in the term "projective plane" arises in connection with the concept of projection mapping. For example, in the Euclidean plane, let L and M be intersecting lines and let p be a point not on L or M (see Figure 1). The mapping that takes q on L to r on M in such a way that p, q, and r lie on a line is called a projection mapping from L to M. If p is considered to be a light source, L is considered to be a set of points, and M is considered to be a screen, then r is the shadow cast by q on M. Physically, r would be a shadow only if q were between p and the screen. This mapping is deficient in two respects: there exists a point on L that has no image on M, and there exists a point on M that cannot receive an image from L. These two points are s and t, respectively: s is the intersection of M' and L where M' is the unique line passing through p and parallel to M, and t is the intersection of L' and M where L' is the unique line passing through

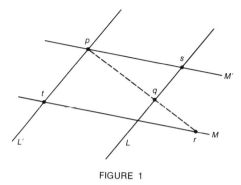

FIGURE 1

p and parallel to L. In the extended Euclidean plane, the parallel lines L and L' and M and M' intersect, so the projection mapping becomes a function with domain L and range M. Such mappings are of prime importance in the study of projective planes; indeed, the projective plane can be defined as the study of the properties that are held invariant under projection mappings as defined here.

Projective geometry is the general study of properties held invariant under projection mappings in two-space and higher dimensional spaces. The first major results in the subject, developed in the early seventeenth century, were due in large part to outstanding mathematicians: France's Poncelet (1788–1869) and Brianchon (1785–1864), Switzerland's Steiner (1796–1863), and Germany's von Staudt (1798–1867). Today projective geometry is an established area of mathematics that is well suited to serve as a general system from which other geometries can be developed and that is also of great interest in its own right. This text explores the special case of projective geometry in two dimensions — the projective plane.

1.2 TERMINOLOGY AND NOTATION

The following terminology and notation are used in this text: Points are denoted by lower-case letters (usually p,q,r,s,a,b,c,d); lines are denoted by capital letters (usually L, M, N); sets $\mathscr{P}$ and $\mathscr{L}$ denote the sets of points and lines, respectively; $\mathscr{I}$ denotes the incidence relation on points and lines (therefore, $\mathscr{I} \subseteq \mathscr{P} \times \mathscr{L}$); and a plane is represented by the triple $(\mathscr{P}, \mathscr{L}, \mathscr{I})$. A basic assumption throughout is that no point is a line and no line is a point (that is, $\mathscr{P} \cap \mathscr{L} = \varnothing$). If $(p,L) \in \mathscr{I}$, then p is "on" L or L "passes through" p. If two or more points lie on the same line, the points are *collinear*. If two or more lines pass through the same point, the lines are *concurrent*. If p, q, r, and s are four points no three of which are collinear, then the quadruple p,q,r,s is called a *complete four-point* or simply a *four-point*. The order of points p, q, r, and s of a four-point is important for reasons that will become clear later. Thus, unless otherwise specified, a four-point is an ordered set of four points. When the text refers to "two points" (lines) or "n points" (lines), the points (lines) are assumed to be distinct unless otherwise noted.

This text departs from the standard textbook notation for points and lines. Usually, capital letters denote points, lower-case letters denote lines. The notation here is consistent with Dembowski (1968) and also has a set theoretical basis since lines are considered to be sets of points.

Finally, the well-known abbreviation "iff" for "if and only if" is used in proofs.

1.3   PLANES

This book focuses almost entirely on affine planes and projective planes. Nevertheless, it is convenient to begin by introducing two weaker planes, the partial plane and the primitive plane. First, however, the term "plane" should be defined.

DEFINITION 1.3.1   *A* plane (incidence plane) *is a triple* $(\mathcal{P}, \mathcal{L}, \mathcal{I})$ *such that* $\mathcal{P}$, $\mathcal{L}$, *and* $\mathcal{I}$ *are sets,* $\mathcal{P} \cap \mathcal{L} = \varnothing$, $\mathcal{P} \cup \mathcal{L} \neq \varnothing$, *and* $\mathcal{I} \subseteq \mathcal{P} \times \mathcal{L}$.

Notation.   Planes are denoted by the letter Σ.

DEFINITION 1.3.2   *A* partial plane *is a plane* $(\mathcal{P}, \mathcal{L}, \mathcal{I})$ *satisfying*

Ax1:   *At most one line passes through two points.*

THEOREM 1.3.3   *If L and M are distinct lines in a partial plane Σ, then there is at most one point incident with both lines.*

Proof.   Suppose that there exist two distinct points p and q on both of the lines. Then $(p, L), (p, M), (q, L), (q, M) \in \mathcal{I}$, and $L \neq M$. This contradicts Ax1.

In a partial plane there is no guarantee that lines pass through any points at all (see Example 1.3.9(i)). Amends are made for this strange state of affairs by the introduction of a new axiom.

DEFINITION 1.3.4 *A* primitive plane *is a partial plane* $(\mathscr{P}, \mathscr{L}, \mathscr{I})$ *satisfying*

  Ax2: *Every line passes through at least two points.*

Primitive planes need not bear a close resemblance to our concept of a plane (see Example 1.3.9(ii) and (iii)). Two important properties are missing: Every two points lie on some line; and there exist three noncollinear points. Both the affine and the projective planes satisfy these conditions.

DEFINITION 1.3.5 *An* affine plane *is a plane* $(\mathscr{P}, \mathscr{L}, \mathscr{I})$ *such that*

  Af1: *Every two points lie on exactly one line.*
  Af2: *If p and L are a given point and line such that p is not on L, then there exists exactly one line M that passes through p and is parallel to L. (M does not intersect L.)*
  Af3: *There exist three noncollinear points.*

Notation. An affine plane is generally denoted by α.

THEOREM 1.3.6 *An affine plane is a primitive plane.*

Proof. The proof is left as an exercise.

DEFINITION 1.3.7 *A* projective plane *is a plane* $(\mathscr{P}, \mathscr{L}, \mathscr{I})$ *such that*

  Pj1: *Every two points lie on exactly one line.*
  Pj2: *Every two lines pass through exactly one point.*
  Pj3: *There exists a four-point.*

Notation. A projective plane is generally denoted by π.

THEOREM 1.3.8 *A projective plane is a primitive plane.*

Proof. The proof is left as an exercise.

EXAMPLE 1.3.9   Let $\mathcal{P}$, $\mathcal{L}$, and $\mathcal{I}$ be as follows:

  i. $\mathcal{P} = \{p\}$, $\mathcal{L} = \{L\}$, $\mathcal{I} = \varnothing$.
 ii. $\mathcal{P} = \{p,q,r,s\}$, $\mathcal{L} = \varnothing$, $\mathcal{I} = \varnothing$.
iii. $\mathcal{P} = \{p,q\}$, $\mathcal{L} = \{L\}$, $\mathcal{I} = \{(p,L),(q,L)\}$.
 iv. $\mathcal{P} = \{p,q,r\}$, $\mathcal{L} = \{L,M,N\}$.
     $\mathcal{I} = \{(p,L),(q,L),(p,M),(r,M),(q,N),(r,N)\}$.
  v. $\mathcal{P} = \{(x,y): x, y$ are real numbers$\}$.
     $\mathcal{L} = \{\{(x,y): ax + by + c = 0\}$ for some real numbers a, b, and c
     where $a = b = c = 0$ is not true$\}$.
     $\mathcal{I} = \{(p,L): p \in L\}$.
 vi. $\mathcal{P} = \{[x,y,z]: x, y,$ and z are real numbers and $x = y = z = 0$ is
     not true$\}$.
     $\mathcal{L} = \{\langle a,b,c \rangle: a, b,$ and c are real numbers and $a = b = c = 0$ is
     not true$\}$.
     $\mathcal{I} = \{([x,y,z],\langle a,b,c \rangle): ax + by + cz = 0\}$.

Here $[x,y,z]$ denotes the class of all triples (rx,ry,rz) where $r \neq 0$, and $\langle a,b,c \rangle$ denotes the class of all triples (ra,rb,rc) where $r \neq 0$.

It is not difficult to verify that: (i)–(vi) are partial planes; (ii)–(vi) are primitive planes; (v) is an affine plane; and (vi) is a projective plane.

DEFINITION 1.3.10   *The plane of Example 1.3.9(v) is called the* real affine plane; *the plane of Example 1.3.9(vi) is called the* real projective plane.

Notation.   The real affine plane is denoted by α_R and the real projective plane is denoted by π_R.

It should be pointed out that from the standpoint of set theory, the point $[x,y,z]$ and the line $\langle x,y,z \rangle$ in π_R are identical. Since $\mathcal{P} \cap \mathcal{L} = \varnothing$ must be true for planes, $[x,y,z]$ and $\langle x,y,z \rangle$ are different objects. Therefore, in this text, the triples in question are used as labels rather than as sets. If a set theoretic interpretation were desired, an exact copy of this plane could easily be provided such that $\mathcal{P} \cap \mathcal{L} = \varnothing$ (see the vector space representation of π_R at the end of this section). Throughout the book the notation "[ ]" will represent points and "$\langle$ $\rangle$" will represent lines.

The definition of $\mathcal{I}$ in π_R might cause confusion because the points and lines of π_R allow for multiple labelings. For example, $[x,y,z]$ can be denoted equally well by $[2x,2y,2z]$ or, in general, by $[rx,ry,rz]$ where $r \neq 0$; furthermore, $\langle a,b,c \rangle$ can be denoted by $\langle sa,sb,sc \rangle$ where $s \neq 0$. The incidence relation, however, is defined in terms of a particular labeling: $([x,y,z], \langle a,b,c \rangle) \in \mathscr{I}$ if and only if $ax + by + cz = 0$. The definition is not confusing if the student realizes that the incidence relation is compatible with the different labelings because $([rx,ry,rz],$ $\langle sa,sb,sc \rangle) \in \mathscr{I}$ if and only if $sarx + sbry + scrz = 0$ if and only if $ax + by + cz = 0$ if and only if $([x,y,z], \langle a,b,c \rangle) \in \mathscr{I}$.

It is clear that α_R is the familiar Euclidean plane represented by Cartesian coordinates, but it is not at all clear that π_R is the same real projective plane that was constructed in Section 1.1. To see that it is, note first that lines in α_R can be represented as triples (a,b,c) where (a,b,c) denotes the line $\{(x,y): ax + by + c = 0\}$. However, since the triple (ra,rb,rc) represents the same line as (a,b,c) (because $ax + by + c = 0$ if and only if $rax + rby + rc = 0$), lines in α_R should, strictly speaking, be denoted by sets of triples of the form (ra,rb,rc) where $r \neq 0$ and $a = b = 0$ is not true. Under this representation the lines in α_R and π_R are the same except that π_R contains the line $\langle 0,0,1 \rangle$ and α_R does not. This extra line in π_R can be thought of as the ideal line. The points in α_R can also be expressed as sets of triples. The point (x,y) can be written as the set $\{(rx,ry,r): r \neq 0\}$. In this setting the points of α_R and π_R are the same except that π_R contains points of the form $[x,y,0]$ and α_R does not. These points can be considered to be ideal points. Notice that all the ideal points lie on the ideal line and that every nonideal line in π_R contains only one such point. Note also that in changing the representation of points and lines of the Cartesian plane to the ordered triple notation, the concept of incidence under the two representations has been preserved.

Some mathematical niceties have been omitted from the preceding discussion, but they will be given a more formal presentation in subsequent sections. For example, different representations of a plane will be discussed formally in Section 1.5, the extension of a plane will be considered in Section 1.7, and the relationship between such planes as α_R and π_R will be developed in Section 2.1.

This section is concluded with a further discussion of the real projective plane. The geometer perhaps thinks of the plane as it is described in Section 1.1, that is, an extension of the Euclidean plane. The algebraist might analyze it in the following way. Let V denote the

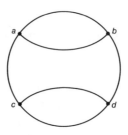

FIGURE 2

three-dimensional vector space excluding the null vector; in other words, $\{(x,y,z): x, y,$ and z are real numbers and $x = y = z = 0$ is not true$\}$. An equivalence relation $\sim$ on V is defined as follows: $(x,y,z) \sim (x',y',z')$ if and only if there exists a nonzero real number r such that $x' = rx$, $y' = ry$, and $z' = rz$. The set of points of π_R is the set of equivalence classes $V/\sim$. Such classes (for example, those in Example 1.3.9(vi)) are denoted by $[x,y,z]$. A line in π_R is defined as $\{[x,y,z]: ax + by + cz = 0\}$ for some real numbers a, b, and c, not all of which are 0. This line is denoted by $\langle a,b,c \rangle$. First we must check that this line is well defined, that is, that if $[x,y,z]$ is on $\langle a,b,c \rangle$ and $(x',y',z') \in [x,y,z]$, then $[x',y',z']$ is also on $\langle a,b,c \rangle$. This is done as in the preceding paragraph: $ax + by + cz = 0$, and $x' = rx$, $y' = ry$, $z' = rz$; therefore, $ax' + by' + cz' = 0$. Next we observe that $\langle a,b,c \rangle = \langle a',b',c' \rangle$ if and only if there exists a real number $s \neq 0$ such that $a' = sa$, $b' = sb$, and $c' = sc$. This also is easily checked. Thus we have the same plane as in Example 1.3.9(vi).

The algebraic topologist would visualize the real projective plane in the following way. Each point in π_R can be considered to be a line passing through the origin in Euclidean three-space (because $\{(rx,ry,rz): x, y,$ and z are fixed and r is any real number$\}$ is such a line). Each of the lines intersects the unit sphere in two points called *antipodal points*. Thus π_R can be considered to be a sphere in three-space with its antipodal points identified (or joined together). If we were to cut from a to b in Figure 2 and remove a circle from around the north pole, a circle would automatically be cut (from d to c) and removed from the south pole. What we have left is a Möbius strip, generated by identifying sides ac and bd of rectangle $abcd$ in Figure 3 in such a way that a is identical to d; c is identical to b, and, in general, x is identical to y where the distance from x to a equals the distance from y to d. We may join together the edge of the Möbius strip and thus regenerate an object like the sphere with the identified antipodal points by identifying s

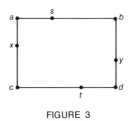

FIGURE 3

with t (see Figure 3), where the distance from a to s equals the distance from t to d. Thus the real projective plane may be generated from a rectangle by appropriate joining together of opposite sides. This step cannot actually be done in three-space and thus is difficult to conceptualize.

EXERCISES

1. Prove Theorem 1.3.6.

2. Prove Theorem 1.3.8.

3. Show that every affine plane satisfies Pj3.

4. a. In π_R find the line joining $[5,2,3]$ and $[6,4,1]$; joining $[x,y,z]$ and $[(x',y',z')]$.
   b. Let $\times$ denote the cross product of vectors in R_3 (the set of triples of real numbers). If $(x,y,z) \times (x',y',z') = (a,b,c)$, show that $\langle a,b,c \rangle$ is the line passing through $[x,y,z]$ and $[x',y',z']$.
   c. Show that the same method can be used to find the intersection of two lines.

5. a. Using your knowledge of Euclidean three-space E, show that the following model is a projective plane. Let p be a given point in space and let
   $$\mathscr{P} = \{L: L \text{ is a line passing through } p \text{ in } E\}$$
   $$\mathscr{L} = \{\gamma: \gamma \text{ is a plane passing through } p\}$$
   $$\mathscr{I} = \{(L,\gamma): L \text{ is in the plane } \gamma\}$$
   b. Is this plane another formulation of π_R?

6. a. Let Q be the field of rational numbers. Let $\mathscr{P}_Q = \{[x,y,z]: x,y,z \in Q$, not all of x, y, and z are 0, and $[x,y,z] = [x',y',z']$ if and only if $x = rx'$, $y = ry'$, $z = rz'$ for some $r \neq 0\}$. Let $\mathscr{L}_Q = \{\langle a,b,c \rangle: a,b,c \in Q$, not all of a,b,c are 0, and $\langle a,b,c \rangle = \langle a',b',c' \rangle$ if and only if $a = sa', b = sb'$, $c = sc'$ for some $s \neq 0\}$. Let $\mathscr{I}_Q = \{([x,y,z], \langle a,b,c \rangle): ax + by + cz = 0\}$. Show that $(\mathscr{P}_Q, \mathscr{L}_Q, \mathscr{I}_Q)$ is a projective plane.

   b. Replace Q by the complex field C and show that $(\mathscr{P}_C, \mathscr{L}_C, \mathscr{I}_C)$ is a
      projective plane.

   c. If you replace Q by the integers Z, can you generate a projective
      plane? Explain your answer.

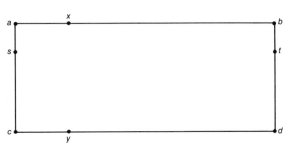

FIGURE 4

7. A torus can be formed by identifying the sides of a rectangle as follows
   (see Figure 4): Identify a with c, b with d, and x with y where the distance from a to x equals the distance from c to y. The resulting cylinder
   is then altered by identifying sides ac and bd so that s is identical with t
   where the distance from a to s and from b to t is equal. Now consider
   Figure 5, which is based on Figure 4. Joining the rectangle results in a
   torus with seven points and seven triangles identified by p_i, T_i, $i = 1$,
   ..., 7. Let T_i be "lines" and let $(p_i, T_j) \in \mathscr{I}$ if and only if p_i is a vertex
   of T_j. Show that $(\mathscr{P}, \mathscr{L}, \mathscr{I})$ is a projective plane.

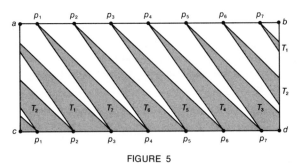

FIGURE 5

8. Consider the following plane proposed by F. R. Moulton (1902):

   $\mathscr{P}$ is the set of points of α_R

   $\mathscr{L} = \{(x,y): x = k$ or $y = mx + k$ where $m \geq 0\}$ $\cup$

      $\{(x,y): y = mx + k$ if $x \geq 0; y = \dfrac{mx}{2} + k$ if $x < 0$ where $m < 0\}$

   $\mathscr{I} = \in$

   Notice that the lines of $\mathscr{L}$ are the lines of α_R if $x = k$ or if $y = mx + k$ and
   $m > 0$. The "lines" of negative slope are bent like refracted light rays at
   the y-axis. Show that $(\mathscr{P}, \mathscr{L}, \mathscr{I})$ is an affine plane.

9. The following expanded set of axioms may be used to define a projective plane:

    A1   There exist at least one point and one line.

    A2   On any two distinct points there is at least one line.

    A3   On any two distinct points there is at most one line.

    A4   On two distinct lines there is at least one point.

    A5   There are at least three distinct points on any line.

    A6   Not all points are on the same line.

An axiom is said to be independent within an axiom system if the system without the axiom in question has a model. For example, A6 is independent because the following model satisfies the other five axioms but does not satisfy A6:

$$\text{Let } \mathscr{P} = \{p,q,r\}; \ \mathscr{L} = \{\{p,q,r\}\}; \ \mathscr{I} = \in.$$

Show that each of the other five axioms is independent.

1.4   DUALITY

One of the most simple and elegant principles in the field of mathematics is the principle of duality. It arises in several areas of study but is perhaps best known for its applications in geometry. The principle applies to classes of structures rather than to individual structures; in geometry, for example, the principle may apply to classes of planes. It is in this setting that the principle is explained.

Let γ be a class of planes and suppose that T is a theorem that holds for every plane in γ. Let T^d denote the statement that results from replacing the word "point" in T by the word "line" and the word "line" by the word "point." This means that such a phrase as "lies on" must be replaced by "pass through" and "pass through" must be replaced by "lies on"; it also means that words like "collinear" and "concurrent" must exchange roles. Statement T^d is called the *dual statement of T*.

THE PRINCIPLE OF DUALITY.   *The principle of duality holds for a class of planes γ if and only if γ has the property that for every theorem T of γ, T^d is also a theorem of γ.*

DEFINITION 1.4.1   *Let $\Sigma = (\mathscr{P}, \mathscr{L}, \mathscr{I})$ be a plane. Then the triple $(\mathscr{L}, \mathscr{P}, \mathscr{I}^{-1})$ (where $\mathscr{I}^{-1} = \{(L,p): (p,L) \in \mathscr{I}\}$) is called the* dual plane of Σ.

Notation. The dual plane of Σ is denoted by Σ^d.

THEOREM 1.4.2 *The principle of duality holds in γ if γ has the property that if $\Sigma \in \gamma$, then $\Sigma^d \in \gamma$.*

Proof. Suppose that $\Sigma \in \gamma$ implies that $\Sigma^d \in \gamma$. Let T be an arbitrary theorem of γ and let Σ be an arbitrary plane in γ. Since T holds true for all members of γ, it holds true in particular for Σ^d. Because T holds true for Σ^d, it follows that T^d holds true for $(\Sigma^d)^d$. But $(\Sigma^d)^d = \Sigma$, so T^d holds for Σ. Thus T^d is a theorem for Σ, and since Σ was arbitrary, T^d is a theorem of γ.

THEOREM 1.4.3 *The principle of duality holds for the class of partial planes.*

Proof. Let $\Sigma = (\mathscr{P}, \mathscr{L}, \mathscr{I})$ be a partial plane. Then $\mathscr{L} \cap \mathscr{P} = \varnothing$, $\mathscr{L} \cup \mathscr{P} \neq \varnothing$, and $\mathscr{I}^{-1} \subseteq \mathscr{L} \times \mathscr{P}$. Furthermore, by Theorem 1.3.3, for every two members $L, M \in \mathscr{L}$, there exists at most one member $p \in \mathscr{P}$ such that $(L, p) \in \mathscr{I}^{-1}$ and $(M, p) \in \mathscr{I}^{-1}$. Thus Ax1 holds for Σ^d and Σ^d is a partial plane. It follows from Theorem 1.4.2 that the principle of duality holds for the class of partial planes.

THEOREM 1.4.4 *The principle of duality does not hold for the class of primitive planes.*

Proof. Consider the dual of Ax2: Every point lies on at least two lines. This statement is not true for the plane of Example 1.3.9(ii). Therefore, the class of primitive planes does not satisfy the principle of duality.

THEOREM 1.4.5 *The principle of duality holds for the class of projective planes.*

Proof. The proof is left as an exercise.

THEOREM 1.4.6 *The principle of duality does not hold for the class of affine planes.*

Proof. The proof is left as an exercise.

Mathematically, the consequences of duality are simple. In a class of planes satisfying the principle of duality, the two undefined terms. point and line, are interchangeable and thus share a symmetric relationship. In a more utilitarian vein it can be observed that with duality we always get two theorems for the proof of one. For those who do not like to think of point and line as undefined terms, the principle of duality is confusing. Conceptually, a point and line are entirely different objects and it does not make sense to consider a point as a line or a line as a point. Perhaps the next section will be of some assistance in this regard, because it explains that for primitive planes the concepts of point and line may be accurately pictured as a dot and a set of dots, respectively.

EXERCISES

1. Prove Theorem 1.4.5.

2. Prove Theorem 1.4.6.

3. Let Σ be a primitive plane satisfying Ax3: Every point has two lines through it. Does the principle of duality hold for the class of primitive planes satisfying Ax3? Explain your answer.

1.5   ISOMORPHISM

The concept of isomorphism has the same meaning in plane geometry that it has in other mathematical disciplines. Thus two planes are said to be isomorphic if they are alike in every mathematical respect.

DEFINITION 1.5.1   *Partial planes* $\Sigma = \{\mathscr{P}, \mathscr{L}, \mathscr{I}\}$ *and* $\Sigma' = \{\mathscr{P}', \mathscr{L}', \mathscr{I}'\}$ *are* isomorphic *if and only if there exist functions f and F such that f is a one-to-one mapping from* $\mathscr{P}$ *onto* $\mathscr{P}'$, *F is a one-to-one mapping from* $\mathscr{L}$ *onto* $\mathscr{L}'$, *and* $(p, L) \in \mathscr{I}$ *if and only if* $(f(p), F(L)) \in \mathscr{I}'$.

Notation.   If Σ and Σ' are isomorphic, the notation used is $\Sigma \sim \Sigma'$. The isomorphism is denoted by the ordered pair (f, F).

EXAMPLE 1.5.2

   i. Define Σ as follows: $\mathscr{P} = \{p, q, r\}$, $\mathscr{L} = \{L, M, N\}$, $\mathscr{I} = \{(p, L),$

$(q,L),(p,M),(r,M),(q,N),(r,N)$}. (Notice that this plane is the same as that in Example 1.3.9(iv)). To show that $\Sigma \sim \Sigma^d$, consider the mappings f and F where $f: p \rightarrow N, q \rightarrow M, r \rightarrow L$, and $F: L \rightarrow r, M \rightarrow q, N \rightarrow p$. That $(x,Y) \in \mathscr{I}$ if and only if $(f(x),F(Y)) \in \mathscr{I}^{-1}$ is easily checked for points x and lines Y in Σ. For example, if $x = p$ and $Y = L$, then $(x,Y) \in \mathscr{I}$. Since $f(p) = N, F(L) = r$, and $(N,r) \in \mathscr{I}^{-1}$, it follows that $(r,N) \in \mathscr{I}$. Thus $(f(x),F(Y)) \in \mathscr{I}^{-1}$.

ii. Let $\Sigma = \alpha_R$ (as in Example 1.3.9(v)) and let $\Sigma' = (\mathscr{P}', \mathscr{L}', \mathscr{I}')$ be defined as in the discussion of π_R—that is, $\mathscr{P}' = \{[x,y,1]: x$ and y are real numbers$\}$, $\mathscr{L}' = \{\langle a,b,c \rangle: a, b,$ and c are real numbers and $a = b = 0$ is not true$\}$, and $\mathscr{I}' = \{([x,y,1], \langle a,b,c \rangle): ax + by + c = 0\}$. Here $[x,y,1] = \{(rx,ry,r): r \neq 0\}$, $\langle a,b,c \rangle = \{(sa,sb,sc): s \neq 0$ and $a = b = 0$ is not true$\}$. Let f and F be defined as follows: $f: (x,y) \rightarrow [x,y,1]$; $F: \{(x,y): ax + by + c = 0\} \rightarrow \langle a,b,c \rangle$. It is easily checked that (f,F) is an isomorphism.

An equivalent definition of isomorphism may be formulated for primitive planes using only one function f mapping points of Σ onto points of Σ'.

THEOREM 1.5.3   *Let Σ and Σ' be primitive planes. Then $\Sigma \sim \Sigma'$ if and only if there exists a one-to-one function f mapping $\mathscr{P}$ onto $\mathscr{P}'$ such that f maps collinear points into collinear points.*

Proof.   First suppose that $\Sigma \sim \Sigma'$ and then seek a function f satisfying the necessary conditions. Since $\Sigma \sim \Sigma'$, there exist functions g and G such that g maps $\mathscr{P}$ onto $\mathscr{P}'$, G maps $\mathscr{L}$ onto $\mathscr{L}'$, and $(p,L) \in \mathscr{I}$ iff $(g(p),G(L)) \in \mathscr{I}'$. Let S be a set of collinear points on line L. If $p \in S$, then $g(p) \in G(L)$, so all points in S under the mapping g lie on $G(L)$ and thus are collinear. The function g then satisfies the necessary conditions.

Next suppose that there exists a function f from $\mathscr{P}$ to $\mathscr{P}'$ preserving collinearity. We can construct a function F from $\mathscr{L}$ to $\mathscr{L}'$ in the following way: Let $L \in \mathscr{L}$ and let p and q denote two points of L; then define $F(L)$ as the line passing through $f(p)$ and $f(q)$. Such a line exists because $f(p)$ and $f(q)$ must be collinear since p and q are collinear. $F(L)$ is unique because of Ax1. Furthermore, F is well defined because other choices r and s on L would result in a line passing through $f(r)$

and $f(s)$; and since p, q, r, and s are collinear, so are their images under f. To show that F is one-to-one, suppose that $L \neq M$ and let p be a point on L but not on M. Let q be another point on L and let r and s be points on M. (Points r and s may be identical with q.) As a result, $F(L)$ is the line passing through $f(p)$ and $f(q)$, $F(M)$ is the line passing through $f(r)$ and $f(s)$, and $F(L) \neq F(M)$ because $f(p)$ is not collinear with $f(r)$ and $f(s)$. The function F is onto since any line $L' \in \mathscr{L}'$ has two points p and q on it, and F maps the line determined by $f^{-1}(p)$, $f^{-1}(q)$ onto L'. Finally, $(p,L) \in \mathscr{I}$ if and only if $(f(p),F(L)) \in \mathscr{I}$ because of the definition of F. Thus $\Sigma \sim \Sigma'$.

DEFINITION 1.5.4 *Let $\Sigma = (\mathscr{P}, \mathscr{L}, \mathscr{I})$ and $\Sigma' = (\mathscr{P}', \mathscr{L}', \mathscr{I}')$ be primitive planes. If f is a one-to-one mapping from $\mathscr{P}$ onto $\mathscr{P}'$ preserving collinearity, then f is an* isomorphism *from Σ onto Σ'.*

EXAMPLE 1.5.5 Let Σ be the plane α_R. Let $g: (x,y) \to (ax + by + c,$ $dx + ey + f)$, where a,b,c,d,e,f are real numbers such that

$$\begin{vmatrix} a & d & 0 \\ b & e & 0 \\ c & f & 1 \end{vmatrix} \neq 0$$

To show that g is an isomorphism of Σ onto itself, first notice that if we define a new plane Σ' letting points be triples $(x,y,1)$ and letting lines be sets of points satisfying equations of the form $ax + by + c = 0$, we have a plane that is isomorphic to Σ, the isomorphism being $j: (x,y)$ $\to (x,y,1)$. The set of points $\mathscr{P}$ of Σ' form a subset S of the three-dimensional vector space V_3. The map in Σ', associated with g, is a restriction to S of the nonsingular linear transformation h in V_3 representable by the matrix

$$\begin{pmatrix} a & d & 0 \\ b & e & 0 \\ c & f & 1 \end{pmatrix}$$

This statement is verified by noting that

$$(x,y,1)\begin{pmatrix} a & d & 0 \\ b & e & 0 \\ c & f & 1 \end{pmatrix} = (ax + by + c,\ dx + ey + f,\ 1)$$

First we observe that $h \mid S$ (h restricted to S) is a one-to-one map from S onto itself preserving collinearity. Clearly, it is one-to-one because

h itself is one-to-one. Let $(x,y,1) \in S$ and note that h^{-1} is also representable in the form

$$\begin{pmatrix} a' & d' & 0 \\ b' & e' & 0 \\ c' & f' & 1 \end{pmatrix}$$

(the student should verify this) and thus that $h^{-1}(x,y,1) \in S$. Therefore, $h(h^{-1}(x,y,1)) = (x,y,1)$ and $h|S$ is onto. Recall that $(x_1,y_1,1)$, $(x_2,y_2,1)$, and $(x_3,y_3,1)$ are collinear if and only if

$$\begin{vmatrix} x_1 & y_1 & 1 \\ x_2 & y_2 & 1 \\ x_3 & y_3 & 1 \end{vmatrix} = 0$$

But

$$\begin{vmatrix} x_1 & y_1 & 1 \\ x_2 & y_2 & 1 \\ x_3 & y_3 & 1 \end{vmatrix} \begin{vmatrix} a & d & 0 \\ b & e & 0 \\ c & f & 1 \end{vmatrix} = \begin{vmatrix} ax_1 + by_1 + c & dx_1 + ey_1 + f & 1 \\ ax_2 + by_2 + c & dx_2 + ey_2 + f & 1 \\ ax_3 + by_3 + c & dx_3 + ey_3 + f & 1 \end{vmatrix}$$

and since

$$\begin{vmatrix} a & d & 0 \\ b & e & 0 \\ c & f & 1 \end{vmatrix} \neq 0$$

the determinant

$$\begin{vmatrix} x_1 & y_1 & 1 \\ x_2 & y_2 & 1 \\ x_3 & y_3 & 1 \end{vmatrix} = 0$$

exactly when

$$\begin{vmatrix} ax_1 + by_1 + c & dx_1 + ey_1 + f & 1 \\ ax_2 + by_2 + c & dx_2 + ey_2 + f & 1 \\ ax_3 + by_3 + c & dx_3 + ey_3 + f & 1 \end{vmatrix} = 0$$

This means that the three points are collinear exactly when their images under $h|S$ are collinear. Thus $h|S$ is an isomorphism. It follows easily that g is also an isomorphism. More precisely, $g = j^{-1} \circ h|S \circ j$, the composition of three isomorphisms. (Functional composition is read from right to left.)

For any primitive plane Σ, we can construct an isomorphic copy having the property that lines are made up of points.

CONSTRUCTION 1.5.6 Let $\Sigma = (\mathscr{P}, \mathscr{L}, \mathscr{I})$ be given. For each $L \in \mathscr{L}$, define L' as $\{p \in \mathscr{P}: (p,L) \in \mathscr{I}\}$. Let $\mathscr{L}' = \{L': L \in \mathscr{L}\}$ and let $\mathscr{I}' = \{(p,L'): p \in L'\}$. Denote $(\mathscr{P}, \mathscr{L}', \mathscr{I}')$ by Σ_s.

THEOREM 1.5.7 Σ_s *is a primitive plane and* $\Sigma_s \sim \Sigma$.

Proof. It is clear that Σ_s is a primitive plane because $(p,L) \in \mathscr{I}$ iff $p \in L'$ iff $(p,L') \in \mathscr{I}'$. The identity map from points of Σ to points of Σ_s preserves collinearity and so, by Theorem 1.5.3, $\Sigma_s \sim \Sigma$.

Theorem 1.5.7 tells us, in effect, that the lines of a primitive plane can be considered to be sets of points and that the incidence relation is simply set membership. The first few sections of the text will be concerned with our intuitive notions of point and line. Although both "point" and "line" are undefined terms, it would be convenient to retain a mental picture that has some appeal both intuitively and mathematically. Conceptually the point is just a dot. A line, however, has dimension and so we visualize it as a path—usually straight, infinitely long in both directions, and made up of closely spaced points. We cannot expect projective or affine planes to correspond precisely to our intuitive ideas of a plane, but an attempt will be made here to preserve as many familiar notions as possible. For example, in the partial plane of Example 1.3.9(i) there is a line without points; however, Ax2 assures that this phenomenon does not occur in primitive planes. In Section 1.4 a dual plane was defined in which points become lines and lines become points. Theorem 1.5.7 tells us that, although points and lines may be considered in different ways, for primitive planes we may always think of points as dots and lines as sets of dots.

Since in this text lines will often be considered to be sets of points and incidence will often be considered to be set membership, it will be convenient to state definitions and theorems in this context. Accordingly, the following notation is introduced.

Notation. For primitive planes, Σ is denoted by $(\mathscr{P}, \mathscr{L}, \in)$, or alternatively, by $\Sigma = (\mathscr{P}, \mathscr{L}, \mathscr{I})$ where $\mathscr{I} = \in$. If p lies on both L and M, the notation is $p = L \cap M$. Notice that $L \cap M$ represents exactly one point in a projective plane and may be the null set in an affine plane; $L \cap M = \varnothing$ is written $L \parallel M$. If a line L lies on both p and q, then L is denoted by pq. Notice that pq denotes a unique line in any partial plane.

EXERCISES

1. Find an example of a partial plane Σ such that
   a. $\Sigma \sim \Sigma^d$.
   b. $\Sigma \not\sim \Sigma^d$.

2. Show that Theorem 1.5.3 does not hold if Σ and Σ' are partial planes rather than primitive planes.

3. Consider the finite geometry consisting of points and lines that satisfy the following axioms:

   There is at least one point.
   Every line is a set of exactly two points.
   Every point lies on exactly two lines.
   There exist exactly three parallel lines that do not intersect a given line.

   Show that:
   a. If two lines intersect, they do so at just one point.
   b. There is at least one line.
   c. To a given line there are exactly two nonparallel lines.
   d. There are exactly six lines.
   e. There are exactly six points.

   Show also that there are exactly two nonisomorphic geometries that satisfy this system.

4. Show that $\pi_R \sim \pi_R{}^d$. Does the same hold for π_Q? for π_C?

5. Let Σ be a primitive plane satisfying Ax3 (see exercise 1.4.3). Formulate a plane Σ_S using a construction dual to Construction 1.5.6. Is it true that $\Sigma_S \sim \Sigma$?

1.6   CONFIGURATIONS

The study of plane geometry is usually associated with such figures as the square, the rectangle, the triangle, and the circle. In projective and affine geometry, there are no circles and no right angles; however, triangles and quadrilaterals do play an important role. The figures commonly used to represent triangles and quadrilaterals, $\triangle$ and $\square$, are not strictly accurate since the concepts of betweenness and line segment are not available.

DEFINITION 1.6.1   *Let Σ be a given partial plane. A pair of sets A and B, where $A \subseteq \mathscr{P}$ and $B \subseteq \mathscr{L}$, is a* complete triangle *in Σ if and only if A is a set of three noncollinear points p,q,r, and B is the set of three* *lines L, M, N joining the three possible distinct pairs of points. The points are called* vertices; *the lines are called* sides.

EXAMPLE 1.6.2

  i. $\mathscr{P} = \{p,q,r\}$
    $\mathscr{L} = \{\{p,q\},\{p,r\},\{q,r\}\}$
    $\mathscr{I} = \in$

  ii. Σ is the plane α_R.

$$p = (0,0) \qquad L: x = 0$$
$$q = (1,0) \qquad M: y = 0$$
$$r = (0,1) \qquad N: y = 1 - x$$

DEFINITION 1.6.3   *A pair of sets A, B, where $A \subseteq \mathscr{P}$ and $B \subseteq \mathscr{L}$, is a* complete quadrangle *if and only if A is a set of four points p, q, r, and s such that p, q, r, s is a four-point and B is the set of six lines L_i: i = 1, . . . , 6 that join the six possible distinct pairs of points. The points are called* vertices; *the lines are called* sides. *If two sides do not meet in a vertex, they are called* opposite *sides; thus there are three pairs of opposite sides. If the opposite sides meet in a point, it is called a* diagonal point.

EXAMPLE 1.6.4

  i. $\mathscr{P} = \{p_i: i = 1,2,3,4\}$
    $\mathscr{L} = \{\{p_i,p_j\}, i \neq j\}$
    $\mathscr{I} = \in$

    The three pairs of opposite sides are:

$$\{p_1,p_2\},\ \{p_3,p_4\}$$
$$\{p_1,p_3\},\ \{p_2,p_4\}$$
$$\{p_1,p_4\},\ \{p_2,p_3\}$$

    There are no diagonal points.

  ii. Σ is the plane α_R.

$$p_1 = (0,0) \qquad L_1: x = 0$$
$$p_2 = (0,1) \qquad L_2: y = 0$$
$$p_3 = (1,0) \qquad L_3: x = 1$$
$$p_4 = (1,1) \qquad L_4: y = 1$$
$$L_5: x = y$$
$$L_6: y = 1 - x$$

The opposite pairs of sides are: L_1, L_3; L_2, L_4; L_5, L_6. There is one diagonal point: $(1/2, 1/2)$.

DEFINITION 1.6.5 *A finite partial plane (that is, $\mathscr{P} \cup \mathscr{L}$ is finite) is called a* configuration.

DEFINITION 1.6.6 *A configuration Σ is called a* confined *configuration if and only if each line passes through at least three points and each point lies on at least three lines.*

EXAMPLE 1.6.7

i. Examples 1.6.2(i) and 1.6.4(i) are configurations that are not confined.

ii. $\mathscr{P} = \{p_i: i = 1, \ldots, 7\}$
$\mathscr{L} = \{\{p_1, p_2, p_4\}, \{p_2, p_3, p_5\}, \{p_3, p_4, p_6\}, \{p_4, p_5, p_7\}, \{p_5, p_6, p_1\}, \{p_6, p_7, p_2\}, \{p_7, p_1, p_3\}\}$
$\mathscr{I} = \in$

Figure 6 is an example of a confined configuration that is somewhat confusing because of the dashed circle, which, like the six line segments, passes through three collinear points of the configuration. It is unfortunate that points p_2, p_3, and p_5, which comprise a line of the configuration, do not lie on a straight line in the Euclidean plane as do all the other collinear triples. Nevertheless, collinearity is conveniently indicated by joining the points with a dashed curve. This problem of diagraming configurations is discussed again at the end of this section and is analyzed in the next section.

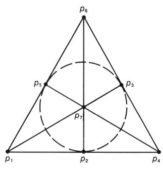

FIGURE 6

DEFINITION 1.6.8   *If Σ is a configuration, Σ^d is called the* dual configuration *of Σ.*

DEFINITION 1.6.9   *A* quadrilateral *is the dual configuration of a quadrangle. Pairs of points that do not lie on a side are called* opposite points. *A line passing through a pair of opposite points is called a* diagonal line.

DEFINITION 1.6.10   *If $\Sigma \sim \Sigma^d$, then Σ is* self-dual.

EXAMPLE 1.6.11

   i. Example 1.6.2(i) is a self-dual configuration; proof of this fact is left as an exercise.

   ii. Example 1.6.7(ii) is also self-dual, as evidenced by the following proof. Label the seven lines of the example as follows:

$$L_1 = \{p_7, p_1, p_3\} \qquad L_5 = \{p_3, p_4, p_6\}$$
$$L_2 = \{p_6, p_7, p_2\} \qquad L_6 = \{p_2, p_3, p_5\}$$
$$L_3 = \{p_5, p_6, p_1\} \qquad L_7 = \{p_1, p_2, p_4\}$$
$$L_4 = \{p_4, p_5, p_7\}$$

Let f map the points of Σ to the points of Σ^d (the lines of Σ) in the following way: $f(p_i) = L_i, i = 1, \ldots, 7$. We must merely verify that collinearity is preserved. Consider points p_3, p_4, and p_6; their images under f are L_3, L_4, and L_6, respectively. These points of Σ^d are collinear because the lines themselves are concurrent (at p_5). (Notice that the incidence relation for Σ^d is not $\in$, it is $\in^{-1}$.)

Now let us focus our attention on a symmetric type of configuration that has m points and n lines such that each point has c lines through it and each line has d points on it. Such a configuration is denoted by $\Sigma(m_c, n_d)$. Clearly, if $c, d \geq 3$, then $\Sigma(m_c, n_d)$ is a confined configuration. If $m = n$ and $c = d$, then we simply write $\Sigma(m_c)$. So Example 1.6.2(i) can be denoted as $\Sigma(3_2)$; Example 1.6.4(i) can be denoted as $\Sigma(4_3, 6_2)$; and Example 1.6.7(ii) can be denoted as $\Sigma(7_3)$. A quadrilateral is of the form $\Sigma(6_2, 4_3)$. It should be pointed out that $\Sigma(m_c, n_d)$ does not necessarily denote a unique plane. ("Unique" means unique up to isomorphism.) The configurations $\Sigma(4_3, 6_2)$, $\Sigma(7_3)$, and $\Sigma(8_3)$ are unique; but there are three nonisomorphic planes of the form $\Sigma(9_3)$ and ten distinct planes of the form $\Sigma(10_3)$.

Incidence tables having entries of only 0, representing nonincidence, and 1, representing incidence, provide a neat conceptual way of representing configurations. Following are three examples of incidence tables.

EXAMPLE 1.6.12

    i. This incidence table represents a configuration of the form $\Sigma(7_3)$, called the *Fano configuration* (see Figure 7).

	L_1	L_2	L_3	L_4	L_5	L_6	L_7
p_1	1	0	1	0	0	0	1
p_2	0	1	0	0	0	1	1
p_3	1	0	0	0	1	1	0
p_4	0	0	0	1	1	0	1
p_5	0	0	1	1	0	1	0
p_6	0	1	1	0	1	0	0
p_7	1	1	0	1	0	0	0

    ii. This incidence table represents a configuration of the form $\Sigma(9_3)$, called the *Pappus configuration* (see Figure 8).

	L_1	L_2	L_3	L_4	L_5	L_6	L_7	L_8	L_9
p_1	1	1	1	0	0	0	0	0	0
p_2	1	0	0	1	1	0	0	0	0
p_3	1	0	0	0	0	1	1	0	0
p_4	0	1	0	1	0	0	0	1	0
p_5	0	1	0	0	0	1	0	0	1
p_6	0	0	1	0	0	0	1	1	0
p_7	0	0	1	0	1	0	0	0	1
p_8	0	0	0	1	0	0	1	0	1
p_9	0	0	0	0	1	1	0	1	0

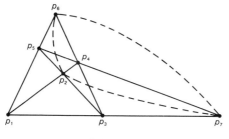

FIGURE 7

iii. This incidence table represents a configuration of the form $\Sigma(10_3)$, called the *Desargues configuration* (see Figure 9).

	L_1	L_2	L_3	L_4	L_5	L_6	L_7	L_8	L_9	L_{10}
p_1	1	1	1	0	0	0	0	0	0	0
p_2	1	0	0	1	1	0	0	0	0	0
p_3	1	0	0	0	0	1	1	0	0	0
p_4	0	1	0	1	0	0	0	1	0	0
p_5	0	1	0	0	0	1	0	0	1	0
p_6	0	0	1	0	1	0	0	1	0	0
p_7	0	0	1	0	0	0	1	0	1	0
p_8	0	0	0	1	0	1	0	0	0	1
p_9	0	0	0	0	1	0	1	0	0	1
p_{10}	0	0	0	0	0	0	0	1	1	1

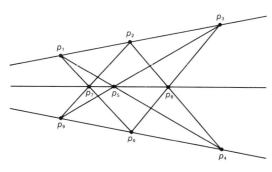

FIGURE 8

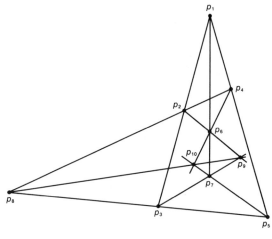

FIGURE 9

These three configurations play an important role in projective geometry. This role is examined briefly in Section 2.8 and is analyzed more fully in Chapters 5 and 6.

In our continuing study of the intuitive nature of point and line, it is relevant to consider the importance of figures, since they will be used as visual aids throughout the book. Certainly the figures for Examples 1.6.12(ii) and (iii) are more revealing than their corresponding incidence tables. This situation is not true for $\Sigma(7_3)$, nor would it be true for $\Sigma(8_3)$ or such other configurations as $\Sigma(13_4)$ and $\Sigma(21_5)$, because the lines in these configurations have the property that their points cannot be arranged to lie on a Euclidean straight line. This phenomenon is analyzed in the next section.

EXERCISES

1. Prove that $\Sigma(3_2)$ is a self-dual configuration.

2. Find an isomorphism between the planes of Examples 1.6.7(ii) and 1.6.12(i); between the plane of Example 1.6.7(ii) and the plane of exercise 1.3.7.

3. Show that every configuration of the form $\Sigma(7_3)$ is isomorphic.

4. a. Show that $P_i = [x_i, y_i, z_i]$ where $i = 1,2,3$ are noncollinear in π_R if and only if

$$\begin{vmatrix} x_1 \ y_1 \ z_1 \\ x_2 \ y_2 \ z_2 \\ x_3 \ y_3 \ z_3 \end{vmatrix} \neq 0$$

b. Let $P_i = [x_i, y_i, z_i]$ where $i = 1, \ldots, 4$. With a 4×4 determinant, formulate a way to determine whether P_1, P_2, P_3, and P_4 form a complete four-point.

5. Construct incidence tables and draw figures with as many straight lines as you can for the configurations $\Sigma(8_3)$ and $\Sigma(13_4)$.

6. There are three nonisomorphic configurations of the form $\Sigma(9_3)$ and ten nonisomorphic configurations of the form $\Sigma(10_3)$. The non-Pappus configurations are labeled $\Sigma_1(9_3)$ and $\Sigma_2(9_3)$. Figure 10 depicts the former; Figure 11 depicts the latter.

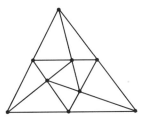

FIGURE 10

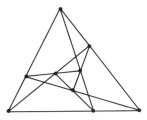

FIGURE 11

a. Show that $\Sigma(9_3)$, $\Sigma_1(9_3)$, and $\Sigma_2(9_3)$ are not isomorphic. Hint: Observe that for each point p in the configuration, exactly six of the eight remaining points are connected to it by a line. List the two nonconnected points, say, q_p and r_p, of a given point p. Next list the nonconnected points of q_p. Continue this listing procedure for all three configurations.

b. Find one configuration of the form $\Sigma(10_3)$ that is not isomorphic to the Desargues configuration.

7. Show that if π is a finite projective plane, then for some number n, every line contains n points and every point lies on n lines. Deduce the

possible cardinality of $\mathscr{P}$ and $\mathscr{L}$ in finite projective and affine planes. Deduce the possible configurations of a finite projective plane and a finite affine plane.

1.7   EMBEDDING

The concept of "embedding" has the same meaning in projective geometry that it has in other areas of mathematics. A stronger form of embedding, called restriction, is also used in the familiar mathematical sense.

DEFINITION 1.7.1   *A primitive plane* $\Sigma = (\mathscr{P}, \mathscr{L}, \mathscr{I})$ *is* embedded *in a primitive plane* $\Sigma' = (\mathscr{P}', \mathscr{L}', \mathscr{I}')$ *if and only if* $\mathscr{P} \subseteq \mathscr{P}'$; *and if* $L \in \mathscr{L}$, *then* $L = L' \cap \mathscr{P}$ *for some* $L' \in \mathscr{L}'$.

Notation.   If Σ is embedded in Σ', the notation used is $\Sigma \subseteq \Sigma'$.

DEFINITION 1.7.2   *The plane* Σ *is a* restriction *of* Σ' *if and only if* $\mathscr{P} \subseteq \mathscr{P}'$; *and* $L \in \mathscr{L}$ *exactly when* $L = L' \cap \mathscr{P}$ *for some* $L' \in \mathscr{L}'$. *If* Σ *is a restriction of* Σ', *then* Σ *is a* subplane *of* Σ'.

Notation.   If Σ is a restriction of Σ', Σ is denoted by $\Sigma' | \mathscr{P}$.

DEFINITION 1.7.3   *The plane* Σ' *is an* extension *of* Σ *if and only if* Σ *is a restriction of* Σ'.

DEFINITION 1.7.4   *The plane* Σ *is a* principal restriction *of* Σ' *if and only if* $\Sigma' = \Sigma | \mathscr{P}' - L'$ *for some* $L' \in \mathscr{L}'$. *($\mathscr{P}' - L'$ means the set theoretic complement of* L' *in* $\mathscr{P}'$.) *If* Σ *is a principal restriction of* Σ', *then* Σ' *is a* principal extension *of* Σ.

EXAMPLE 1.7.5

   i. Let Σ be a plane with only the following four noncollinear points from the Cartesian plane: (0,0), (0,1), (1,0), and (1,1). Here Σ is embedded in the Cartesian plane but the embedding is not a restriction.

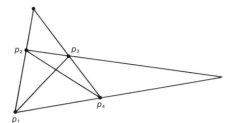

FIGURE 12

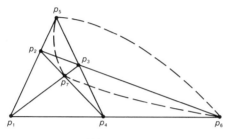

FIGURE 13

ii. Let Σ be the $\Sigma(4_3, 6_2)$ plane depicted in Figure 12. Thus:

$$\mathscr{P} = \{p_i: i = 1, \ldots, 4\}$$
$$\mathscr{L} = \{\{p_i, p_j\}: i \neq j\}$$

Let Σ' be the $\Sigma(7_3)$ plane depicted in Figure 13. Thus:

$$\mathscr{P} = \{p_i: i = 1, \ldots, 7\}$$

$\mathscr{L}$ consists of the following sets:

$\{p_1, p_2, p_5\}$	$\{p_2, p_4, p_7\}$
$\{p_1, p_3, p_7\}$	$\{p_3, p_4, p_5\}$
$\{p_1, p_4, p_6\}$	$\{p_5, p_6, p_7\}$
$\{p_2, p_3, p_6\}$	

Here Σ is a principal restriction of Σ'; in particular, $\Sigma = \Sigma' \,|\, \mathscr{P}'$ $- L'$, where $L' = \{p_5, p_6, p_7\}$.

iii. Let $\Sigma = \pi_R$ and let $\Sigma' = \alpha_R$. The fact that α_R is a principal restriction of π_R should come as no surprise. We know from Example 1.5.2(ii) and the discussion at the end of Section 1.3 that the addition of the line at infinity ($\langle 0,0,1 \rangle$) to α_R yields π_R, or, equivalently, that the deletion of the line at infinity from π_R yields α_R. This fact can be proved in a slightly different way. Consider the plane π_R with the line $\langle 0,0,1 \rangle$ deleted. First observe that every point not on $\langle 0,0,1 \rangle$ is of the form $[x,y,z]$ where $z \neq 0$. Every point of this form can be represented as $[x',y',1]$, since $[x,y,z] = [x/z, \, y/z, \, z/z]$. Let f map these points into points of α_R as follows: $f: [x,y,1] \rightarrow (x,y)$. Clearly, this is a one-to-one, onto mapping. Also, f preserves collinearity by the following argument: $[x_i,y_i,1]$, $i = 1,2,3$, are collinear if and only if there exists a,b,c such that $ax_i + by_i + c = 0$ for all i; this argument holds if and only if (x_i,y_i), $i = 1,2,3$, are collinear. Thus f is an isomorphism.

What is true in Examples 1.7.5(ii) and (iii) is true in general — that is, that every principal restriction of a projective plane is an affine plane. The proof is straightforward and is left as an exercise.

THEOREM 1.7.6   *A principal restriction $\pi | \mathscr{P} - L$ of a projective plane π is an affine plane.*

Notation.   This affine plane is denoted by $\alpha(\pi_L^-)$.

It is also possible to reverse this process, that is, to form a projective plane by constructing a principal extension of an affine plane. The procedure is analogous to the construction of an extension of the Euclidean plane as described in Section 1.1. Each line in the affine plane is granted an extra point, called an *ideal point*; parallel lines are given the same ideal point, and all the ideal points are placed on a new line called the *ideal line*. Formally, the procedure is as follows.

CONSTRUCTION 1.7.7   Suppose that α is an affine plane with points $\mathscr{P}$ and lines $\mathscr{L}$. For convenience, assume that lines are made up of points (in other words, that $\mathscr{I} = \in$). First, observe that parallelism is an equivalence relation on $\mathscr{L}$. The set of equivalence classes $\mathscr{L} / \parallel$ is denoted by X, and the members of X are denoted by $[L]$. Thus $[L]$ equals the set of all lines parallel to L. For simplicity, the following notation is also used: $\mathscr{P}^+ = \mathscr{P} \cup X$, $L' = L \cup \{[L]\}$, $\mathscr{L}' = \{L' : L \in \mathscr{L}\}$, and $\mathscr{L}^+ = \mathscr{L}' \cup \{X\}$. It is easily checked that $(\mathscr{P}^+, \mathscr{L}^+, \in)$ is eligible for study as a plane.

THEOREM 1.7.8   *The plane $(\mathscr{P}^+, \mathscr{L}^+, \in)$ is a projective plane.*

This theorem is established by showing that axioms Pj1, Pj2, and Pj3 hold in the plane.

Proof.

Pj1.   Let p and q be two distinct points.

Case 1. $p, q \in \mathscr{P}$. By Af1 there exists exactly one line joining p and q.

Case 2. $p \in \mathscr{P}, q \in X$. Let $q = [L]$. By Af3, let M be the unique line through p parallel to L. Thus $[L] = [M]$ and p and q lie on M'. If $p \in X$ and $q \in \mathscr{P}$, a similar argument is used.

Case 3. $p, q \in X$. Clearly, X is the unique line joining p and q.

Pj2.   Let L and M be distinct lines.

Case 1. $L, M \neq X$ and $L \not\parallel M$. Then L and M intersect in a unique point in the affine plane α.

Case 2. $L \parallel M$. Then L intersects M at point $[L]$.

Case 3. $L = X$. Then L intersects M at point $[M]$. (If $M = X$, then L intersects M at $[L]$.)

Pj3.   Proof that this axiom holds is left as an exercise.

Notation.   This projective plane is denoted by $\pi(\alpha^+)$.

DEFINITION 1.7.9   *The points $[L]$ that have been added here are called* ideal points *or* points at infinity. *Line X is called the* ideal line *or the* line at infinity.

Notation.   Line X is often denoted by L_∞.

The projective plane constructed in the proof of Theorem 1.7.8 is really a principal extension of the original affine plane. This fact follows by direct application of the appropriate definitions.

THEOREM 1.7.10   *The principal restriction $\pi(\alpha^+) | \mathscr{P} - L_\infty$ is the plane α, that is, $\alpha(\pi(\alpha^+)_{L\infty^-}) = \alpha$.*

Two key questions remain to be answered concerning the relationship of projective planes and affine planes. First: Is there only one projective extension for a given affine plane? Stating the question another way, if we begin with a projective plane, form an arbitrary restriction by deleting one line, and then construct the principal extension of Construction 1.7.7, is the resulting plane isomorphic to the original one? The answer to this question is yes, as shown in the following theorem.

THEOREM 1.7.11 *If π_0 is a projective plane, then there exists an isomorphism f from π_0 to $\pi(\alpha((\pi_0)_L^-)^+)$ and $f\colon L \to L_\infty$.*

Proof. Let $\mathscr{P}$ denote the set of points and let $\mathscr{L}$ denote the set of lines of π_0. Thus the points of $\alpha((\pi_0)_L^-)$ are in the set $\mathscr{P} - L$ and the lines of $\alpha((\pi_0)_L^-)$ are of the form $M - (M \cap L)$ for those lines $M \in \mathscr{L}$ such that $M \neq L$. Let $\mathscr{P}'$ denote the points of $\pi(\alpha((\pi_0)_L^-)^+)$. Then $\mathscr{P}' = (\mathscr{P} - L) \cup \{[M]\colon M \in \mathscr{L}; M \neq L\}$. A line in $\pi(\alpha((\pi_0)_L^-)^+)$ has the form $(M - (M \cap L)) \cup \{[M]\}$.

Define the function f as follows: If $p \notin L$, then $f(p) = p$; if $p \in L$, then $f(p) = [M]$ where M is a line such that $M \cap L = p$. It is easily checked that f is well defined, is one-to-one, and preserves collinearity and that $f_i\colon L \to L_\infty$. Thus f is the isomorphism we are seeking.

COROLLARY 1.7.12 *If π and π' are principal extensions of α, then $\pi \sim \pi'$.*

The second question is analogous to the first: Is every affine restriction of a given projective plane isomorphic? In other words, if we begin with an affine plane, form its unique projective extension, and then impose an arbitrary affine restriction, is the resulting plane isomorphic to the original one? The answer to this question is no, as will be proved in Section 2.7.

This section indicates that a close relationship exists between projective and affine planes. The study of projective and affine planes might be termed essentially the study of the consequences of one axiom system. This is not quite true, of course; and although this text considers mainly the axiom system for projective planes, considerable detail also will be supplied for the affine case. Generally, however, affine analogies will be left to the student.

As suggested in Section 1.6, $\Sigma(7_3)$ cannot be drawn with straight lines in the Euclidean plane. This fact will be proved in the remainder of this section.

THEOREM 1.7.13 *The plane $\Sigma(7_3)$ cannot be embedded in α_R.*

Proof. Consider a quadrangle with four points, $p = (0,0)$, $q = (1,0)$, $r = (0,1)$, and $s = (a,b)$, such that all three diagonal points exist. This condition requires only that $a,b \neq 0,1$ and $a + b \neq 0,1$. Then the diagonals are:

$$rs \cap qp = \left(\frac{-a}{b-1}, 0 \right)$$

$$qs \cap pr = \left(0, \frac{-b}{a-1} \right)$$

$$ps \cap qr = \left(\frac{a}{a+b}, \frac{b}{a+b} \right)$$

From analytic geometry we know that three points, (x_1,y_1), (x_2,y_2), and (x_3,y_3), are collinear iff

$$\begin{vmatrix} x_1 & y_1 & 1 \\ x_2 & y_2 & 1 \\ x_3 & y_3 & 1 \end{vmatrix} = 0$$

If the diagonal points are collinear, then

$$\begin{vmatrix} \dfrac{-a}{b-1} & 0 & 1 \\ 0 & \dfrac{-b}{a-1} & 1 \\ \dfrac{a}{a+b} & \dfrac{b}{a+b} & 1 \end{vmatrix} = 0$$

and thus $a + b = 1$. But qr is the line $x + y = 1$, and so s must be on qr, which contradicts the hypothesis that p,q,r,s is a four-point. Thus the diagonal points of this particular quadrangle are not collinear.

The remainder of the proof shows that for any quadrangle with vertices p', q', r', s' having the property that all three diagonal points exist, we can, for a suitable point s, construct an isomorphism of the plane onto itself that maps p, q, r, s onto p', q', r', s', respectively. This will complete the argument because if a plane $\Sigma(7_3)$ were to be found in the Cartesian plane, we could choose any four points of an embedded quadrangle of $\Sigma(7_3)$ and map them with an isomorphism onto p, q, r, s. We could thus deduce that p, q, r, s and its three diagonal points form a $\Sigma(7_3)$ plane. But this conclusion would contradict the preceding one.

Let $p' = (x_1, y_1)$, $q' = (x_2, y_2)$, and $r' = (x_3, y_3)$. Then the map $g: (x, y) \rightarrow (ax_1 + by_1 + c, dx_1 + ey_1 + f)$, where $a = x_2 - x_1$, $b = x_3 - x_1$, $c = x_1$, $d = y_2 - y_1$, $e = y_3 - y_1$, $f = y_1$, maps p, q, r, s onto p', q', r', s' where $s = g^{-1}(s')$. By Example 1.5.5, g is an isomorphism, so the proof is complete.

Theorem 1.7.13 is important for two reasons. First, the embedding of configurations in projective planes (as well as in affine planes) is generally an important tool in analyzing those planes. The three most important configurations in this regard are the Fano configuration, the Desargues configuration, and the Pappus configuration. Planes that allow for Fano configurations in all permissible situations—for example, $\Sigma(7_3)$ and $\Sigma(21_5)$—are called Fano planes.

DEFINITION 1.7.14   *A projective plane π is a* Fano plane *if and only if the diagonal points of every quadrangle in that plane are collinear.*

Similarly, planes that allow for Desargues (or Pappus) configurations in all permissible situations are called Desarguesian (or Pappian) planes. Desarguesian and Pappian planes are defined and examined in Chapters 5 and 6, respectively. Clearly, π_R is not a Fano plane. In fact, Theorem 1.7.13 indicates that a stronger statement can be made: $\Sigma(7_3)$ cannot be embedded anywhere in π_R. Curiously enough, a plane satisfying this property is said to satisfy Fano's Axiom:

FANO'S AXIOM   *The diagonals of every quadrangle in a projective plane are noncollinear.*

Thus a plane satisfying Fano's Axiom is not a Fano plane and a Fano plane does not satisfy Fano's Axiom.

THEOREM 1.7.15   *The plane π_R satisfies Fano's Axiom.*

Proof.   The proof is left as an exercise.

The other reason that Theorem 1.7.13 is important is an informal one concerning the nature of point and line. A primitive plane can be embedded in α_R if and only if it can be diagramed so that its points are dots and its lines are sets of points lying on a straight line in the Cartesian plane. The fact that $\Sigma(7_3)$ cannot be embedded in α_R suggests that drawings of finite planes may display invalid relationships. Figures are valuable aids in understanding definitions, theorems, and proofs, but figures often may be inadequate and sometimes may even be misleading.

EXERCISES

1. Show that the affine plane $\Sigma(4_3, 6_2)$ is a subplane of every affine plane α.

2. Show that every principal restriction of a projective plane is an affine plane.

3. Which of the following configurations can you embed in $\Sigma(13_4)$?
   a. $\Sigma(7_3)$.
   b. $\Sigma(8_3)$.
   c. $\Sigma(9_3)$, the Pappus configuration.
   d. $\Sigma(10_3)$, the Desargues configuration.

4. Consider the incidence table for $\Sigma(21_5)$ displayed in Figure 14. This table is the same as those in Example 1.6.12 except that black squares have been substituted for the 1's and blank squares have been substituted for the 0's. Horizontal numbers indicate subscripts of the twenty-one lines; vertical numbers indicate subscripts of the twenty-one points. Attempt to find the following configurations in $\Sigma(21_5)$:
   a. $\Sigma(7_3)$.
   b. $\Sigma(8_3)$.
   c. $\Sigma(9_3)$, the Pappus configuration.
   d. $\Sigma(10_3)$, the Desargues configuration.

5. Which of the following configurations can you embed in π_Q? in π_R? in π_C?
   a. $\Sigma(9_3)$, the Pappus configuration.
   b. $\Sigma(10_3)$, the Desargues configuration.
   c. $\Sigma(7_3)$.
   d. $\Sigma(8_3)$.

6. Prove Theorem 1.7.15.

7. Show that if $\alpha \sim \alpha'$, then there exists an isomorphism f mapping $\pi(\alpha^+)$ onto $\pi((\alpha')^+)$ such that $f: L_\infty \to M_\infty$ where M_∞ is the line at infinity in $\pi((\alpha')^+)$.

8. Show that all affine restrictions of a given plane π are isomorphic if and

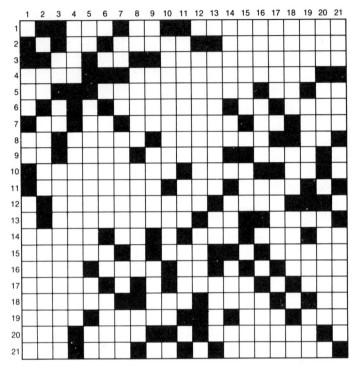

FIGURE 14

only if there exists an isomorphism from π onto itself that maps L to M for any lines L and M.

9. a. Prove or disprove the following statement: If α and α' are affine planes such that $\alpha \subseteq \alpha'$, then $\pi(\alpha^+) \subseteq \pi((\alpha')^+)$.

b. Show that the following statement is false: If $\pi(\alpha^+) \subseteq \pi((\alpha')^+)$, then $\alpha \subseteq \alpha'$.

c. Add appropriate restrictions to the statement in 9(b) to make a tenable theorem and prove your theorem.

CHAPTER 2

EXAMPLES AND
ELEMENTARY PROPERTIES

It will be convenient to introduce several more examples of projective and affine planes before undertaking a theoretical analysis. These examples will not only serve to illustrate concepts in the following sections but will also help to reveal some basic problems that might have been overlooked had sophisticated techniques been developed first. In this chapter five different representations are given for projective and affine planes—a sparse listing, as can be seen from a brief glance at Part 3. The focus here will be on the variety of representations, the assistance that the disciplines of algebra and number theory can offer, and the task of finding the geometric invariants that will help in distinguishing between planes of the same cardinality.

2.1   PLANES DEFINED BY FIELDS

Section 1.3 included a definition of the real projective plane π_R, in which both points and lines were representable as equivalence classes of ordered triples of real numbers. A projective plane can be constructed in the same way if the field of real numbers is replaced by an arbitrary field. (Actually, the field can be replaced by an arbitrary division ring, a procedure that is discussed more fully in Chapter 7.) This section begins with a brief review of the concept of "field."

DEFINITION 2.1.1   *A set S with two binary operations, +, ·, is called a field if and only if the following axioms hold:*

1. *If $a,b \in S$, then $a + b \in S$ and $a \cdot b \in S$.*

2. *If $a,b,c \in S$, then $a + (b + c) = (a + b) + c$ and $a \cdot (b \cdot c) = (a \cdot b) \cdot c$.*

3. *There exists $o,e \in S$, $o \neq e$, such that $a + o = o + a = a$ for all $a \in S$, and $a \cdot e = e \cdot a = a$ for all $a \in S$.*

4. *For all $a \in S$, there exists $b \in S$ (depending on a) such that $a + b = b + a = o$. For all $a \neq o \in S$, there exists $c \in S$ (depending on a) such that $a \cdot c = c \cdot a = e$.*

5. *For all $a,b \in S$, $a + b = b + a$ and $a \cdot b = b \cdot a$.*

6. *For all $a,b,c \in S$, $a \cdot (b + c) = a \cdot b + a \cdot c$.*

Notation.   The field $(S, +, \cdot)$ is denoted by F.

In this text a will loosely be said to be an element of the field F; in other words, "$a \in F$" will be written when what is strictly meant is $a \in S$. Also, "ab" will be used as a shorthand notation to denote $a \cdot b$. Many different kinds of algebras will be introduced in this book, and to preserve distinctions it will be necessary to use several different symbols to indicate addition and multiplication. For example, $\oplus$ and $\boxplus$ will also be used for addition and $\circ$, $\odot$, $\circledcirc$, $\boxdot$, and $\otimes$ will be used for multiplication. The multidot ($\cdot$) is the only multiplication symbol that will be omitted in multiplication.

EXAMPLE 2.1.2

i. The real numbers, with ordinary addition and multiplication, form a field. This field is denoted by R.

ii. The complex numbers, with ordinary addition and multiplication, form a field. This field is denoted by C.

iii. The rational numbers, with the ordinary operations, form a field. This field is denoted by Q.

iv. Let p be any prime number and let the set of elements of $S = \{[0],[1], \ldots, [p-1]\}$ where $[i] = \{i+pk: k \in J\}$ and where J denotes the set of all integers. Define addition, $\oplus$, and multiplication, $\circ$, as follows: $[i] \oplus [j] = [i+j]$, $[i] \circ [j] = [i \cdot j]$. It is a rather simple task to check that S, in terms of this addition and multiplication, is a field. This field is denoted by J/p. The notation is read, "the field of integers modulo p." The integers modulo p may be thought of simply as the set $\{0,1, \ldots, p-1\}$ with addition and multiplication modulo p, that is, $i \oplus j = k$, and $i \circ j = r$, where k and r are the remainders of $i+j$ and $i \cdot j$, respectively, when divided by p.

v. Consider the addition and multiplication tables represented by Tables 1 and 2. These tables can easily be shown to represent fields of four and nine elements, respectively. Such fields will be denoted by GF(4) and GF(9), respectively. The letters G and F stand for Galois Field: GF(4) is the Galois Field of order 4,

TABLE 1

Let $S = \{0,1,a,1+a\}$.
Let $b = 1+a$.

+	0	1	a	b
0	0	1	a	b
1	1	0	b	a
a	a	b	0	1
b	b	a	1	0

·	0	1	a	b
0	0	0	0	0
1	0	1	a	b
a	0	a	b	1
b	0	b	1	a

TABLE 2

Let $S = \{0,1,2,a,1+a,2+a,2a,1+2a,2+2a\}$.
Let $b = 1+a$, $c = 2+a$, $d = 2a = a+a$, $e = 1+2a$, $f = 2+2a$.

+	0	1	2	a	b	c	d	e	f
0	0	1	2	a	b	c	d	e	f
1	1	2	0	b	c	a	e	f	d
2	2	0	1	c	a	b	f	d	e
a	a	b	c	d	e	f	0	1	2
b	b	c	a	e	f	d	1	2	0
c	c	a	b	f	d	e	2	0	1
d	d	e	f	0	1	2	a	b	c
e	e	f	d	1	2	0	b	c	a
f	f	d	e	2	0	1	c	a	b

·	0	1	2	a	b	c	d	e	f
0	0	0	0	0	0	0	0	0	0
1	0	1	2	a	b	c	d	e	f
2	0	2	1	d	f	e	a	c	b
a	0	a	d	e	1	b	c	f	2
b	0	b	f	1	c	d	2	a	e
c	0	c	e	b	d	2	f	1	a
d	0	d	a	c	2	f	e	b	1
e	0	e	c	f	a	1	b	2	d
f	0	f	b	2	e	a	1	d	c

GF(9) is the Galois Field of order 9. Galois Fields are studied in greater detail in Chapter 7.

DEFINITION 2.1.3   *A field has* characteristic n *if and only if n is the least integer such that* $\underbrace{1 + 1 + \ldots + 1}_{n \text{ times}} = 0$. *If no such integer exists, the field has* characteristic 0.

EXAMPLE 2.1.4

   i. The fields Q, R, C have characteristic 0.

   ii. The fields J/p have characteristic p.

   iii. The field GF(4) has characteristic 2.

   iv. The field GF(9) has characteristic 3.

A general theorem about finite characteristics is the following:

THEOREM 2.1.5   *If the field F has characteristic* $n \neq 0$, *then n is a prime number.*

Proof.   The proof is left as an exercise.

A projective plane can be constructed over an arbitrary field F in the same manner as the real projective plane was constructed over R in Example 1.3.9(vi).

CONSTRUCTION 2.1.6   Let F be an arbitrary field. Form the plane $\pi_F = (\mathscr{P}, \mathscr{L}, \mathscr{I})$ as follows. Let $\mathscr{P} = \{[x,y,z]: x,y,z \in F$ and $x = y = z = 0$ is not true$\}$ where $[x,y,z] = [x',y',z']$ if and only if there exists $r \in F$, $r \neq 0$, such that $x = rx'$, $y = ry'$, $z = rz'$. Let $\mathscr{L} = \{\langle a,b,c \rangle: a,b,c \in F$ and $a = b = c = 0$ is not true$\}$ where $\langle a,b,c \rangle = \langle a',b',c' \rangle$ if and only if there exists $s \in F$, $s \neq 0$, such that $a = sa'$, $b = sb'$, $c = sc'$. Let $\mathscr{I}$ be defined as follows: $([x,y,z], \langle a,b,c \rangle) \in \mathscr{I}$ if and only if $ax + by + cz = 0$.

THEOREM 2.1.7   *The plane* π_F *is a projective plane.*

Proof.   The first step is to prove Pj1: Every two points lie on exactly one line. Let $[x,y,z]$ and $[x',y',z']$ be distinct points. It is easily checked that the line $\langle yz' - y'z, zx' - z'x, xy' - x'y \rangle$ passes through both points.

We must check, however, that this triple is not $\langle 0,0,0 \rangle$. If the triple is $\langle 0,0,0 \rangle$, then $yz' = y'z$, $xz' = x'z$, $xy' = x'y$. Since one of $x',y',z' \neq 0$, we may assume without loss of generality that $x' \neq 0$; thus $x = rx'$, $y = ry'$, and $z = rz'$ where $r = x/x'$, contradicting our assumption that the two points are distinct.

Suppose that there are two lines through the given points. Let $\langle a,b,c \rangle$, where $a = yz' - y'z$, $b = zx' - z'x$, and $c = xy' - x'y$ denote one line and let $\langle a',b',c' \rangle$ denote the other line. Maintaining the assumption of the preceding paragraph that $x' \neq 0$, it may be observed that the two equations $x'y - xy' = 0$ and $zx' - z'x = 0$ cannot hold simultaneously because these two equations together imply that $x = rx'$, $y = ry'$, and $z = rz'$ where $r = x/x'$, again contradicting the distinctness of the two points. Without loss of generality we may assume that $x'y - xy' \neq 0$.

Since $\langle a',b',c' \rangle$ passes through the two points, we obtain the following two equations:

$$a'x + b'y + c'z = 0$$
$$a'x' + b'y' + c'z' = 0$$

Appropriate multiplication of these two equations yields:

$$a'xx' + b'yx' + c'zx' = 0$$
$$a'x'x + b'y'x + c'z'x = 0$$

and

$$a'xy' + b'yy' + c'zy' = 0$$
$$a'x'y + b'y'y + c'z'y = 0$$

Solving each pair of equations separately, we get

$$b'(x'y - xy') + c'(zx' - z'x) = 0$$
$$a'(x'y - xy') + c'(z'y - zy') = 0$$

Since $x'y - xy' \neq 0$, these equations may be rewritten:

$$b' = \frac{(z'x - zx')c'}{x'y - xy'}$$

$$a' = \frac{(y'z - yz')c'}{x'y - xy'}$$

From the definitions of a, b, and c we obtain

$$b = \frac{(z'x - zx')c}{x'y - xy'}$$

$$a = \frac{(y'z - yz')c}{x'y - xy'}$$

It follows that $bc' = b'c$ and $ac' = a'c$, and therefore $a = ra'$, $b = rb'$, and $c = rc'$ where $r = c/c'$. (Notice that $c' \neq 0$ because the equations containing it imply that if $c' = 0$, then $a' = b' = 0$, which is impossible.) Thus the two lines under consideration are identical.

The proof of Pj2 follows by the same argument. The points $[1,0,0]$, $[0,1,0]$, $[0,0,1]$, and $[1,1,1]$ form a four-point, satisfying Pj3.

EXAMPLE 2.1.8

   i. Let $F = J/2$. Then π_F is as follows:

$\mathscr{P} = \{[1,0,0], [0,1,0], [0,0,1], [1,1,1], [1,1,0], [0,1,1], [1,0,1]\}$
$\mathscr{L} = \{\langle 1,0,0\rangle, \langle 0,1,0\rangle, \langle 0,0,1\rangle, \langle 1,1,1\rangle, \langle 1,1,0\rangle, \langle 0,1,1\rangle, \langle 1,0,1\rangle\}$

Figure 15 displays the incidence relationships.

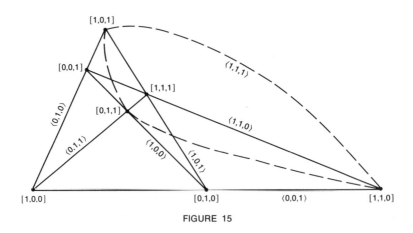

FIGURE 15

   ii. Let $F = J/3$. π_F is written out as follows:

$\mathscr{P} = \{[1,0,0], [0,1,0], [0,0,1], [1,1,1], [1,1,0], [1,0,1], [0,1,1],$
$[2,0,1], [0,2,1], [2,1,1], [1,2,1], [2,2,1], [2,1,0]\}$

The lines and their points are:

$$\langle 1,0,0 \rangle = \{[0,1,0], [0,1,1], [0,0,1], [0,2,1]\}$$
$$\langle 0,1,0 \rangle = \{[1,0,0], [1,0,1], [0,0,1], [2,0,1]\}$$
$$\langle 0,0,1 \rangle = \{[1,0,0], [0,1,0], [1,1,0], [2,1,0]\}$$
$$\langle 1,1,1 \rangle = \{[1,1,1], [2,0,1], [0,2,1], [2,0,1]\}$$
$$\langle 1,1,0 \rangle = \{[0,0,1], [2,1,0], [2,1,1], [1,2,1]\}$$
$$\langle 1,0,1 \rangle = \{[0,1,0], [2,0,1], [2,1,1], [2,2,1]\}$$
$$\langle 0,1,1 \rangle = \{[1,0,0], [0,2,1], [1,2,1], [2,2,1]\}$$
$$\langle 2,0,1 \rangle = \{[0,1,0], [1,0,1], [1,1,1], [1,2,1]\}$$
$$\langle 0,2,1 \rangle = \{[1,0,0], [0,1,1], [1,1,1], [2,1,1]\}$$
$$\langle 2,1,1 \rangle = \{[1,1,0], [0,2,1], [2,1,1], [1,0,1]\}$$
$$\langle 1,2,1 \rangle = \{[1,1,0], [0,1,1], [1,2,1], [2,0,1]\}$$
$$\langle 2,2,1 \rangle = \{[2,1,0], [1,2,0], [1,0,1], [0,1,1]\}$$
$$\langle 2,1,0 \rangle = \{[1,1,0], [1,1,1], [2,2,1], [0,0,1]\}$$

It is possible to form an affine plane over an arbitrary field F. The construction is analogous to that of α_R, the Cartesian plane.

CONSTRUCTION 2.1.9   Let F be an arbitrary field. The plane $\alpha_F = (\mathscr{P}, \mathscr{L}, \mathscr{I})$ is formed as follows. Let

$$\mathscr{P} = \{(x,y): x,y \in F\}$$
$$\mathscr{L} = \{L(a,b,c): a,b,c \in F\} \text{ where } L(a,b,c)$$
$$= \{(x,y): ax + by + c = 0\}, \text{ and at}$$
$$\text{least one of } a,b,c \text{ is not } 0$$
$$\mathscr{I} = \in$$

The relationship between π_F and α_F is the same as that between π_R and α_R and can be stated formally as a theorem.

THEOREM 2.1.10   *The plane $\alpha((\pi_F)_{(0,0,1)}^-)$ is isomorphic to α_F and the plane $\pi((\alpha_F)^+)$ is isomorphic to π_F.*

Proof.   The proof is left as an exercise.

EXERCISES

1. Show that π_F is self-dual.

2. Show that if F' is a subfield of F, then $\pi_{F'}$ is isomorphic to a subplane of π_F.

3. Prove Theorem 2.1.5.

4. Prove Theorem 2.1.10.

5. a. Show that if π_F is a Fano plane, then F is of characteristic 2.
   b. Show that if π_F satisfies Fano's Axiom, then F is not of characteristic 2.
   c. Prove the converse of 5(a) and (b).

6. An alternate way of coordinating a plane is to use a Cartesian group. A system $(R,+,\cdot)$ is a Cartesian group if and only if the following six conditions are satisfied:

   (1) $(R,+)$ is a group.
   (2) The equations $ax = b$ and $xa = b$ have a unique solution for all $a,b \in R - \{0\}$.
   (3) There exists $e \in R$ such that $ex = xe = x$ for all $x \in R$.
   (4) Let o denote the additive identity; then $ox = xo = o$ for all $x \in R$.
   (5) Given $a,b,c,d \in R$ such that $a \neq c$, there exists a unique $x \in R$ such that $xa + b = xc + d$.
   (6) Given $a,b,c,d \in R$ such that $a \neq c$, there exists a unique pair $(x,y) \in R \times R$ such that $ax + y = b$ and $cx + y = d$.

   Algebras M and A, which follow, are examples of Cartesian groups.

   Let $M = (R,+,\circ)$ where R denotes the set of reals, $+$ denotes ordinary addition on the reals, and $\circ$ is defined as follows:

   $$x \circ y = xy \text{ if either } x \text{ or } y \text{ or both} \geq 0$$
   $$x \circ y = \frac{xy}{2} \text{ if } x \text{ and } y < 0$$

   Let $A = (R,+,\circ)$ where R is the set of elements of $GF(9)$, $+$ denotes the addition on $GF(9)$, and $\circ$ is defined as in Table 3.

   TABLE 3

$\circ$	1	2	a	b	c	d	e	f
1	1	2	a	b	c	d	e	f
2	2	1	d	f	e	a	c	b
a	a	d	b	1	f	c	2	e
b	b	f	e	c	1	2	d	a
c	c	e	1	d	a	f	b	2
d	d	a	f	2	b	e	1	c
e	e	c	2	a	d	b	f	1
f	f	b	c	e	2	1	a	d

   a. Show that if $(R,+,\cdot)$ is a field, then it is a Cartesian group.
   b. Verify that M and A are Cartesian groups.

    c. Verify that M and A are not fields. Find as many field axioms as you can that M and A fail to satisfy.

    d. Let $(R,+,\cdot)$ be a Cartesian group, let 0 and 1 denote the additive and multiplicative identities, respectively, and construct $\mathscr{P}$, $\mathscr{L}$, and $\mathscr{I}$ as follows:

$$\mathscr{P} = \{[x,y,1]: x,y \in R\} \cup \{[1,x,0]: x \in R\} \cup \{[0,1,0]\}$$
$$\mathscr{L} = \{\langle m,1,k\rangle: m,k \in R\} \cup \{\langle 1,0,k\rangle: k \in R\} \cup \{\langle 0,0,1\rangle\}$$
$$\mathscr{I} = \{([x,y,z], \langle m,n,k\rangle): xm + yn + zk = 0\}$$

Show that $(\mathscr{P},\mathscr{L},\mathscr{I})$ is a projective plane.

    e. If F is a field, show that π_F is isomorphic to the plane $(\mathscr{P},\mathscr{L},\mathscr{I})$ constructed in 6(d).

    f. Let π_M denote the plane constructed from the Cartesian group M. Show that $\alpha((\pi_M)_{\langle 0,0,1\rangle^-})$ is isomorphic to the plane of exercise 1.3.8. The plane π_M is called the *Moulton plane*.

7. Let $P = (R,+,\cdot)$ and $P' = (R',\oplus,\circ)$ be Cartesian groups. The mapping γ from R onto R' is an *isomorphism* if and only if $\gamma(x + y) = \gamma(x) \oplus \gamma(y)$ and $\gamma(x \cdot y) = \gamma(x) \circ \gamma(y)$.

    a. Find an example of an isomorphism from P onto itself where:

       (1) $P = GF(4)$.

       (2) $P = GF(9)$.

       (3) $P = C$, the field of complex numbers.

       (4) $P = A$ where A is the Cartesian group of exercise 6.

    b. Let F and F' be fields. Show that if there exists an isomorphism from F to F', then π_F is isomorphic to $\pi_{F'}$.

    c. Let π_P and $\pi_{P'}$ denote the planes constructed over the Cartesian groups P and P', respectively. Show that if there exists an isomorphism from P onto P', then π_P and $\pi_{P'}$ are isomorphic.

2.2   FINITE AFFINE PLANES AND LATIN SQUARES

This section and the two subsequent sections examine examples of finite planes, that is, planes $(\mathscr{P},\mathscr{L},\mathscr{I})$ such that $\mathscr{P} \cup \mathscr{L}$ is finite. The area of finite planes is rich in exciting problems and elegant solutions and has the status of a separate and distinct discipline of mathematics (see Dembowski, 1968). Its study naturally leads to the study of finite fields (as Section 2.1 has suggested), Latin squares, difference sets, incidence matrices, and finite algebraic structures (as Sections 2.2, 2.3, and 2.4 will show), and generally to the study of number theory, group theory, and combinatorial theory.

THEOREM 2.2.1 *If α is a finite affine plane, then every line has n points on it and every point has n + 1 lines passing through it, for some $n \geq 2$.*

Proof. Suppose that $L = \{p_1, p_2, \ldots, p_n\}$. Clearly, $n \geq 2$. Let q be any point not on L; such a point exists by Af3. By Af1 and Af2, there are $n + 1$ lines through q: qp_i: $i = 1, \ldots, n$, including the unique line parallel to L through q. Let M be any line distinct from L in α. It is easily verified (see exercise 1.3.3) that $\mathscr{P} \neq L \cup M$, so there exists a point q that is not on L or M. The $n + 1$ lines through q meet M in n points, since exactly one of the $n + 1$ lines is parallel to M. Any additional points on M would result in additional lines through q; thus M has exactly n points on it. The proof that every point on L has $n + 1$ lines through it is left as an exercise.

DEFINITION 2.2.2 *If α has n points on every line, then α is of order n.*

Finite projective planes also have the property that all lines within such planes have the same number of points. The order of a projective plane is defined in the next section. More complete coverage of cardinality is given in Section 2.6.

It follows easily from Theorem 2.2.1 that a finite plane with n points on a line has a total of n^2 points and $n^2 + n$ lines.

THEOREM 2.2.3 *If α is an affine plane of order n, then α has n^2 points and $n^2 + n$ lines.*

Proof. Let p be a given point of α; p has $n + 1$ lines through it and all the lines contain $n - 1$ points, excluding p. Thus there are $(n + 1)(n - 1) + 1 = n^2$ points, including p. Proof that there are $n^2 + n$ lines is also relatively simple.

Affine planes whose points were represented by ordered pairs of elements from a field were introduced in Section 2.1. Because all finite affine planes have n^2 points for some n, the question naturally arises whether, for any affine plane of order n, a field of order n exists from which we may draw ordered pairs to represent the points and linear combinations of points to represent the lines. The answer is no.

Certainly we may always represent points as ordered pairs of elements from some set of n elements, but the hope that this set is a field, complete with all the interplay of algebra and geometry mentioned previously, is not realized. (This failure will become evident in Section 2.4 and will be formally proved in Chapter 5.) There are, however, other algebraic structures of n elements that may be substituted for the role of a field in the representation of a finite plane. An example of such a structure is given in Section 2.4 and a general study of such structures is included in Part 3. This section will pursue a somewhat different approach—one that does not require advanced algebraic skills and yet reveals the fundamental character of the finite affine plane equally well. This approach involves the use of Latin squares.

DEFINITION 2.2.4 *A Latin square of order n, $n \geq 1$, is a square array of numbers such that for every i and j, a_{ij} is an integer and $1 \leq a_{ij} \leq n$. Furthermore, every row and column of the array contains each number from 1 to n exactly once.*

$$\begin{bmatrix} a_{11} & a_{12} & \cdots & a_{1n} \\ a_{21} & a_{22} & \cdots & a_{2n} \\ \cdot & & & \cdot \\ \cdot & & & \cdot \\ \cdot & & & \cdot \\ a_{n1} & a_{n2} & \cdots & a_{nn} \end{bmatrix}$$

EXAMPLE 2.2.5 Following are two different examples of Latin squares of order 3:

$$\begin{bmatrix} 1 & 2 & 3 \\ 2 & 3 & 1 \\ 3 & 1 & 2 \end{bmatrix} \qquad \begin{bmatrix} 3 & 2 & 1 \\ 1 & 3 & 2 \\ 2 & 1 & 3 \end{bmatrix}$$

DEFINITION 2.2.6 *The following two Latin squares of order n*

$$A = \begin{bmatrix} a_{11} & \cdots & a_{1n} \\ \cdot & & \cdot \\ \cdot & & \cdot \\ \cdot & & \cdot \\ a_{n1} & \cdots & a_{nn} \end{bmatrix} \qquad B = \begin{bmatrix} b_{11} & \cdots & b_{1n} \\ \cdot & & \cdot \\ \cdot & & \cdot \\ \cdot & & \cdot \\ b_{n1} & \cdots & b_{nn} \end{bmatrix}$$

are orthogonal *if and only if the square array*

$$C = \begin{bmatrix} c_{11} & \cdots & c_{1n} \\ \cdot & & \cdot \\ \cdot & & \cdot \\ \cdot & & \cdot \\ c_{n1} & \cdots & c_{nn} \end{bmatrix}$$

defined by $c_{ij} = (a_{ij}, b_{ij})$ *contains all* n^2 *different ordered pairs. Intuitively, we may think of C as the array obtained by superimposing A on B.*

EXAMPLE 2.2.7   The pair of Latin squares in Example 2.2.5 are orthogonal Latin squares of order 3 because

$$\begin{bmatrix} (1,3) & (2,2) & (3,1) \\ (2,1) & (3,3) & (1,2) \\ (3,2) & (1,1) & (2,3) \end{bmatrix}$$

contains all nine different ordered pairs.

DEFINITION 2.2.8   *A set of* $n - 1$ *mutually orthogonal Latin squares of order n is called a* complete orthogonal system *of Latin squares.*

EXAMPLE 2.2.9   The Latin squares in Example 2.2.5 form a complete orthogonal system of Latin squares of order 3. This follows directly from Definition 2.2.8.

Now we are ready for the theorem that combines the finite affine plane with a system of orthogonal Latin squares.

THEOREM 2.2.10   *There exists an affine plane of order n if and only if there exists a complete orthogonal system of Latin squares of order n.*

Proof.   Let $\alpha(n)$ denote a given affine plane of order n. As we observed in Construction 1.7.7, parallelism is an equivalence relation on the lines of an affine plane. For $\alpha(n)$, parallelism partitions the lines into $n + 1$ equivalence classes, each containing n mutually parallel lines. Let β and γ be two such classes and number the lines in each class from 1 to n (see Figure 16). Each point p lies on exactly one line of β, say line

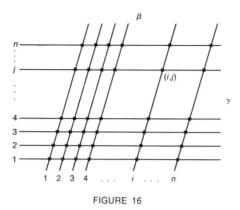

FIGURE 16

number i, and exactly one line from γ, say line number j. Thus we may attach coordinates (i,j) to each point p. Line number 1 from γ may be thought of as the x-axis and line number 1 from β may be thought of as the y-axis.

There are $n - 1$ remaining classes of parallel lines. Suppose that the lines from a fixed arbitrary class δ have been numbered from 1 to n. We may construct a square array of numbers of order n as follows: For each position (i,j) in the array, enter the number of the line in δ that passes through the point with coordinates (i,j). The next step is to show that this array is a Latin square. By Definition 2.2.4, every entry in column i in the array is a number from the list of numbers 1 to n. If two numbers are the same, then one line L of δ passes through two distinct points p and q whose coordinates are (i,j) and (i,k), respectively. Thus line $L = pq$ and line L is in β. But β and δ are disjoint. This contradiction implies that every entry in row i is distinct. A similar argument shows that every entry in row j comes from the list of numbers 1 to n and is distinct. Thus the array is a Latin square.

It is clear that there is an associated Latin square for each class. To show that the resulting $n - 1$ Latin squares are mutually orthogonal, let S and S' be two such squares arising from δ and δ', respectively. Suppose that the (i,j)th and the (k,m)th positions have the same ordered pair (a,b) when the squares S and S' have been superimposed. This means that both line number a from δ and line number b from δ' pass through point p and point q, where p has coordinates (i,j) and q has coordinates (k,m). But δ and δ' are disjoint. This contradiction implies that each entry in the square array of ordered pairs is different.

Thus S and S' are orthogonal. Since S and S' were arbitrary, the set of $n - 1$ Latin squares is mutually orthogonal. Proof of the converse is left as an exercise.

EXERCISES

1. Show that, in Theorem 2.2.1, every point on L has $n + 1$ lines through it.

2. Suppose that $\mathscr{P}$, $\mathscr{L}$, and $\mathscr{I}$ satisfy the following four properties:
   $\mathscr{P}$ has n^2 members.
   $\mathscr{L}$ has $n^2 + n$ members.
   Each line contains n points.
   Every two points lie on a unique line.
   Show that $(\mathscr{P}, \mathscr{L}, \mathscr{I})$ is an affine plane.

3. Complete the proof of Theorem 2.2.10.

4. Find three mutually orthogonal Latin squares of order 4.

5. Consider the following Latin squares:
$$\begin{bmatrix} 0 & 1 & 2 \\ 1 & 2 & 0 \\ 2 & 0 & 1 \end{bmatrix} \quad \begin{bmatrix} 2 & 1 & 0 \\ 0 & 2 & 1 \\ 1 & 0 & 2 \end{bmatrix}$$
   Let β, γ, δ, and ϵ be the four classes of parallel lines as described in the proof of Theorem 2.2.10. Label the lines $L_\alpha{}^i$ where $i = 0,1,2$ and $\alpha = \beta, \gamma, \delta, \epsilon$.
   a. List these twelve lines and their associated points. (Recall that the points are of the form (x,y) where $x,y = 0,1,2$.)
   b. Are these lines of the form $x = k$ or $y = mx + k$ where $m,k = 0,1,2$ mod 3?

2.3 FINITE CYCLIC PLANES AND DIFFERENCE SETS

This section introduces a new way of representing finite planes, both projective and affine. First, perfect difference sets are used to represent finite projective planes (see Hall, 1947). The section begins with a cardinality theorem.

THEOREM 2.3.1 *If π is a finite projective plane, then every line has n points on it and every point has n lines through it for some $n \geq 3$.*

Proof. This theorem is really a corollary of Theorem 2.2.1. Let $\pi = \pi(\alpha^+)$ for some affine restriction α. Now α has $n - 1$ points on every line and n lines through every point for some $n \geq 3$. Since $\pi(\alpha^+)$ adds one point to each line of α, the proof is complete.

COROLLARY 2.3.2 *If π is a projective plane of order n, then π has $n^2 + n + 1$ points and $n^2 + n + 1$ lines.*

Proof. The proof is left as an exercise.

DEFINITION 2.3.3 *If π has $n + 1$ points on every line, π is of order n.*

Thus the order of a finite projective plane π is the same number as the order of any of its affine restrictions $\alpha(\pi_L{}^-)$. This number does not equal the number of points on a line in π, but it does reflect the number of elements in the field F if $\pi = \pi_F$. We shall see that it also corresponds to the number of members of a representative perfect difference set.

DEFINITION 2.3.4 *Let $S = \{a_1, a_2, \ldots, a_{n+1}\}$ where $n \geq 2$ and $0 \leq a_i < k$ for $i = 1, \ldots, n + 1$. Then S is called a* perfect difference *set of order n (with respect to k) if and only if, for every $m \neq 0 \mod k$, there exists a unique ordered pair (a_i, a_j) such that $a_i, a_j \in S$ and $a_i - a_j = m \mod k$.*

EXAMPLE 2.3.5 $S = \{0, 1, 3\}$ is a perfect difference set of order 2 (with respect to 7) because

$$1 = 1 - 0 \mod 7 \qquad 4 = 0 - 3 \mod 7$$
$$2 = 3 - 1 \mod 7 \qquad 5 = 1 - 3 \mod 7$$
$$3 = 3 - 0 \mod 7 \qquad 6 = 0 - 1 \mod 7$$

If S is a perfect difference set of order n with respect to k, then $k = n^2 + n + 1$. This is easily explained. There are $(n + 1)(n)$ ordered pairs in a set of $n + 1$ elements; since these pairs must generate a unique element of J/k (the integers modulo k) and since all elements except 0 are generated, $k = (n + 1)n + 1 = n^2 + n + 1$. So we may say, without ambiguity, that S is a perfect difference set of order n. Notice that perfect difference sets of a given order are not unique. For example, $\{1, 2, 4\}$ and $\{2, 3, 5\}$ are also perfect difference sets of order 2. In fact, the sets $\{i, 1 + i, 3 + i\}$, where $i = 0, \ldots, 6$, and addition is modulo 7, are all perfect difference sets of order 2. Furthermore, two numbers do not uniquely determine a perfect difference set, as examples $\{0,1,3\}$ and $\{0,1,5\}$ show.

Notation. A perfect difference set of order n is denoted by S_n.

DEFINITION 2.3.6 *Let $S_n = \{a_1, \ldots, a_{n+1}\}$ be a perfect difference set of order n. Let $\mathscr{P} = \{p_0, p_1, \ldots, p_{n^2+n}\}$, $\mathscr{L} = \{L_0, L_1, \ldots, L_{n^2+n}\}$, and $\mathscr{I} = \{(p_i, L_j): i+j \in S_n\}$; then $(\mathscr{P}, \mathscr{L}, \mathscr{I})$ is called a* cyclic plane *of order n generated by S_n.*

EXAMPLE 2.3.7

i. Let $S_2 = \{0,1,3\}$ be a perfect difference set of order 2. The associated cyclic plane is illustrated in Table 4, in which the letters i and j stand for subscripts of the points and lines, respectively. Subscripts of points on line L_j can be read vertically from the table (for example, L_2 contains p_5, p_6, and p_1).

TABLE 4

					j			
		0	1	2	3	4	5	6
		0	6	5	4	3	2	1
i		1	0	6	5	4	3	2
		3	2	1	0	6	5	4

ii. Let $S_3 = \{0,1,3,9\}$ be a perfect difference set for $n = 3$; the associated plane is illustrated in Table 5.

TABLE 5

							j						
		0	1	2	3	4	5	6	7	8	9	10	11	12
		0	12	11	10	9	8	7	6	5	4	3	2	1
		1	0	12	11	10	9	8	7	6	5	4	3	2
i		3	2	1	0	12	11	10	9	8	7	6	5	4
		9	8	7	6	5	4	3	2	1	0	12	11	10

iii. Let $S_4 = \{0,1,4,14,16\}$ be a perfect difference set for $n = 4$; the associated plane is shown in Table 6.

TABLE 6

	j																				
	0	1	2	3	4	5	6	7	8	9	10	11	12	13	14	15	16	17	18	19	20
	0	20	19	18	17	16	15	14	13	12	11	10	9	8	7	6	5	4	3	2	1
	1	0	20	19	18	17	16	15	14	13	12	11	10	9	8	7	6	5	4	3	2
i	4	3	2	1	0	20	19	18	17	16	15	14	13	12	11	10	9	8	7	6	5
	14	13	12	11	10	9	8	7	6	5	4	3	2	1	0	20	19	18	17	16	15
	16	15	14	13	12	11	10	9	8	7	6	5	4	3	2	1	0	20	19	18	17

It is easily seen that these cyclic planes of order 2, 3, and 4 are projective planes of order 2, 3, and 4, respectively. In fact, every cyclic plane is a projective plane, as indicated by the following theorem.

THEOREM 2.3.8 *A cyclic plane of order n is a projective plane of order n.*

Proof. The first step is to show that every line contains exactly $n + 1$ points. Let L_j be given. Let $S_n = \{a_1, \dots, a_{n+1}\}$ be a perfect difference set of order n and let i_k for $1 \le k \le n + 1$ be defined by $i_k = a_k - j$ mod $n^2 + n + 1$. Such numbers i_k exist and are unique for each k because $J/n^2 + n + 1$ is a group under addition. Then the $n + 1$ points p_{i_k} are on L_j because $i_k + j = a_k \in S_n$. A dual argument shows that every point has exactly $n + 1$ lines passing through it.

To show that at most one line L_k can pass through two distinct points p_i and p_j, suppose that both L_k and L_m pass through p_i and p_j. Then $i + k = a_r$, $j + k = a_s$, $i + m = a_t$, and $j + m = a_u$ for some $a_r, a_s, a_t, a_u \in S_n$. As a result $(i + k) - (j + k) = a_r - a_s$, and $(i + m) - (j + m) = a_t - a_u$; thus $i - j = a_r - a_s = a_t - a_u$. Since $i - j \ne 0$, $i - j$ is generated by the difference of a unique pair of elements of S_n. Therefore, $a_r = a_t$ and $a_s = a_u$, that is, $k = m$. Thus $L_k = L_m$, and so only one line passes through p_i and p_j. A dual argument shows that two lines may intersect at only one point.

The next step is to show that every two points p and q must lie on a line. There are $n + 1$ lines through p, each containing n points excluding p. Thus $n(n + 1) + 1$ or $n^2 + n + 1$ points are contained in this set of lines passing through p. But there are only $n^2 + n + 1$ points in the whole plane, so q must be on one of these lines. A dual argument shows that every two lines must intersect.

The final step is to show that there exist four points, no three of which are collinear. Let $p = p_{a_1}$ and $q = p_{a_2}$. Thus line pq is $\{p_i : i \in S_n\}$. Line pq has $n + 1$ points on it, and since $n^2 + n + 1 > n + 1$, we may choose a point r that is not on pq. The lines pq, pr, and qr contain $3n$ points collectively. Since $n \geq 2$, we know that $n^2 + n + 1 > 3n$, so a fourth point s may be chosen that is not on any of these three lines. Thus p,q,r,s is a four-point.

This concludes the proof that the cyclic plane is a projective plane. Clearly, this projective plane is of order n.

Notation. The cyclic plane generated by S_n is denoted by π_{S_n}.

For future reference, a short list of perfect difference sets is given in Table 7.

TABLE 7

n	Perfect Difference Set
2	$\{0,1,3\}$
3	$\{0,1,3,9\}$
4	$\{0,1,4,14,16\}$
5	$\{0,1,3,8,12,18\}$
7	$\{0,1,3,13,32,36,43,52\}$
8	$\{0,1,3,7,15,31,36,54,63\}$
9	$\{0,1,3,9,27,49,56,61,77,81\}$
11	$\{0,1,3,12,20,34,38,81,88,94,104,109\}$
13	$\{0,1,3,16,23,28,42,76,82,86,119,137,154,175\}$
16	$\{0,1,3,7,15,31,63,90,116,127,136,181,194,204,233,238,255\}$
17	$\{0,1,4,38,40,85,92,110,115,163,179,189,246,252,265,279,$ $287,290\}$
19	$\{0,1,4,22,47,61,81,111,116,147,149,160,189,201,252,284,$ $310,358,366,375\}$
23	$\{0,1,3,17,36,42,64,93,131,149,161,193,204,214,219,227,$ $264,273,313,400,448,452,472,479\}$
25	$\{0,1,47,135,186,195,200,202,219,311,331,361,392,430,448,$ $471,474,482,519,544,548,554,576,597,612,639\}$

A word of explanation is needed here regarding the omission from this table of difference sets of order $n = 6,10,12,14,15,18,20,21,22,24$. It has been shown that there are no perfect difference sets of these orders. In fact, for $n \leq 3600$, every perfect difference set of order n is of prime power order. (This discussion is expanded more fully in Section 2.6.) It should also be pointed out that not all finite projective planes can be represented by perfect difference sets. (This statement is not proven until Part 2, but examples in support of it are given in Section 2.4.)

The affine analogue of the cyclic plane and the perfect difference set is not obvious, so there will be a brief discussion of it here (see Bose, 1942; Hoffman, 1952).

DEFINITION 2.3.9 *Let* $R = \{a_1, a_2, \ldots, a_n\}$ *where* $n \geq 2$ *and let* $0 \leq a_i < n^2 - 1$ *for* $i = 1, \ldots, n$. *Then* R *is called an* affine difference set *of order* n *if and only if, for every* $m \neq 0$ mod $n + 1$, *there exists a unique pair* (a_i, a_j) *such that* $a_i, a_j \in R$ *and* $a_i - a_j = m$ mod $n^2 - 1$.

Notation. An affine difference set of order n is denoted by R_n.

EXAMPLE 2.3.10 The set $R_3 = \{0,1,3\}$ is an affine difference set of order 3 because

$$1 = 1 - 0 \text{ mod } 8 \qquad 5 = 0 - 3 \text{ mod } 8$$
$$2 = 3 - 1 \text{ mod } 8 \qquad 6 = 1 - 3 \text{ mod } 8$$
$$3 = 3 - 0 \text{ mod } 8 \qquad 7 = 0 - 1 \text{ mod } 8$$

Since $n + 1 = 4$, it is not necessary to generate 4 in the subtraction. A short list of affine difference sets is given in Table 8.

TABLE 8

n	Affine Difference Set
2	$\{0,1\}$
3	$\{0,1,3\}$
4	$\{0,1,3,7\}$
5	$\{0,1,4,9,11\}$
7	$\{0,1,4,6,13,23,34\}$
8	$\{0,1,3,7,15,20,31,41\}$

CONSTRUCTION 2.3.11   Let $R_n = \{a_1, \ldots, a_n\}$ be an affine difference set of order n. Define $\mathcal{P}, \mathcal{L}, \mathcal{I}$ as follows:

$\mathcal{P} = \{p_i: i = 0, \ldots, n^2 - 2\} \cup \{q\}$ where q is any
object distinct from p_i and the elements of $\mathcal{L}$ and
$\mathcal{I}$ as given here

$\mathcal{L} = \{L_i: i = 0, \ldots, n^2 - 2\} \cup \{M_i: i = 0, \ldots, n\}$

$\mathcal{I} = \{(p_i, L_j): i + j \in R_n\} \cup \{(p_i, M_j): j = i \bmod n + 1\}$
$\cup \{(q, M_j): j = 0, \ldots, n\}$

EXAMPLE 2.3.12   Let $R_4 = \{0,1,3,7\}$. $(\mathcal{P}, \mathcal{L}, \mathcal{I})$ is then described as in Table 9. For convenience of notation, let $q = p_x$. Clearly, there are sixteen points and twenty lines in this plane. If we check Af1 and Af2 for one case each, we find that L_7 is the unique line containing p_8 and p_{11}; L_3 is the unique line that contains p_4 and that is parallel to L_{13}. Points p_0, p_1, q are three noncollinear points satisfying Af3.

TABLE 9

	j for L_j															j for M_j				
	0	1	2	3	4	5	6	7	8	9	10	11	12	13	14	0	1	2	3	4
	0	14	13	12	11	10	9	8	7	6	5	4	3	2	1	0	1	2	3	4
i	1	0	14	13	12	11	10	9	8	7	6	5	4	3	2	5	6	7	8	9
	3	2	1	0	14	13	12	11	10	9	8	7	6	5	4	10	11	12	13	14
	7	6	5	4	3	2	1	0	14	13	12	11	10	9	8	x	x	x	x	x

The following theorem offers more formal proof that $(\mathcal{P}, \mathcal{L}, \mathcal{I})$ is an affine plane.

THEOREM 2.3.13   *The triple* $(\mathcal{P}, \mathcal{L}, \mathcal{I})$ *of Construction 2.3.11 is an affine plane.*

Proof.   Let $(\mathcal{P}, \mathcal{L}, \mathcal{I})$ be constructed from $R_n = \{a_1, \ldots, a_n\}$. The first step is to establish cardinality properties. Clearly, there are n^2 points in $\mathcal{P}$ and $n(n + 1)$ lines in $\mathcal{L}$. Each line contains n points by the following argument.

Case 1.   Line L_j contains exactly those points p_i such that

$i+j \in R_n$. There are n such points; they are of the form p_{i_k}, where $i_k = j - a_k$: $k = 1, \ldots, n$.

Case 2.   Line M_j contains q and exactly those points p_i such that $j - i = 0$ mod n: $i = 0, \ldots, n^2 - 2$. There are $n - 1$ such numbers i. Thus M_j contains n points.

A similar argument by cases can be used to demonstrate that each point has $n + 1$ lines passing through it.

Case 1.   This statement holds for q by definition. The lines are M_j: $j = 0, \ldots, n$.

Case 2.   There are exactly n lines of the form L_j passing through point p_i. This follows from a dual argument to that of Case 1 for lines. Also, $p_i \in M_j$ where $j = i$ mod $n + 1$. Thus there are $n + 1$ lines through p_i.

The next step is to prove Af1.

Case 1.   Suppose that the two points are p_i and q. The only lines that may pass through them are of the form M_j. If M_j and M_k pass through p_i, then $j = i$ mod $n + 1$ and $k = i$ mod $n + 1$. Thus $j = k$ and the lines are the same.

Case 2.   Let p_i and p_j be the two distinct points under consideration. There are three subcases.

Subcase A.   Points p_i and p_j lie on M_r and M_s. Thus $r = s$ by the argument of Case 1 for Af1.

Subcase B.   Points p_i and p_j lie on L_k and M_r; because they lie on L_k, we have $i + k \in R_n$ and $j + k \in R_n$. Thus $(i + k) - (j + k)$ is not divisible by $n + 1$ because differences of elements of R_n do not generate multiples of $n + 1$. But since points p_i and p_j lie on M_r, we have $r = i$ mod $n + 1$ and $r = j$ mod $n + 1$. This implies that $i - j$ is divisible by $n + 1$, a contradiction.

Subcase C.   Points p_i and p_j lie on L_k and L_m. Thus $i + k = a_r$, $j + k = a_s$, $i + m = a_t$, $j + m = a_u$. Therefore, $i - j = a_r - a_s$ and $i - j = a_t - a_u$. But differences are unique in R_n, so $a_r = a_t$ and $a_s = a_u$. Thus $k = m$ and $L_k = L_m$.

Now Af2 follows easily. Consider any point p and line L not incident with p. Since there are $n + 1$ lines passing through p and only n points on L, there must be exactly one line through p that is parallel to L.

Points p_0, p_1, and q are always noncollinear, so Af3 holds. Thus $(\mathscr{P}, \mathscr{L}, \mathscr{I})$ is an affine plane.

DEFINITION 2.3.14 *The plane of Construction 2.3.11 is called a* cyclic affine plane *generated by* R_n.

Notation. The cyclic affine plane generated by R_n is denoted by α_{R_n}.

The relationship between cyclic affine planes and cyclic projective planes is not completely determined. For all known examples, the principal extension of a cyclic affine plane is a cyclic projective plane, and any principal restriction of a cyclic projective plane is a cyclic affine plane.

EXERCISES

1. Prove Corollary 2.3.2.

2. Suppose that $(\mathscr{P}, \mathscr{L}, \mathscr{I})$ is a plane such that both $\mathscr{P}$ and $\mathscr{L}$ have $n^2 + n + 1$ members, each line has $n + 1$ points, and two lines meet in exactly one point. Show that $(\mathscr{P}, \mathscr{L}, \mathscr{I})$ is a projective plane.

3. Show that every cyclic plane is self-dual.

4. If n is an order of some perfect difference set, there are in general many perfect difference sets of this order. An equivalence relation between perfect difference sets of a fixed order n may be defined as follows: $S_n = \{a_0, \ldots, a_n\}$ is equivalent to $S_n' = \{b_0, \ldots, b_n\}$ if and only if $b_i = a_i + j \bmod m$ for some $j < m$ where $m = n^2 + n + 1$. For example, $\{0,1,3\}$ and $\{5,6,1\}$ are equivalent difference sets of order two where $j = 5$. It has been found that for all known examples $n = p^k$, there are $\phi(m)/3k$ nonequivalent perfect difference sets of order n where ϕ is the Euler phi function. The Euler phi function at m counts the number of integers n such that $1 \le n < m$ and n is relatively prime to m. For example, $\phi(4) = 2$ because 1 and 3 are relatively prime to 4 and 2 is not.
   a. Find as many nonequivalent perfect difference sets of order 2, 3, and 4 as you can.
   b. Prove that π_{S_n} is isomorphic to $\pi_{S_{n'}}$ if S_n is equivalent to S_n'.
   c. For nonequivalent perfect difference sets S_n and T_n in general, it is unknown whether $\pi_{S_n} \sim \pi_{T_n}$. However, no instance has been found

where $\pi_{S_n} \not\sim \pi_{T_n}$. For the sets that you found in 4(a), show that $\pi_{S_n} \sim \pi_{T_n}$.

5. Are π_{S_n} and $\pi_{GF(n)}$ isomorphic where $n = 2,3,4$?

6. Let $S_n = \{a_0, \ldots, a_n\}$ be a perfect difference set and let $T_m = \{b_0, \ldots, b_m\}$ be a subset of S_n. The set T_m is a *subdifference set* of S_n if and only if the set $\{(b_i - b_j)\bmod N: i, j = 0, \ldots, m\}$ equals the set $\{kp: k = 0, \ldots, N/p - 1\}$ for some number p where $N = n^2 + n + 1$ and p divides N.

The sets T_2 and T'_2 are subdifference sets of S_4 and S'_4, respectively, where $S_4 = \{0,1,4,14,16\}$, $T_2 = \{1,4,16\}$, $S'_4 = \{0,1,6,8,18\}$, and $T'_2 = \{0,6,18\}$.

   a. Find a subdifference set for the following perfect difference sets displayed in Table 7: S_9; S_{16}; S_{25}.

   b. Show that if T_m is a subdifference set of S_n, then $m^2 + m + 1$ must divide $n^2 + n + 1$.

   c. Place restrictions on k so that if $m^k = n$, then $m^2 + m + 1$ divides $n^2 + n + 1$.

   d. Show that if T_m is a subdifference set of S_n, then π_{T_m} is a subplane of π_{S_n}.

   e. Find a cyclic subplane for the following cyclic planes: π_{S_4}; π_{S_9}; $\pi_{S_{16}}$. (Find a subplane of order 2 and one of order 4.)

2.4 TWO PLANES OF ORDER NINE

Sections 2.1 and 2.3 included a discussion of different types of representations of projective and affine planes. The specific information given in these sections will facilitate representation of planes of order 2,3,4,5,7, and 9 in two different ways. As will be shown later, the corresponding planes π_F and π_{S_n} are isomorphic for a given order n, where $n = 2,3,4,5,7,9$. This section gives two more representations for projective planes of order 9. The planes defined here are not isomorphic to the plane π_{S_9}; nor, in fact, are they isomorphic to each other.

   The first plane will be represented by notation similar to that used for cyclic planes.

CONSTRUCTION 2.4.1   The plane $(\mathscr{P}, \mathscr{L}, \mathscr{I})$ is constructed as follows:

$$\mathscr{P} = \{a_i, b_i, c_i, d_i, e_i, f_i, g_i: i = 0, \ldots, 12\}$$
$$\mathscr{L} = \{L_i: i = 0, \ldots, 90\} \text{ where:}$$
$$L_{7i} = \{a_i, a_{i+3}, a_{i+4}, a_{i+11}, b_i, c_i, d_i, e_i, f_i, g_i\}$$

$$L_{7i+1} = \{a_i, b_{i+1}, b_{i+6}, b_{i+12}, e_{i+4}, e_{i+5}, f_{i+3}, f_{i+7}, g_{i+8}, g_{i+11}\}$$
$$L_{7i+2} = \{a_i, c_{i+1}, c_{i+6}, c_{i+12}, g_{i+4}, g_{i+5}, e_{i+3}, e_{i+7}, f_{i+8}, f_{i+11}\}$$
$$L_{7i+3} = \{a_i, d_{i+1}, d_{i+6}, d_{i+12}, f_{i+4}, f_{i+5}, g_{i+3}, g_{i+7}, e_{i+8}, e_{i+11}\}$$
$$L_{7i+4} = \{a_i, e_{i+1}, e_{i+6}, e_{i+12}, b_{i+4}, b_{i+5}, c_{i+3}, c_{i+7}, d_{i+8}, d_{i+11}\}$$
$$L_{7i+5} = \{a_i, f_{i+1}, f_{i+6}, f_{i+12}, d_{i+4}, d_{i+5}, b_{i+3}, b_{i+7}, c_{i+8}, c_{i+11}\}$$
$$L_{7i+6} = \{a_i, g_{i+1}, g_{i+6}, g_{i+12}, c_{i+4}, c_{i+5}, d_{i+3}, d_{i+7}, b_{i+8}, b_{i+11}\}$$
$$\mathscr{I} = \in$$

Clearly, there are ninety-one points and ninety-one lines in this plane. The task of verifying that this is a projective plane is extremely tedious, but a few observations will help to make this fact clear. Notice that two lines of the form L_{7i} intersect in exactly one point, a point of the form a_i, because the subscripts of a_i on L_0 (namely, 0,3,4,11) form a perfect difference set of order 3. Two lines of the form L_{7i+j}, for a fixed j, $j = 1, \ldots, 6$, intersect at exactly one point, a point of the form $b_i, \ldots, g_i$, because the various subscripts of $b_i, \ldots, g_i$ are members of exactly one of the sets $\{1,6,12\}$, $\{4,5\}$, $\{3,7\}$, $\{8,11\}$; the respective differences modulo 13 of elements taken from each set are unique and exhaust the set $\{1, \ldots, 12\}$. If the student is imaginative enough to devise a clever way of showing that two lines of the form L_{7i+j} and $L_{7i'+j'}$, for $i \neq i'$ and $j \neq j'$, intersect in only one point, he will be able to show that such a plane is projective since the plane is self-dual (as will be shown in Section 2.8).

Notation. This projective plane of order 9 is denoted by $\pi_H(9)$. It is sometimes referred to as the "Hughes plane" of order 9; hence the use of *H* in the notation.

This plane will be considered in greater detail in the next few sections; a fairly thorough discussion of it is included in Chapter 9. Nevertheless, the student can conduct his own analysis of $\pi_H(9)$ here. The foldout in this section is an incidence table for $\pi_H(9)$ that displays the geometric properties of this plane.

The second plane will be represented by triples of elements from an algebraic structure of order 9 called a *right nearfield*. A study of nearfields and other such algebraic structures will be delayed until Part 3; it will be sufficient to state here that a right nearfield $(N,+,\cdot)$ satisfies all the field axioms except left distributivity and commutativity of multiplication.

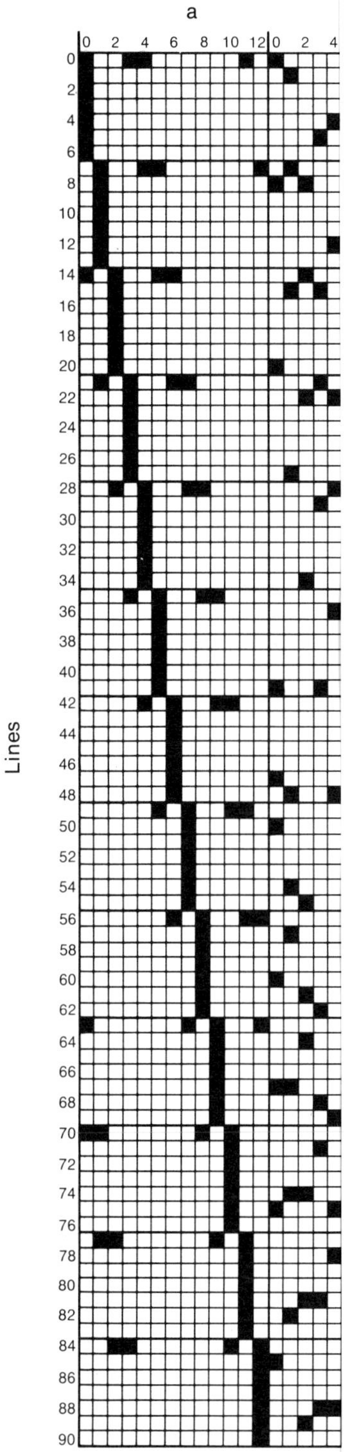
a
0 2 4 6 8 10 12 0 2 4
0 2 4 6 8 10 12 14 16 18 20 22 24 26 28 30 32 34 36 38 40 42 44 46 48 50 52 54 56 58 60 62 64 66 68 70 72 74 76 78 80 82 84 86 88 90
Lines

CONSTRUCTION 2.4.2   Let $N = \{0,1,2,a,b,c,d,e,f\}$. Addition in N is the same as it is in GF(9); thus $b = a + 1$, $c = a + 2$, $d = a + a$, $e = d + 1$, $f = d + 2$, and $1 + 2 = a + d = 0$. Multiplication is as follows: $x \cdot 0 = 0 \cdot x = 0$, and $x \cdot y$ is shown in Table 10.

TABLE 10

·	1	2	a	b	c	d	e	f
1	1	2	a	b	c	d	e	f
2	2	1	d	f	e	a	c	b
a	a	d	2	e	b	1	f	c
b	b	f	c	2	d	e	a	1
c	c	e	f	a	2	b	1	d
d	d	a	1	c	f	2	b	e
e	e	c	b	d	1	f	2	a
f	f	b	e	1	a	c	d	2

Notation.   This nearfield is denoted by $N(9)$.

The projective plane over N is constructed in the following manner.

CONSTRUCTION 2.4.3   The sets $\mathscr{P}$, $\mathscr{L}$, and $\mathscr{I}$ are defined as follows:

$$\mathscr{P} = \{[x,y,1]: x,y \in N\} \cup \{[1,x,0]: x \in N\} \cup \{[0,1,0]\}$$
$$\mathscr{L} = \{\langle m,1,k\rangle: m,k \in N\} \cup \{\langle 1,0,k\rangle: k \in N\} \cup \{\langle 0,0,1\rangle\}$$
$$\mathscr{I} = \{([x,y,z],\langle m,n,k\rangle): xm + yn + zk = 0\}$$

There are ninety-one points and ninety-one lines in this plane also. As with $\pi_{II}(9)$, a detailed proof that this plane is projective will be delayed until Part 3. The proof is merely begun here to get the ambitious student on the right track; it is continued in exercise 2.4.1.

Four cases may be used to show that every two points lie on at least one line.

Case 1.   Suppose that $p = [x,y,1]$, $p' = [x',y',1]$, and $x \neq x'$; then the line $L = \langle -(x-x')^{-1}(y-y'), 1, x(x-x')^{-1}(y-y') - y\rangle$ passes through p and p'. Clearly, $p \in L$. And $p' \in L$ because

$$-x'((x-x')^{-1}(y-y')) + y' + x((x-x')^{-1}(y-y')) - y$$
$$= -(x'-x)((x-x')^{-1}(y-y')) + y' - y$$

(using right distributivity)

$$= ((x - x')(x - x')^{-1})(y - y') + y' - y$$

(using associativity of multiplication
and the fact that $-(a - b) = b - a$)

$$= (y - y') - (y - y') = 0$$

Case 2. Suppose that $p = [x,y,1]$ and that $p' = [x,y',1]$. It is easily seen that $\langle 1,0,-x \rangle$ contains p and p'.

Case 3. Suppose that $p = [x,y,1]$ and that $q = [1,z,0]$. Then the line $L = \langle -z, 1, xz - y \rangle$ contains p and q.

Case 4. Suppose that $p = [1,x,0]$ and that $p' = [1,x',0]$. Then the line $\langle 0,0,1 \rangle$ contains p and p'.

Notation. This projective plane of order 9 is denoted by $\pi_N(9)$. The N in the notation comes from the representative nearfield N.

This plane is occasionally referred to as the Veblen-Wedderburn plane of order 9 since the structure $(N,+,\cdot)$ is a Veblen-Wedderburn system. It is curious to note that the plane $\pi_H(9)$ was first displayed by Veblen and Wedderburn; yet their names are associated with the plane $\pi_N(9)$ (see Veblen, Wedderburn, 1907). This plane will be considered in some detail in the following sections and will be analyzed thoroughly in Chapter 12.

It is natural to consider one affine restriction of $\pi_N(9)$, namely, the plane resulting from the deletion of $\langle 0,0,1 \rangle$. An isomorphic copy of this plane can be constructed using the familiar notation of α_F:

$$\mathcal{P} = \{(x,y)\colon x,y \in N\}$$
$$\mathcal{L} = \{\{(x,y)\colon y = xm + k \text{ or } x = k\}\colon m,k \in N\}$$
$$\mathcal{I} = \in$$

Notation. This particular affine restriction of $\pi_N(9)$ is denoted by $\alpha_N(9)$.

EXERCISES

1. Prove the following three statements to establish that $\pi_N(9)$ is a projective plane:

   a. $\pi_N(9)$ has ninety-one points and ninety-one lines.

   b. Every line contains ten points and every point lies on ten lines.

   c. $\pi_N(9)$ is a projective plane. (Use 1(a) and (b) and the result proved in the text that every two points have at least one line passing through them.)

2. Consider the plane $(\mathscr{P},\mathscr{L},\mathscr{I})$, constructed as follows:

$$\mathscr{P} = \{a_i,b_i,c_i,d_i,e_i,f_i,g_i: i = 0, \ldots, 12\}$$
$$\mathscr{L} = \{L_i: i = 0, \ldots, 90\}: \text{ where}$$
$$L_{7i} = \{a_i,a_{i+7},a_{i+9},a_{i+12},b_i,c_i,d_i,e_i,f_i,g_i\}$$
$$L_{7i+1} = \{a_i,b_{i+3},b_{i+5},b_{i+10},e_{i+2},e_{i+12},f_{i+8},f_{i+9},g_{i+7},g_{i+11}\}$$
$$L_{7i+2} = \{a_i,c_{i+3},c_{i+5},c_{i+10},g_{i+2},g_{i+12},e_{i+8},e_{i+9},f_{i+7},f_{i+11}\}$$
$$L_{7i+3} = \{a_i,d_{i+3},d_{i+5},d_{i+10},f_{i+2},f_{i+12},g_{i+8},g_{i+9},e_{i+7},e_{i+11}\}$$
$$L_{7i+4} = \{a_i,e_{i+3},e_{i+5},e_{i+10},b_{i+2},b_{i+12},c_{i+8},c_{i+9},d_{i+7},d_{i+11}\}$$
$$L_{7i+5} = \{a_i,f_{i+3},f_{i+5},f_{i+10},d_{i+2},d_{i+12},b_{i+8},b_{i+9},c_{i+7},c_{i+11}\}$$
$$L_{7i+6} = \{a_i,g_{i+3},g_{i+5},g_{i+10},c_{i+2},c_{i+12},d_{i+8},d_{i+9},b_{i+7},b_{i+11}\}$$
$$\mathscr{I} = \in$$

This plane is denoted by $\pi_{H'}(9)$.

   a. Provide some evidence that this plane is a projective plane in the same manner that the text does for the plane $\pi_H(9)$.
   b. Construct another plane having the same format as planes $\pi_H(9)$ and $\pi_{H'}(9)$.

3. Recalling the structure $(N,+,\cdot)$ of Construction 2.4.2, define a new operation $\circ$ on N as follows: $a \circ b = b \cdot a$.

   a. Verify that the structure $(N,+,\circ)$ is a left nearfield. (A left nearfield satisfies all the field axioms with the exception of right distributivity and commutativity of multiplication.)
   b. Denote by $\pi_{N'}(9)$ the plane $(\mathscr{P},\mathscr{L},\mathscr{I})$ constructed exactly as in Construction 2.4.3 except that the left nearfield of 3(a) is the underlying algebraic structure. Notice that this change does not alter the form of the points and lines but does affect the incidence relation. Show that $\pi_{N'}(9)$ is isomorphic to the dual plane of $\pi_N(9)$.

4. This exercise assumes familiarity with Table 3 (p. 44) and exercise 2.1.7(e). Consider the Cartesian groups $B = (R,+,\odot)$ and $C = (R,+,\odot)$, where R and $+$ are the same as in A (and therefore identical to $(N,+)$ and $(GF(9),+))$, $\odot$ is defined as in Table 11, and $\odot$ is defined as in Table 12. Notice that B and C are Cartesian groups satisfying the law of right distributivity.

TABLE 11

$\odot$	1	2	a	b	c	d	e	f
1	1	2	a	b	c	d	e	f
2	2	1	d	f	e	a	c	b
a	a	d	e	c	1	f	b	2
b	b	f	1	d	a	c	2	e
c	c	e	b	1	f	2	d	a
d	d	a	c	e	2	b	f	1
e	e	c	f	2	b	1	a	d
f	f	b	2	a	d	e	1	c

TABLE 12

$\odot$	1	2	a	b	c	d	e	f
1	1	2	a	b	c	d	e	f
2	2	1	d	f	e	a	c	b
a	a	e	c	1	d	b	f	2
b	b	d	f	c	2	1	a	e
c	c	f	2	d	b	e	1	a
d	d	c	e	2	a	f	b	1
e	e	b	1	a	f	c	2	d
f	f	a	b	e	1	2	d	c

    a. Verify that neither B nor C satisfies the following laws: commutativity of multiplication; associativity of multiplication; left distributivity.

    b. Show that B and C are not isomorphic. (Recall that isomorphism was defined in exercise 2.1.7.)

    c. Let $P = (R,+,\otimes)$ be a Cartesian group satisfying right distributivity (where R and $+$ are the same as in A) and suppose that there exists an element $z \in R$ such that for all $x, m \in R$, the equation $x \otimes (m \otimes z) = (x \circ m) \otimes z$ is true (where $\circ$ is the multiplication of A). Let π_P denote the plane constructed over P (as explained in exercise 2.1.6) and let f be a mapping from the points of π_A to the points of π_P defined as follows:

$$f\colon [x,y,1] \to [x,y \otimes z,1]$$
$$[1,x,0] \to [1,x \otimes z,0]$$
$$[0,1,0] \to [0,1,0]$$

Construct a mapping F from the lines of π_A to the lines of π_P and show that (f,F) is an isomorphism.

    d. Find an isomorphism from π_A to π where $\pi = \pi_B$; where $\pi = \pi_C$.

  *5. In the plane $\pi_H(9)$ show that two lines of the form L_{7i+j} and $L_{7i'+j'}$, where $i \neq i'$ and $j \neq j'$, have a unique intersection.

2.5   EXTENSION PLANES

This section introduces a method for generating new planes from given planes. Suppose that Σ is a primitive plane that is not projective. Thus Σ fails to satisfy at least one of the axioms Pj1, Pj2, Pj3; in other words, either two points have no line joining them, two lines fail to intersect, or Σ does not contain a four-point. To make Σ a projective plane, we would have to add either the missing line or the missing point or points. Such addition may introduce nonintersecting pairs of lines or noncollinear pairs of points, so further addition is necessary. This process is continued until a projective plane is constructed and can take place either within a larger plane, as Example 2.5.1 shows, or without a plane, as Construction 2.5.5 explains.

EXAMPLE 2.5.1   Let $\pi = \pi_{S_4}$, as shown in Table 6. Let $\Sigma = (\mathcal{P}, \mathcal{L}, \mathcal{I})$ where $\mathcal{P} = \{p_0, p_1, p_2, p_3\}$, $\mathcal{L} = \varnothing$, $\mathcal{I} = \varnothing$. If points p_i and p_j are connected with the line $\{p_i, p_j\}$ for $i,j = 0,1,2,3$, and $i \neq j$, there will be six new lines that may be labeled $L_{i,j}$ where $L_{i,j} = \{p_i, p_j\}$. Three pairs of these lines do not intersect: $L_{0,1}, L_{2,3}$; $L_{0,2}, L_{1,3}$; $L_{0,3}, L_{1,2}$. If we label these intersections $L_{0,1} \cap L_{2,3} = q_1$, $L_{0,2} \cap L_{1,3} = q_2$, and $L_{0,3} \cap L_{1,2} = q_3$, there will be three new points and thus three new pairs of non-collinear points. These points may be connected with three new lines $\{q_i, q_j\}$, $i,j = 1,2,3$, and $i \neq j$, and the lines labeled $M_{i,j}$ where $M_{i,j} = \{q_i, q_j\}$. Since we are working within the plane π_{S_4}, $q_1 = p_{16}$, $q_2 = p_8$, and $q_3 = p_{15}$ (see Table 6). Furthermore, these three points are collinear in π_{S_4}, so the lines $M_{i,j}$ should be one and the same line, namely, $\{q_1, q_2, q_3\}$ or $\{p_{16}, p_8, p_{15}\}$. The result of the addition of points and lines at this stage is seven points and seven lines in the configuration $\Sigma(7_3)$. Thus there are no more lines to intersect and no more points to join; the extension process is complete.

It may be said that $\Sigma(7_3)$ is generated by Σ in π_{S_4} or that $\Sigma(7_3)$ is the projective extension of Σ in π_{S_4}. A formal definition of these terms will be given in an equivalent but different manner. First a lemma is required.

LEMMA 2.5.2 *Let $\pi = (\mathscr{P}, \mathscr{L}, \mathscr{I})$ be a projective plane, let Σ be a subplane containing a four-point, and let $\gamma = \{\mathscr{P}': \mathscr{P}'$ is the point set of a projective plane $\pi' \subseteq \pi$ and $\Sigma \subseteq \mathscr{P}'\}$. Then $\pi_o = \pi | \cap \{\mathscr{P}': \mathscr{P}' \in \gamma\}$ is a projective plane.*

Proof. Let $\pi_o = (\mathscr{P}_o, \mathscr{L}_o, \mathscr{I}_o)$; thus $\mathscr{P}_o = \cap \gamma$. ($\cap \gamma$ is shorthand for $\cap \{\mathscr{P}': \mathscr{P}' \in \gamma\}$. We must show that π_o is a projective plane. Let $p, q \in \mathscr{P}_o$; then the unique line $pq \cap \mathscr{P}_o$ joins p and q in π_o, and thus Pj1 is satisfied. Let L_o and M_o be two lines of $\mathscr{L}_o$. Thus $L_o = L \cap \mathscr{P}_o$ and $M_o = M \cap \mathscr{P}_o$ for some lines $L, M \in \mathscr{L}$. Let $p = L \cap M$. Now $p \in \mathscr{P}'$ for any $\mathscr{P}' \in \gamma$ because the lines $L \cap \mathscr{P}'$ and $M \cap \mathscr{P}'$ of π' must intersect in π'. Thus $p \in \mathscr{P}'$ for all $\mathscr{P}' \in \gamma$. It follows that $p \in \cap \gamma$ and so L_o and M_o have a unique intersection in π_o. Thus Pj2 is satisfied. Furthermore, Pj3 is satisfied because the points of Σ are contained in $\mathscr{P}'$ for all $\mathscr{P}'$, and Σ itself contains a four-point.

DEFINITION 2.5.3 *The plane π_o defined in Lemma 2.5.2 is called the least subplane containing Σ.*

DEFINITION 2.5.4 *Let π be a projective plane and let Σ be a sub-plane of π containing a four-point. Then the projective plane π_o is* *the* projective extension *of* Σ *if and only if* π_o *is the least subplane of* π *that contains* Σ. *We also say that* Σ *generates* π_o *in* π.

The extension process of Example 2.5.1 came to an end after three stages of adjunction. In the first stage we adjoined doubleton sets of points called lines. In the second stage we adjoined objects called points, which were determined by pairs of lines. In the third stage we again adjoined doubleton sets of points; however, since the process was taking place inside the plane π_{S_4}, we were subject to the restriction that diagonal points of a quadrangle are collinear. Without this restriction, we could have continued the process with seven points and nine lines, three of which were new. Following is a formal definition of an extension process that is free from the restrictions imposed by superplanes.

CONSTRUCTION 2.5.5   Let $\Sigma_o = (\mathscr{P}_o, \mathscr{L}_o, \in)$ be the given primitive plane. Define Σ_1 as $(\mathscr{P}_1, \mathscr{L}_1, \in)$ where:

$$\mathscr{P}_1 = \mathscr{P}_o$$
$$\mathscr{L}_1 = \mathscr{L}_o \cup \{\{p,q\}: \text{no line in } \mathscr{L}_o \text{ joins } p \text{ and } q\}$$

Thus $\Sigma_o \subseteq \Sigma_1$.

Define Σ_2 as $(\mathscr{P}_2, \mathscr{L}_2)$ where

$$\mathscr{P}_2 = \mathscr{P}_1 \cup \{\{L,M\}: L,M \in \mathscr{L}_1 \text{ and } L \parallel M\}$$
$$\mathscr{L}_2 = \{N \cup \{\{L,M\}\}: N = L \text{ or } M \text{ and } L,M \in \mathscr{L} \text{ and } L \parallel M\}$$
$$\cup \{L \in \mathscr{L}_1: \text{no line is parallel to } L\}$$

Thus $\Sigma_1 \subseteq \Sigma_2$.
   In general, for $2k + 1$, Σ_{2k+1} is formed from Σ_{2k} as follows:

$$\mathscr{P}_{2k+1} = \mathscr{P}_{2k}$$
$$\mathscr{L}_{2k+1} = \mathscr{L}_{2k} \cup \{\{p,q\}: \text{no line joins } p \text{ and } q\}$$

Thus, for Σ_{2k+1}, every two points are joined by a unique line.
   Σ_{2k+2} is formed from Σ_{2k+1} as follows:

$$\mathscr{P}_{2k+2} = \mathscr{P}_{2k+1} \cup \{\{L,M\}: L,M \in \mathscr{L}_{2k+1} \text{ and } L \parallel M\}$$
$$\mathscr{L}_{2k+2} = \{N \cup \{\{L,M\}\}: N = L \text{ or } M; L,M \in \mathscr{L}_{2k+1}, L \parallel M\}$$
$$\cup \{L: L \in \mathscr{L}_{2k+1} \text{ and } L \nparallel M \text{ for any } M \in \mathscr{L}_{2k+1}\}$$

Thus for Σ_{2k+2}, any two lines intersect in a unique point.

Now let $\Sigma_0^+ = (\mathcal{P}^+, \mathcal{L}^+, \in)$ where $\mathcal{P} = \overset{\infty}{\underset{i=0}{\cup}} \mathcal{P}_i$ and $\mathcal{L} = \{L:$ for some $k_0, L \cap \mathcal{P}_k \in \mathcal{L}_k$ for all $k > k_0\}$.

It is easily checked that Σ_0^+ is a primitive plane such that (1) every two points are joined by a unique line; (2) every two lines intersect in a unique point; and (3) $\Sigma_i \subseteq \Sigma_0$ for any $i \geq 0$.

DEFINITION 2.5.6 *The plane Σ_0^+ constructed here is called the* free extension *of Σ_0.*

Note that Σ_0^+ is not an extension of Σ_0 in the sense of Definition 1.7.3. Note also that this "extension" procedure is unique; that is, if $\Sigma_0^+ \sim \Sigma_1^+$, then $\Sigma_0 \sim \Sigma_1$.

THEOREM 2.5.7 *If Σ is a primitive plane containing a four-point, then the free extension Σ^+ of Σ is a projective plane.*

Proof. As previously mentioned, Pj1 and Pj2 are satisfied in Σ^+. The added condition requiring the existence of a four-point yields Pj3.

EXAMPLE 2.5.8

  i. Let $\Sigma_0 = (\mathcal{P}, \mathcal{L}, \in)$ where $\mathcal{P} = \{p, q, r\}$ and $\mathcal{L} = \varnothing$. Then $\Sigma_0^+ = \Sigma_i = \Sigma(3_2)$.

  ii. Let $\Sigma_0 = (\mathcal{P}, \mathcal{L}, \in)$ where $\mathcal{P} = \{p, q, r, s\}$ and $\mathcal{L} = \varnothing$. Then Σ_0^+ is a projective plane and has denumerably many points and lines.

  iii. Let $\Sigma_0 = (\mathcal{P}, \mathcal{L}, \in)$ where $(\mathcal{P} - \{p\}, \mathcal{L}) = \Sigma(7_3)$, that is, Σ_0 is the plane $\Sigma(7_3)$ with an additional point. Here Σ_0^+ is a projective plane and also has denumerably many points and lines.

The study of extension planes will be continued in the remaining sections of Chapter 2.

EXERCISES

1. a. Show that every four-point Σ in the projective plane π of order 2 generates π.
   b. Is this true of the projective plane of order 3?

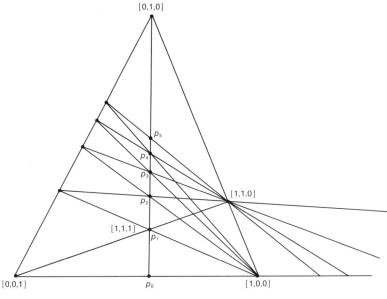

FIGURE 17

2. In $\pi_{J/5}$ find the projective extension of $\Sigma = \{[1,0,0], [0,1,0], [0,0,1], [1,1,1]\}$ with the help of Figure 17. Observe that $p_1 = [1,1,1]$ and show that $p_0 = [1,0,1]$, $p_2 = [1,2,1]$, $p_3 = [1,3,1]$, $p_4 = [1,4,1]$, and $p_5 = [1,5,1] = [1,0,1]$.

3. Let π_F be a plane of order p^k where $F = \mathrm{GF}(p^k)$. Following the idea of exercise 2, show that the four-point $\Sigma = \{[0,0,1],[0,1,0],[1,0,0],[1,1,1]\}$ generates a plane $\pi_{F'}$ where $F' = J/p$.

4. In $\pi_N(9)$ find the plane generated by
   a. $\Sigma = \{[0,0,1],[0,1,0],[1,0,0],[1,1,1]\}$.
   b. $\Sigma = \{[0,0,1],[2,0,1],[b,1,1],[1,e,1]\}$.

5. Let Σ_o be a finite primitive plane. State necessary and sufficient conditions for $\Sigma_o{}^+$ to contain infinitely many points and lines.

2.6 CARDINALITY

The last three sections analyze some elementary properties of projective and affine planes. The property of cardinality is not strictly a geometric property but it is certainly basic, especially to the present study of finite planes. As shown in Theorem 2.3.1, for finite projective planes π, every line contains n points and every point has n lines passing through it for some $n \geq 3$. The cardinality properties of a finite projective plane may be summarized in the following theorem.

THEOREM 2.6.1 π *is a finite projective plane if and only if* $\pi = \Sigma(m_n)$ *where* $n \geq 3$ *and* $m = n^2 - n + 1$.

Proof.   The proof is left as an exercise.

THEOREM 2.6.2 α *is a finite affine plane if and only if* $\alpha = \Sigma(m_c, n_d)$ *where* $c \geq 3$, $d = c - 1$, $m = (c - 1)^2$, *and* $n = c(c - 1)$.

Proof.   The proof is left as an exercise.

If π has infinitely many points on a line, a theorem analogous to Theorem 2.3.1 may be stated and proved.

THEOREM 2.6.3   *Let* π *be an arbitrary projective plane. Then any two lines considered as point sets are in one-to-one correspondence. There is also a one-to-one correspondence between the set of points on any line and the set of lines through any point.*

Proof.   Let L and M be distinct lines and let p be a point not on L or M. Furthermore, let $S = \{pq \cap M: q \in L\}$. It is clear that $S = M$ and that S is in one-to-one correspondence with L. Thus every two lines are in one-to-one correspondence. Let p and L be an arbitrary point and line, respectively, such that $p \notin L$. Also let $R = \{N \in \mathscr{L}: p \in N\}$ and $T = \{N \cap L: N \in R\}$. Clearly, $T = L$ and R is in one-to-one correspondence with T. On the basis of this result and the preceding result in this proof, it follows that the set of points on any line and the set of lines through any point are in one-to-one correspondence.

COROLLARY 2.6.4   *Any two lines of an affine plane are in one-to-one correspondence.*

These theorems may be used to define the concept of order for infinite planes.

DEFINITION 2.6.5 *The* order *of an infinite projective plane or infinite affine plane is the cardinality of the set of points on any line of the plane.*

Note that for infinite planes, the cardinality of line L equals the cardinality of $L - \{p\}$ for some $p \in L$; in fact, the cardinality of L equals the cardinality of $\mathscr{P}$. Obviously a theorem for infinite planes analogous to Theorem 2.6.1 cannot be given because configurations are finite by definition.

There are two natural questions to ask concerning the cardinality of planes. First, given a cardinal number, is there a projective plane of this order? Second, if there is a projective plane of a given order, is it unique?

For infinite cardinals, the answer to the first question is affirmative. Since there exist fields F of any infinite cardinality, there must exist planes π_F and α_F of any infinite order inasmuch as the cardinality of F and the order of π_F and α_F are the same.

For finite cardinals the answer is not known. Section 2.1 has shown that finite planes exist at least for those orders n for which finite fields of order n exist. Section 2.2 has shown that finite planes exist precisely for those orders n for which complete systems of mutually orthogonal sets of Latin squares of order n exist. Section 2.3 has shown that finite planes of order n exist at least for those n for which perfect difference sets or affine difference sets of order n exist.

Finite fields are known to exist for all prime power orders, that is, for all orders n such that $n = p^k$ for some prime p and positive integer k. Thus, there exist finite planes of all prime power orders. It also follows that there are complete systems of mutually orthogonal Latin squares of order n for $n = p^k$. Tarry (1901) showed that there is no complete system of mutually orthogonal Latin squares of order 6. Tarry enumerated all possible Latin squares of order 6 and found that not only are there not five mutually orthogonal Latin squares of order 6, there are not even two such orthogonal Latin squares. Actually, Euler had claimed to know this fact in 1772 and furthermore had hypothesized that for all orders $n = 4k + 2$, there are no orthogonal Latin squares of order n. This hypothesis remained a conjecture for nearly two centuries until Bose, Shrikhande, and Parker (1960) showed it to be false for $n > 6$. Unfortunately, they failed to solve the geometric problem of existence of planes, since the existence of two orthogonal squares does not imply the existence of $n - 1$ orthogonal squares. It is interesting to note that a lower bound has been found on the number of orthogonal Latin squares of order n. If $n = p_1^{r_1} \cdot p_2^{r_2} \cdot \ldots \cdot p_k^{r_k}$, then a lower bound $b = \min(p_i^{r_i} - 1)$, $i = 1,2, \ldots, k$. A general discussion in this regard is included in Hall (1967). It is also known that perfect difference sets exist for orders $n = p^k$. Moreover, it has been shown (Evans and Mann, 1951; V. H. Keiser, unpublished) that for $n \leq 3600$, there are no other perfect difference sets except those of prime power order.

From the preceding remarks it would appear that there are no finite planes of order n other than for $n = p^k$. Indeed, as of this writing none have been found. The current state of affairs concerning nonexistence may be summarized in the following theorem.

THEOREM 2.6.6   (*Bruck and Ryser, 1949*). *Let n be an integer of the form $4k + 1$ or $4k + 2$ where $k \geq 1$. Then, if $n \neq a^2 + b^2$ for some non-negative integers a and b, there is no finite projective plane of order n.*

Proof.   The proof may be found in Hall, 1959.

There are infinitely many numbers n of the form given in Theorem 2.6.6; the first few are 6,14,21,22,30. There are also infinitely many numbers n for which it is not known whether a projective plane (or affine plane) of order n exists. The number 10 is the first of these; 12,15,18,20,24,26,28 are the next few. Computers have been working long hours in an attempt to settle this question for $n = 10$, but without more theoretical help, the number of steps involved remains far too great. For example, it has been estimated that there are $2^{10,000}$ possible candidates for an incidence matrix for the plane $\Sigma(111_{11})$.

The next question to be considered is that of uniqueness. Uniqueness cannot be expected for planes of infinite order. For example, π_Q and Σ_0^+, where Q is the rational field and Σ_0 is a four-point, are non-isomorphic planes with a denumerable order; π_R and π_C are non-isomorphic planes of the first uncountable order. The uniqueness of finite planes has not been completely determined. For orders $n = 2,3,4,5,7,8$, it is known that the planes of order n are unique. (For $n = 7$, see Hall (1953, 1954b); for $n = 8$, see Hall, Swift, and Killgrove (1959)). Accordingly, the following notation is used here.

Notation. The planes of order 2,3,4,5,7,8 are denoted by $\pi(2)$, $\pi(3)$, $\pi(4)$, $\pi(5)$, $\pi(7)$, and $\pi(8)$, respectively.

There are at least four planes of order 9: π_{S_9}, $\pi_H(9)$, $\pi_N(9)$, and $(\pi_N(9))^d$. The plane π_F, where $F = \mathrm{GF}(9)$, is isomorphic to π_{S_9}, and the planes π_F and $\pi_H(9)$ are self-dual. These planes will be discussed more fully in the following sections. In general, for $n > 8$, projective planes of order p^k, $k > 1$, are not unique. The situation with regard to projective planes of order p is not known; as of this writing the only known plane of order p is π_F where $F = J/p$. There are also cyclic planes of all prime power orders, but it is unknown whether any of the cyclic planes differ from the planes π_F of the same order. In fact, it has not even been proved that all cyclic planes of the same order are isomorphic. It is known that π_{S_n} is unique for orders n, where $n = i$ or i^2, $i = 2,3,4,5,7,8,9$ (Bruck, 1960).

There are many unanswered questions concerning the existence and uniqueness of planes of finite order. Naturally, it is hoped that answers to all the questions will eventually be found; in the meantime, the following conjectures merit the most credence:

1. All cyclic planes of the same order are isomorphic.

2. Every cyclic plane is of prime power order.

3. Every cyclic plane is isomorphic to a plane π_F. (The converse has been proved (Singer, 1938) and will be discussed in Chapter 7.)

4. Every plane of order p is isomorphic to π_F where $F = J/p$.

5. Every finite plane is of order p^k for some prime p and natural number k.

The fifth conjecture is the most important and has received by far the most attention. Very likely it will be the most difficult to solve.

EXERCISES

1. Prove Theorem 2.6.1.

2. Prove Theorem 2.6.2.

3. Define $I(m_c)$ to be an $m \times m$ matrix with entries of only 0 and 1 such that each row and column contains c entries of 1 and $m - c$ entries of 0.

Clearly, a plane $\Sigma = (\mathcal{P}, \mathcal{L}, \mathcal{I})$ may be associated with $I(m_c)$ where $I(m_c)$ represents the incidence table for Σ. If $m = c^2 - c + 1$, is Σ necessarily a projective plane?

4. If $I(m_c)$ is the incidence table for a projective plane π, show that $I(m_c)(I(m_c))^T = J$ where J is the $m \times m$ matrix with each entry of the principal diagonal equal to c and every other entry equal to 1. (Here $(I(m_c))^T$ denotes the transpose of the matrix $I(m_c)$.) Is the converse true?

2.7   SUBPLANES

This section will consider projective subplanes of projective planes. First to be considered are planes of the form π_F.

EXAMPLE 2.7.1   Consider the plane π_F where $F = \mathrm{GF}(4)$. Let the plane $\pi' = \pi_F | \mathcal{P}'$ where $\mathcal{P}' = \{[x,y,z]: x,y,z \in \{0,1\}\}$. A simple calculation shows that π' has seven lines:

$$L_1 = \{[1,0,0], [1,1,0], [0,1,0]\}$$
$$L_2 = \{[1,0,0], [1,0,1], [0,0,1]\}$$
$$L_3 = \{[0,0,1], [0,1,1], [0,1,0]\}$$
$$L_4 = \{[0,1,0], [1,0,1], [1,1,1]\}$$
$$L_5 = \{[1,1,0], [1,0,1], [0,1,1]\}$$
$$L_6 = \{[1,0,0], [0,1,1], [1,1,1]\}$$
$$L_7 = \{[1,1,0], [0,0,1], [1,1,1]\}$$

Clearly, this is the plane $\pi(2)$; it is thus a subplane of π.

Let $F = \mathrm{GF}(4)$ and $F' = J/2$; observe that π', the subplane of π_F in Example 2.7.1, is isomorphic to $\pi_{F'}$. This fact suggests the following theorem.

THEOREM 2.7.2   *If $\pi' = \pi_F | \mathcal{P}$, where $\mathcal{P} = \{[x,y,z]: x,y,z \in F'\}$ and where F' is a subfield of F, then π' is a subplane of π_F isomorphic to $\pi_{F'}$.*

Proof.   Denote points of π' by $[x,y,z]$ and points of $\pi_{F'}$ by $[[x,y,z]]$ where $x,y,z \in F'$. Let $f: [[x,y,z]] \rightarrow [x,y,z]$. The points $p_i = [[x_i,y_i,z_i]]$, $i = 1,2,3$, are collinear iff there exists $a,b,c \in F'$ such that $ax_i + by_i$

$+ cz_i = 0$ for all i. But since $F' \subseteq F$, it follows that $[x_i, y_i, z_i]$ are collinear in π_F. We may conclude that the points $f(p_i)$ are collinear in π' and thus that f is an isomorphism.

COROLLARY 2.7.3   If F' is a subfield of F, then $\pi_{F'}$ is isomorphic to a subplane of π_F.

It will be shown in Chapter 8 that if π' is a subplane of π_F, there exists a subfield F' of F such that $\pi_{F'} \sim \pi'$. This fact, together with Theorem 2.7.2, implies that the class of subfields of F completely determines the class of subplanes of π_F. Thus, for example, π_Q has no proper subplanes, π_R has π_Q as one of its subplanes, and π_C contains π_R and π_Q. For finite fields of order p^n, there exist subfields of order p^k where k divides n; thus the subplanes of π_F where $F = \mathrm{GF}(p^n)$ are isomorphic to the planes $\pi_{F'}$ where $F' = \mathrm{GF}(p^k)$ and k divides n. (Note that the field J/p may also be denoted by $\mathrm{GF}(p)$.)

The next topic to be discussed is subplanes of cyclic planes.

EXAMPLE 2.7.4

  i. Let π be the cyclic plane of order 4 as displayed in Table 6 (p. 53). A cyclic subplane of π may be described as follows. $\{1, 4, 16\} \subset S_4$; therefore, P_0, P_3, and P_{15} are collinear (they are on L_1). But $\{0, 1, 5\}$ is a perfect difference set of order 2, so the set $\{0, 3, 15\}$, together with the operation of subtraction modulo 21, generates the set $\{0, 3, 6, 9, 12, 15, 18\}$. Let $\mathscr{P} = \{p_{3i} : i = 0, \ldots, 6\}$ and $L'_j = L_j \cap \mathscr{P}$; then $\pi | \mathscr{P}$ may be written as follows:

$$
\begin{array}{ll}
L'_1 = \{p_0, p_3, p_{15}\} & L'_{10} = \{p_{12}, p_{15}, p_6\} \\
L'_4 = \{p_{18}, p_0, p_{12}\} & L'_{13} = \{p_9, p_{12}, p_3\} \\
L'_7 = \{p_{15}, p_{18}, p_9\} & L'_{16} = \{p_6, p_9, p_0\} \\
& L'_{19} = \{p_3, p_6, p_{18}\}
\end{array}
$$

Clearly, $\pi | \mathscr{P}$ is the plane of order 2.

  ii. Let $\pi = \pi_{S_9}$. Proceeding in the same way as in step (i), observe that $\{0, 49, 56, 77\} \subset S_9$ and that $p_0, p_{49}, p_{56}, p_{77}$ all lie on L_0. Since $\{0, 7, 8, 11\}$ is a perfect difference set of order 3, the set $\{0, 49, 56, 77\}$, together with the operation of subtraction modulo 91, generates the set $\{7i : i = 0, 1, \ldots, 12\}$. Let $\mathscr{P} = \{p_{7i} : i = 0, \ldots, 12\}$; then $\pi | \mathscr{P}$ may be written as follows:

$$L'_0 = \{p_0, p_{49}, p_{56}, p_{77}\} \qquad L'_{49} = \{p_{42}, p_0, p_7, p_{28}\}$$
$$L'_7 = \{p_{84}, p_{42}, p_{49}, p_{70}\} \qquad L'_{56} = \{p_{35}, p_{84}, p_0, p_{21}\}$$
$$L'_{14} = \{p_{77}, p_{35}, p_{42}, p_{63}\} \qquad L'_{63} = \{p_{28}, p_{77}, p_{84}, p_{14}\}$$
$$L'_{21} = \{p_{70}, p_{28}, p_{35}, p_{56}\} \qquad L'_{70} = \{p_{21}, p_{70}, p_{77}, p_7\}$$
$$L'_{28} = \{p_{63}, p_{21}, p_{28}, p_{49}\} \qquad L'_{77} = \{p_{14}, p_{63}, p_{70}, p_0\}$$
$$L'_{35} = \{p_{56}, p_{14}, p_{21}, p_{42}\} \qquad L'_{84} = \{p_7, p_{56}, p_{63}, p_{84}\}$$
$$L'_{42} = \{p_{49}, p_7, p_{14}, p_{35}\}$$

Clearly, $\pi | \mathcal{P}$ is the plane of order 3.

In these particular examples, construction of the subplane π_{S_m} is based on the fact that $m^2 + m + 1$ divides $n^2 + n + 1$. For example, if $m = 2$ and $n = 4$, then $m^2 + m + 1 = 7$ and $n^2 + n + 1 = 21$; if $m = 3$ and $n = 9$, then $m^2 + m + 1 = 13$ and $n^2 + n + 1 = 91$. This suggests two conjectures. First, if $m^2 + m + 1$ divides $n^2 + n + 1$ and $m \geq 2$, then π_{S_n} contains a subplane of the form π_{S_m}; second, if π_{S_m} is a subplane of π_{S_n}, then $m^2 + m + 1$ divides $n^2 + n + 1$. Both conjectures are easily disproved. To disprove the first, let $m = 2$ and $n = 9$. Now 7 divides 91, but π_{S_2} is not a subplane of π_{S_9}. To disprove the second, let $m = 2$ and $n = 8$. Then π_{S_2} is a subplane of π_{S_8}, but 7 does not divide 73.

Since all known cyclic planes are of the type π_F, we may conjecture that for cyclic planes of order $n = p^k$, all subplanes are of orders p^m, where m divides k.

Consider the planes $\pi_H(9)$ and $\pi_N(9)$. For $\pi_H(9)$ we may easily choose a subplane π_0 of order 3 by letting $\mathcal{P}_0 = \{a_i : i = 0, \ldots, 12\}$ and $\mathcal{L}_0 = \{L_{7i} | \mathcal{P}_0 : i = 0, \ldots, 12\}$. What is significant, however, is that we may also choose a subplane π' of order 2.

EXAMPLE 2.7.5   Let $\mathcal{P}' = \{a_0, c_4, d_1, f_1, b_3, g_1, g_7\}$. Letting $L'_j = L_j | \mathcal{P}'$, we have:

$$L'_5 = \{a_0, b_3, f_1\} \qquad L'_{23} = \{c_4, f_1, g_7\}$$
$$L'_6 = \{a_0, c_4, g_1\} \qquad L'_{68} = \{b_3, c_4, d_1\}$$
$$L'_7 = \{d_1, f_1, g_1\} \qquad L'_{62} = \{b_3, g_1, g_7\}$$
$$L'_3 = \{a_0, d_1, g_7\}$$

This shows that there are finite planes for which the order of the subplane need not divide the order of the plane. The same phenomenon occurs in $\pi_N(9)$. The plane π_0, where $\mathcal{P}_0 = \{[x, y, 1]: x, y = 0, 1, 2\} \cup$

$\{[1,x,0]: x = 0,1,2\} \cup \{[0,1,0]\}$ and $\mathscr{L}_o = \{L|\mathscr{P}_o: L \in \mathscr{L}\}$, is easily seen to be the plane of order 3.

EXAMPLE 2.7.6   The following subplane of $\pi_N(9)$ is of order 2:

$$\langle 0,1,0 \rangle = \{[0,0,1], [a,0,1], [2,0,1]\}$$
$$\langle b,1,0 \rangle = \{[0,0,1], [b,1,1], [e,a,1]\}$$
$$\langle c,1,0 \rangle = \{[0,0,1], [1,e,0], [1,e,1]\}$$
$$\langle 2,1,a \rangle = \{[a,0,1], [b,1,1], [1,e,1]\}$$
$$\langle c,1,f \rangle = \{[a,0,1], [1,e,0], [e,a,1]\}$$
$$\langle c,1,c \rangle = \{[b,1,1], [1,e,0], [2,0,1]\}$$
$$\langle e,1,e \rangle = \{[2,0,1], [1,e,1], [e,a,1]\}$$

Obviously, we cannot hope for a theorem for finite planes that states that the order of a subplane must divide the order of a plane; nevertheless, we can establish a significant cardinality theorem for subplanes of finite planes.

THEOREM 2.7.7   *Let π be a plane of order n. If π' is a subplane of order k, then either $k^2 = n$ or $k^2 + k \leq n$.*

Proof.   Let $\mathscr{P}$ and $\mathscr{L}$ represent the points and lines, respectively, of π, let $\mathscr{P}'$ and $\mathscr{L}'$ represent the points and lines, respectively, of π', let $L' = L \cap \mathscr{P}'$ where $L \in \mathscr{L}$, and let $p \in \mathscr{P}$ such that $p \in L - L'$. Thus we have two cases: (1) every line of π passing through p also passes through a point of π'; (2) there exists a line N of π that passes through p such that $N \cap \mathscr{P}' = \varnothing$.

> *Case 1.*   Let $\gamma = \{pq: q \in \mathscr{P}', q \notin L'\}$. Since there are $k^2 + k + 1$ points in π' and $k + 1$ points on L', there exist k^2 points such that $q \in \mathscr{P}'$ and $q \notin L'$. We can prove that γ contains k^2 lines by the following argument. Suppose that $pq = pr$; thus p, q, and r are collinear. Let M be the line containing p, q, and r and let $M' = M \cap \mathscr{P}'$. Then $M' \cap L' = \varnothing$ because $M \cap L = p$ and $p \notin \mathscr{P}'$. Since π' is a projective plane, $M' \cap L' \in \mathscr{P}'$; hence a contradiction. Thus the members of γ are distinct and γ contains k^2 lines. Through p there are at least $k^2 + 1$ lines, namely, L and the members of γ that intersect $\mathscr{P}'$. It is easily seen that $k^2 + 1$ is also the maximum number of lines that can pass through p as well as through points of $\mathscr{P}'$. There are $n + 1$ lines of π that pass through p,

and since every line passing through p also passes through points of $\mathscr{P}'$, we have $k^2 + 1 = n + 1$. We conclude that $k^2 = n$.

Case 2. Let N be a line of π' such that $p \in N$ and $N \cap \mathscr{P}' = \varnothing$ and let $\beta = \{L \cap N: L \cap \mathscr{P}' = L' \in \mathscr{L}'\}$. Recall that there are $k^2 + k + 1$ lines in π'. The following argument shows that β consists of $k^2 + k + 1$ distinct points. Suppose that $L \cap N = M \cap N$. Since L' and M' are lines in π', they must intersect at a point $q \in \mathscr{P}'$. But $q \in N$, which contradicts the fact that $N \cap \mathscr{P}' = \varnothing$. Thus N contains $k^2 + k + 1$ points from the set β. Since N contains $n + 1$ points in all, we may conclude that $k^2 + k \leq n$.

Theoretically, there are many possible orders for subplanes of a given plane, but it is interesting to note that for all known finite planes, the only subplanes that fail to divide the order of the plane are subplanes of order 2.

This section is concluded with a study of the possible subplanes of a free extension plane. A free extension plane can contain any subplane or combinations of subplanes of any orders; we simply let Σ_0 be any union of disjoint primitive planes. (This union is, itself, a primitive plane.) A natural question to ask at this point is whether any new subplanes become evident during the extension process. The answer is yes, but the only new subplanes are free extension planes — no new finite planes are constructed. In fact, we may prove a stronger statement: No new confined configuration arises during the process.

THEOREM 2.7.8 *If* $\Sigma \subseteq \Sigma_0{}^+$ *and* Σ *is a confined configuration, then* $\Sigma \subseteq \Sigma_0$.

Proof. For points $p \in \mathscr{P}^+$, define the level of p to be the smallest $n \geq 0$ such that $p \in \mathscr{P}_n$. For lines L in $\mathscr{L}^+$, define the level of L to be the smallest $n \geq 0$ such that $L \cap \mathscr{P}_n \in \mathscr{L}_n$.

Since Σ is a configuration, it has a finite number of points and lines. Let n be the maximum level of a point or line in Σ. Suppose that a line has level n. Then $L \cap \mathscr{P}_n$ is a line of $\mathscr{L}_n$ and $L \cap \mathscr{P}_{n-1}$ is not a line of $\mathscr{L}_{n-1}$. If $n > 0$, then L joins two points of Σ_{n-1} that did not lie on a line. But all points of Σ have level $\leq n$, so they are all in $\mathscr{P}_n$, and by construction, L can join only two of them. This contradicts the fact that L must have three points on it. Therefore, n must be 0. A similar argument shows that $n = 0$ if we suppose that a point has a maximum level. Thus $n = 0$ and $\Sigma \subseteq \Sigma_o$.

We may use this theorem and Example 2.5.8(iii) to prove that principal restrictions of projective planes need not be isomorphic. This settles a question that was raised in the discussion following Corollary 1.7.12.

THEOREM 2.7.9 *If α and α' are principal restrictions of π, it is not necessarily true that $\alpha \sim \alpha'$.*

Proof. Let π denote the plane of Example 2.5.8(iii). Let $\alpha = \pi | \mathscr{P} - pr$ and let $\alpha' = \pi | \mathscr{P} - pq$, where r is a point of $\Sigma(7_3) \subseteq \Sigma_o$ and q is not a point of $\Sigma(7_3)$. Such a q exists — for example, it could be the intersection of pr with st, where s and t are points of $\Sigma(7_3)$ not collinear with r. (This intersection would take place in Σ_2.) Now α does not have a confined configuration embedded in it because we have removed a point from $\Sigma(7_3)$, and, by Theorem 2.7.8, $\Sigma(7_3)$ is the only possible confined configuration to be found in π. The removal of the line pq does not affect $\Sigma(7_3)$ because there is no point p' of $\Sigma(7_3)$ collinear with p and q. (Newly constructed lines contain only two points.) Therefore, α' does have $\Sigma(7_3)$ embedded in it and thus is not isomorphic to α.

EXERCISES

1. Let f be an isomorphism from π onto itself and let $\mathscr{P}'$ be the set of all points held fixed by f. Show that if $\mathscr{P}'$ contains a four-point, then $\pi | \mathscr{P}'$ is a subplane of π.

2. a. Find a subplane of order 5 in π_F where $F = GF(25)$.
   b. Find a subplane of order 5 in $\pi_{S_{25}}$.
   c. Find a subplane of order 2 in π_{S_8}.

3. Show that the following statement is not necessarily true: If $\pi(\alpha^+) \subseteq \pi((\alpha')^+)$, then $\alpha \subseteq \alpha'$.

4. The subplane π' of π is a *Baer subplane* if and only if every line of π contains a point of π' and every point of π lies on the extension of some line of π'.

   a. Give an example of a Baer subplane in the following planes:
      (1) $\pi_N(9)$.

(2) $\pi_H(9)$.

(3) π_F where $F = GF(4)$.

(4) π_F where $F = GF(9)$.

(5) π_F where $F = C$, the field of complex numbers.

b. Show that if π has order n *and* π' is a Baer subplane of π, then π' has order m where $m^2 = n$.

5. The plane π is *partitioned* into n subplanes π_i where $i = 1, \ldots , n$ if and only if every point of π is in exactly one of the subplanes π_i.

a. Show that π_{S_4} may be partitioned into three Baer subplanes.

b. Show that π_{S_9}, $\pi_{S_{16}}$, and $\pi_{S_{25}}$ may also be partitioned into Baer subplanes and find the number of such subplanes in each such plane.

c. Into how many subplanes of order 2 can you partition $\pi_{S_{16}}$?

6. a. Suppose that π' is a subplane of π, that π has order n, and that π' has order m.

(1) How many lines in π intersect π' in $m + 1$ points?

(2) How many lines in π intersect π' in one point?

(3) How many lines in π do not intersect π'?

b. Find the number of lines in $\pi_H(9)$ that intersect a subplane $\pi(2)$ in: three points; one point; no points.

c. Find the number of lines in $\pi_H(9)$ that intersect a subplane $\pi(3)$ in: four points; one point; no points.

2.8   COMPARING PROJECTIVE PLANES

This section reviews the criteria used to analyze projective planes. If, for example, we wish to distinguish between two projective planes, we may: (1) compare the cardinalities of the planes, (2) check for self-duality, or (3) analyze the subplane structure (or, more generally, the configurational substructure of each plane).

Because the property of cardinality is not strictly a geometric property, only the second and third criteria will be utilized; the comparison will be restricted to planes of the same cardinality. For example, we are already acquainted with four planes of order 9 (π_{S_9}, $\pi_H(9)$, $\pi_N(9)$, and $\pi_N(9)^d$), two planes of countable cardinality (π_Q and Σ_0^+, where Q is the field of rational numbers and Σ_0^+ is the partial plane of four points and no lines), and two planes of uncountable cardinality (π_R and π_C, where C is the field of complex numbers).

Let us first examine the criterion of self-duality. It is clear from the construction of Σ_0^+ that Σ_0^+ is self-dual; in fact, every free extension

plane that is generated from a partial plane consisting solely of points
is self-dual. The plane $\pi_H(9)$ is also self-dual, as proved in the following
theorem.

THEOREM 2.8.1 $\pi_H(9)$ is self-dual.

Proof.   Refer back to Section 2.4; it is easily checked that the function
$f: \pi_H(9) \rightarrow (\pi_H(9))^d$ is an isomorphism where f is defined as follows:

$$f: a_i \rightarrow L_{-7i} \qquad e_i \rightarrow L_{4-7i}$$
$$b_i \rightarrow L_{1-7i} \qquad f_i \rightarrow L_{5-7i}$$
$$c_i \rightarrow L_{2-7i} \qquad g_i \rightarrow L_{6-7i}$$
$$d_i \rightarrow L_{3-7i}$$

where $i = 0, \ldots , 12$ and where the arithmetic of the subscripts is
modulo 91.

The plane $\pi_N(9)$ is not self-dual; proof of this fact will be presented in
Chapter 12. The plane π_{S_9} is self-dual; self-duality is a property shared
by all cyclic planes.

THEOREM 2.8.2   *The planes π_{S_n} and $(\pi_{S_n})^d$ are isomorphic for any
perfect difference set S_n.*

Proof.   Recall that the points and lines of π_{S_n} are denoted by p_i and L_j
where $i,j = 0,1, \ldots , n$ and $(p_i,L_j) \in \mathscr{I}$ iff $i + j \in S_n$ modulo $n^2 + n$
$+ 1$. Points and lines of $(\pi_{S_n})^d$ are therefore of the form L_i and p_j, respectively, and $(L_i,p_j) \in \mathscr{I}^{-1}$ iff $j + i \in S_n$. Let $f: p_i \rightarrow L_i$ and let $F:$
$L_j \rightarrow p_j$ where $i,j = 0, \ldots , n$. It is easily seen that $(p_i,L_j) \in \mathscr{I}$ iff
$(f(p_i),F(L_j)) \in \mathscr{I}^{-1}$. Thus (f,F) is an isomorphism.

Planes coordinated by fields are also self-dual, as shown by Theorem
2.8.3.

THEOREM 2.8.3   *For any field F, the planes π_F and $(\pi_F)^d$ are isomorphic.*

Proof.   Recall that the points and lines of π_F are of the form $[x,y,z]$
and $\langle a,b,c \rangle$, respectively, where $x,y,z,a,b,c \in F$ and $([x,y,z], \langle a,b,c \rangle)$
$\in \mathscr{I}$ iff $ax + by + cz = 0$. Therefore, points and lines of $(\pi_F)^d$ are of the

form $\langle x,y,z \rangle$ and $[a,b,c]$, respectively, where $(\langle x,y,z \rangle, [a,b,c]) \in \mathscr{I}^{-1}$
iff $xa + yb + zc = 0$. Let $g: [x,y,z] \to \langle x,y,z \rangle$ and $G: \langle a,b,c \rangle \to [a,b,c]$;
it follows that $([x,y,z], \langle a,b,c \rangle) \in \mathscr{I}$ iff $(g([x,y,z]), G(\langle a,b,c \rangle)) \in \mathscr{I}^{-1}$
because $ax + by + cz = 0$ iff $xa + yb + zc = 0$. Thus (g,G) is an iso-
morphism.

Let us now focus on the subplane structure. Theorem 2.7.8 indicates
that there are no proper subplanes of $\Sigma_o{}^+$. In Section 2.7 the statement
was made (but not proved) that the subplanes of π_F correspond to the
subfields of F; for example, π_Q has no proper subplane and π_F, where
$F = \mathrm{GF}(9)$, has only $\pi(3)$ as a proper subplane. Section 2.7 also indicated that the planes $\pi_H(9)$, $\pi_N(9)$, and $\pi_N(9)^d$ have subplanes of orders
2 and 3. Interestingly, there are a number of different subplanes of
orders 2 and 3 in these planes. Killgrove (1965) compiled such information (see also Room and Kirkpatrick, 1971) and found that both
$\pi_N(9)$ and $\pi_N(9)^d$ have 51,840 subplanes of order 2 and 1,080 subplanes of order 3. At first it may seem surprising that two different
planes have the same number of subplanes, but a little thought reveals
that both $\pi(2)$ and $\pi(3)$ are self-dual and therefore exist in the same
numbers in both $\pi_N(9)$ and its dual. The plane $\pi_H(9)$ has 33,696 subplanes of order 2 and 1,080 subplanes of order 3.

To facilitate a more thorough study of these statistics, the concept of
the characteristic of a quadrangle is introduced here.

DEFINITION 2.8.4   *Let Σ be a quadrangle in π and let $\pi(\Sigma)$ denote
the least subplane containing Σ. The* characteristic *of Σ is the order of
$\pi(\Sigma)$ if $\pi(\Sigma)$ is finite; if $\pi(\Sigma)$ is infinite, the characteristic is 0. If the
characteristic of every quadrangle is the same number $n \geq 0$, then n is
the* characteristic *of π.*

EXAMPLE 2.8.5

    i. Consider the plane $\pi_H(9)$. Let Σ_1 contain the points a_0, a_1, a_2, a_3.
    It is easily checked that $\pi(\Sigma_1)$ is the plane of order 3 made up of
    the points $\{a_0, a_1, \ldots, a_{12}\}$. Let Σ_2 contain the points a_0, c_4, d_1, f_1.
    A check of Example 2.7.5 shows that $\pi(\Sigma_2) = \pi(2)$. Let Σ_3 con-
    tain the points a_0, b_1, b_2, c_1. A check of the plane $\pi_H(9)$ shows that
    the extension process yields the following results:

Stage 1:

$$a_0 b_1 = L_1 \qquad b_2 c_1 = L_{81}$$
$$a_0 c_1 = L_2 \qquad b_1 b_2 = L_{74}$$
$$a_0 b_2 = L_{14} \qquad b_1 c_1 = L_7$$

Stage 2:

$$L_1 \cap L_{81} = e_4$$
$$L_2 \cap L_{74} = e_3$$
$$L_{14} \cap L_7 = a_5$$

Stage 3:

$$e_4 e_3 = L_{85}$$
$$e_4 a_5 = L_{89}$$
$$e_3 a_5 = L_{38}$$

Stage 4:

$L_{85} \cap L_1 = e_4$	$L_{39} \cap L_1 = e_4$	$L_{38} \cap L_1 = g_8$
$L_{85} \cap L_2 = e_3$	$L_{39} \cap L_2 = c_{12}$	$L_{38} \cap L_2 = e_3$
$L_{85} \cap L_{14} = f_2$	$L_{39} \cap L_{14} = a_5$	$L_{38} \cap L_{14} = a_5$
$L_{85} \cap L_{81} = e_4$	$L_{39} \cap L_{81} = e_4$	$L_{38} \cap L_{81} = d_6$
$L_{85} \cap L_{74} = e_3$	$L_{39} \cap L_{74} = e_{11}$	$L_{38} \cap L_{74} = e_3$
$L_{85} \cap L_7 = a_{12}$	$L_{39} \cap L_7 = a_5$	$L_{38} \cap L_7 = a_5$

Stage 5:

$$c_1 f_2 = L_{26} \qquad \text{(and so on)}$$

Stage 6:

$$L_{26} \cap L_{39} = b_{10} \qquad \text{(and so on)}$$

This extension can be interrupted because we already have generated fourteen points. By Theorem 2.7.7, the only possible subplanes of $\pi_H(9)$ are $\pi(2)$ and $\pi(3)$, and since $\pi(\Sigma_3)$ cannot be either of these, $\pi(\Sigma_3)$ must be $\pi_H(9)$ itself. Thus there are quadrangles in $\pi_H(9)$ of characteristics 2, 3, and 9.

ii. Consider the plane $\pi_N(9)$. Let Σ_1 contain the points $[0,0,1]$, $[0,1,0]$, $[1,0,0]$, $[1,1,1]$; then clearly, $\pi(\Sigma_1) = \pi_F$ where $F = J/3$.

Let Σ_2 contain the points $[0,0,1]$, $[a,0,1]$, $[b,1,1]$, $[1,e,0]$. Example 2.7.6 shows that $\pi(\Sigma_2) = \pi(2)$. Proof that there exists a quadrangle of characteristic 9 in $\pi_N(9)$ is left as an exercise.

It will be proved in Chapter 8 that π_F has a characteristic that equals the characteristic of the field F. Thus for the plane $\pi_F(9)$ (the plane over GF(9)), every four-point generates $\pi(3)$. It is natural to ask whether planes other than those representable by fields have characteristics. For infinite planes the answer is easily seen to be yes. (For example, the plane Σ_0^+ mentioned previously has characteristic 0.) For finite planes the answer is unknown. Only for $n = 2$ is it known that finite planes of characteristic n (that is, Fano planes) are representable by fields (Gleason, 1956).

This analysis of the characteristics of quadrangles in planes of order 9 is completed by the addition of some further information compiled by Killgrove. Clearly, in a 91-point projective plane, there are

$$\frac{91 \cdot 90 \cdot 81 \cdot 64}{24} = 1,769,040$$

quadrangles. For $\pi_F(9)$, all these quadrangles have characteristic 3. For $\pi_H(9)$, there are 235,872 quadrangles of characteristic 2, 252,720 quadrangles of characteristic 3, and 1,280,448 quadrangles of characteristic 9. For $\pi_N(9)$, there are 362,880 quadrangles of characteristic 2, 252,720 quadrangles of characteristic 3, and 1,153,770 quadrangles of characteristic 9. Since there are seven distinct quadrangles in every plane $\pi(2)$ and 234 distinct quadrangles in every plane $\pi(3)$, we obtain the numbers of subplanes $\pi(2)$ and $\pi(3)$ given previously.

This section is concluded with an analysis of the occurrence of configurations $\Sigma(7_3)$, $\Sigma(8_3)$, $\Sigma(9_3)$ (the Pappus configuration), and $\Sigma(10_3)$ (the Desargues configuration) in projective planes. These configurations have three possible relationships with a projective plane π: (1) they do not occur at all in π, (2) they occur everywhere in π, (3) they occur sometimes but not everywhere in π. We say that the configuration Σ *occurs in* π if and only if Σ can be embedded in π. The phrase "occurs everywhere in π" is clarified for all four configurations in Figures 18 through 21. (Collinearity in these figures is depicted by the connection of appropriate points with lines.)

The configuration $\Sigma(7_3)$ occurs everywhere in π if and only if $\Sigma(7_3)$ occurs in π and the existence of the primitive plane as represented by

Figure 18 implies the collinearity of p_5, p_6, and p_7. In other words, the diagonals of every four-point are collinear (and π is thus a Fano plane).

The configuration $\Sigma(8_3)$ occurs everywhere in π if and only if $\Sigma(8_3)$ occurs in π and the existence of the primitive plane as represented by Figure 19 implies the collinearity of p_6, p_7, and p_8.

The configuration $\Sigma(9_3)$ occurs everywhere in π if and only if $\Sigma(9_3)$ occurs in π and the existence of the primitive plane as represented by Figure 20 implies the collinearity of p_7, p_8, and p_9.

The configuration $\Sigma(10_3)$ occurs everywhere in π if and only if $\Sigma(10_3)$ occurs in π and the existence of the primitive plane as represented by Figure 21 implies the collinearity of p_8, p_9, and p_{10}.

We already know some facts about planes of order 9. We know that $\Sigma(7_3)$ occurs, although not everywhere, in $\pi_H(9)$, $\pi_N(9)$, and $\pi_N(9)^d$, but that it does not occur at all in $\pi_F(9)$. The configuration $\Sigma(8_3)$ is easily found to occur in all four of these planes; some additional effort may be required to show that it does not occur everywhere in any of them. It is easily verified that $\Sigma(9_3)$ and $\Sigma(10_3)$ occur, but not everywhere, in $\pi_H(9)$, $\pi_N(9)$, and $\pi_N(9)^d$. Both of these configurations occur everywhere in $\pi_F(9)$. It follows from Theorem 2.7.8 that none of the four configurations ever occurs in the plane Σ_0^+. As for the occurrence of these four configurations in the other planes under consideration, Theorems 2.8.6 through 2.8.9 apply to planes coordinatized by any field.

THEOREM 2.8.6   In π_F, if $F \neq J/2$, $J/3$, or GF(4), the Desargues configuration $\Sigma(10_3)$ occurs everywhere.

THEOREM 2.8.7   In π_F, where $F \neq J/2$, the Pappus configuration $\Sigma(9_3)$ occurs everywhere.

THEOREM 2.8.8   In π_F, if the equation $x^2 + x + 1 = 0$ has a solution in F, then the configuration $\Sigma(8_3)$ occurs; if the equation has no solution, then the configuration does not occur.

THEOREM 2.8.9   In π_F, if the field F is of characteristic 2, then the configuration $\Sigma(7_3)$ occurs everywhere (that is, π_F is a Fano plane); if F does not have characteristic 2, then the configuration does not occur at all (that is, π_F satisfies Fano's Axiom).

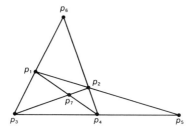

FIGURE 18

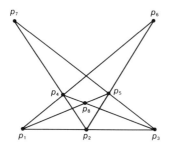

FIGURE 19

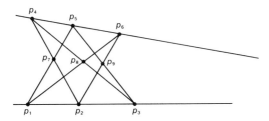

FIGURE 20

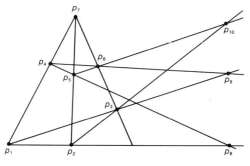

FIGURE 21

Formal, complete proofs are not given for these theorems; such proofs would necessitate extending to all four-points the results proved here for a fixed four-point. If the plane π_F were identical about every four-point the steps outlined below could be expanded into proofs. In the next two chapters it will be shown that for planes π_F, the geometric "lay of the land" about every four-point is identical.

For the partial proof of Theorem 2.8.9, consult Figure 18 and let $p_1 = [1,0,0]$, $p_2 = [0,1,0]$, $p_3 = [0,0,1]$, and $p_4 = [1,1,1]$. It is easily checked that $p_3 p_4 = \langle -1,1,0 \rangle$ and $p_1 p_2 = \langle 0,0,1 \rangle$, so $p_5 = [1,1,0]$; that $p_1 p_3 = \langle 0,1,0 \rangle$ and $p_2 p_4 = \langle -1,0,1 \rangle$, so $p_6 = [1,0,1]$; and that $p_1 p_4 = \langle 0,-1,1 \rangle$ and $p_2 p_3 = \langle 1,0,0 \rangle$, so $p_7 = [0,1,1]$. Since $p_5 p_6 = \langle -1,1,1 \rangle$, it follows that $p_7 \in p_5 p_6$ if and only if $1 + 1 = 0$.

For the partial proof of Theorem 2.8.8, consult Figure 19 and let $p_1 = [1,0,0]$, $p_2 = [1,1,0]$, $p_4 = [0,0,1]$, and $p_8 = [0,1,1]$. Also let p_6 be an arbitrary point on $p_1 p_4$ distinct from p_1 and p_4; thus $p_6 = [x,0,1]$ where $x \neq 0$. It is easily checked that $p_2 p_6 = \langle -1,1,x \rangle$ and $p_1 p_8 = \langle 0,-1,1 \rangle$, so $p_5 = [x+1,1,1]$; that $p_4 p_8 = \langle 1,0,0 \rangle$ and $p_1 p_2 = \langle 0,0,1 \rangle$, so $p_3 = [0,1,0]$; and that $p_2 p_4 = \langle -1,1,0 \rangle$ and $p_3 p_5 = \langle -1,0,x+1 \rangle$, so $p_7 = [x+1, x+1,1]$. Since $p_6 p_8 = \langle 1,x,-x \rangle$, it follows that $p_7 \in p_6 p_8$ if and only if $[x+1,x+1,1] \in \langle 1,x,-x \rangle$. This occurs if and only if $x^2 + x + 1 = 0$.

For the partial proof of Theorem 2.8.7, consult Figure 20 and let $p_1 = [0,1,0]$, $p_2 = [1,1,0]$, $p_4 = [0,0,1]$, and $p_5 = [1,0,1]$. Also let p_3 be an arbitrary point on $p_1 p_2$; thus $p_3 = [x,1,0]$ where $x \neq 0,1$. Similarly, let p_6 be an arbitrary point such that $p_6 \in p_4 p_5$ and $p_6 \notin p_1 p_2$. Thus $p_6 = [y,0,1]$ where $y \neq 0,1$. It is easily checked that $p_1 p_5 = \langle -1,0,1 \rangle$ and $p_2 p_4 = \langle -1,1,0 \rangle$, so $p_7 = [1,1,1]$; $p_1 p_6 = \langle -1,0,y \rangle$ and $p_3 p_4 = \langle -1,x,0 \rangle$, so $p_8 = [xy,y,x]$; and $p_2 p_6 = \langle -1,1,y \rangle$ and $p_3 p_5 = \langle -1,x,1 \rangle$, so $p_9 = [xy-1,y-1,x-1]$. Note that p_7, p_8, and p_9 all lie on the line $\langle x-y, xy-x, y-xy \rangle$.

For the partial proof of Theorem 2.8.6, consult Figure 21 and let $p_1 = [0,0,1]$, $p_2 = [0,1,0]$, $p_4 = [1,0,1]$, $p_5 = [1,1,0]$, and $p_7 = [1,0,0]$. Let p_3 and p_6 be arbitrary points subject to the restrictions that p_3, p_6, and p_7 are distinct and collinear and that neither p_3 nor p_6 lies on a side of the complete quadrangle with vertices p_1, p_2, p_4, and p_5. It is left as an exercise to show that $p_3 = [x,y,1]$ and $p_6 = [z,y,1]$ where $x,y,z \neq 0$; that $x,z \neq 1$; that $x \neq y \neq z \neq x$; and that $x,z \neq y+1$. It is easily checked that $p_1 p_2 = \langle 1,0,0 \rangle$ and $p_4 p_5 = \langle -1,1,1 \rangle$, so $p_8 = [0,-1,1]$; that $p_1 p_3 = \langle -y,x,0 \rangle$ and $p_4 p_6 = \langle -y,z-1,y \rangle$, so $p_9 = [xy,y^2,y+yx-yz]$; and that $p_2 p_3 = \langle -1,0,x \rangle$ and $p_5 p_6 = \langle 1,-1,y-z \rangle$, so $p_{10} = [x,x+y-z,1]$.

Observe that p_8, p_9, and p_{10} all lie on the line $\langle z - y - x - 1, x, x \rangle$. It should be noted that to ensure that the configuration is $\Sigma(10_3)$, we must insist on the restriction that $z \neq x + y + 1$.

EXERCISES

1. Complete the informal proof of Theorem 2.8.6 by showing that $p_3 = [x,y,1]$ and that $p_6 = [z,y,1]$ where x, y, and z satisfy the stated conditions $(x,y,z \neq 0;\ x,z \neq 1;\ x \neq y \neq z \neq x;$ and $x,z \neq y + 1)$.

2. Table 13 is constructed as follows. If the configuration $\Sigma(m_n)$ occurs everywhere in π, number 1 fills the appropriate position; if the configuration occurs sometimes, number 0 is used; if the configuration does not occur at all, number -1 is used. Supply reasons for each entry in Table 13.

TABLE 13

	$\pi(3)$	$\pi(4)$	$\pi(5)$	$\pi(7)$	$\pi_F(9)$	$\pi_N(9)$	$\pi_H(9)$	π_R	π_C
$\Sigma(7_3)$	-1	1	-1	-1	-1	0	0	-1	-1
$\Sigma(8_3)$	0	0	-1	0	0	0	0	-1	0
$\Sigma(9_3)$	1	1	1	1	1	0	0	1	1
$\Sigma(10_3)$	-1	-1	1	1	1	0	0	1	1

3. Find a geometric difference between π_R and π_M, the Moulton plane (see exercises 1.3.8 and 2.1.6(f)).

4. a. Show that if F is a finite field of order 3, 7, 13, or 31, then $\Sigma(8_3)$ occurs in π_F.
   b. Using exercise 4(a), formulate a sufficient condition for the occurrence of $\Sigma(8_3)$ in π_F.

5. To do this exercise, refer back to exercise 1.6.6.
   a. Explain what is meant by "the configuration $\Sigma_1(9_3)$ occurs everywhere." Do the same for $\Sigma_2(9_3)$.
   b. State and informally prove a theorem concerning $\Sigma_1(9_3)$ that is analogous to Theorems 2.8.6–2.8.9. Do the same for $\Sigma_2(9_3)$. Hint: To prove the theorem concerning $\Sigma_1(9_3)$, refer to Figure 22 and let $p_1 = [1,0,0]$, $p_2 = [1,1,1]$, $p_3 = [0,1,0]$, $p_4 = [0,0,1]$, $p_5 = [y,1,1]$, and $p_6 = [0,x,1]$ where $x,y \neq 0,1$.
   c. Using the numbers 1, 0, and -1 as in Table 13, construct a table relating the configurations $\Sigma_1(9_3)$ and $\Sigma_2(9_3)$ to the planes $\pi(3)$, $\pi(4)$, $\pi(5)$, $\pi(7)$, $\pi_F(9)$, π_Q, and π_R.

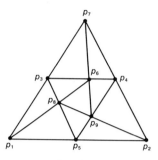

FIGURE 22

6. Several planes of order 9 were introduced in the exercises following Sections 2.1 and 2.4: specifically, π_A (exercise 2.1.6), $\pi_{H'}(9)$ (exercise 2.4.2), and π_B and π_C (exercise 2.4.4). Since there are only four distinct planes of order 9, each of these planes must be isomorphic to one of the four planes $\pi_F(9)$, $\pi_H(9)$, $\pi_N(9)$, and $\pi_N(9)^d$.

   a. Show that $\pi_H(9)$ and $\pi_{H'}(9)$ are isomorphic. Hint: Let f map the points of $\pi_H(9)$ to the points of $\pi_{H'}(9)$ as follows: $f\colon x_i \to x_{3i}$ where $x = a, \ldots, g$. Find the induced map F from the lines of $\pi_H(9)$ to the lines of $\pi_{H'}(9)$ and show that (f,F) is an isomorphism.

   b. Show that $\pi_N(9)$, π_A, π_B, and π_C are isomorphic. Hint: Since exercise 2.4.4 establishes that π_A, π_B, and π_C are isomorphic, it is sufficient to exhibit an isomorphism (f,F) from $\pi_N(9)$ to π_A. First develop f on points of the form $[x,y,1]$ and F on lines of the form $\langle m,1,k \rangle$ by observing that $x \cdot m = x \circ (m - 1) + x$ for all $x, m \in R$ where $\cdot$ is the multiplication in N and $\circ$ is the multiplication in A.

CHAPTER 3

TRANSFORMATIONS

The first two chapters have left us with an abundance of planes and a formidable list of unsolved problems, but only limited means for solving them. This discouraging situation will be ameliorated in this chapter and the next. The most fruitful avenue of approach is through the analysis of transformation groups. This chapter will acquaint the student with the definitions, examples, and elementary properties of the important transformations. Techniques will be developed in the next chapter to analyze significantly the planes of Parts 2 and 3.

3.1 COLLINEATIONS

The most important kind of transformation is the isomorphism of a plane onto itself.

DEFINITION 3.1.1 *An isomorphism of a plane* Σ *onto itself is called a* collineation. *If* Σ *is an affine plane, the term* affine collineation *is often used.*

Notation. As with isomorphisms, two alternate notations are used for collineations. Letting $\Sigma = (\mathcal{P}, \mathcal{L}, \mathcal{I})$, a collineation may be denoted by the pair (f, F) where f is a one-to-one mapping from $\mathcal{P}$ onto $\mathcal{P}$ and F is a one-to-one mapping from $\mathcal{L}$ onto $\mathcal{L}$ such that $(p.L) \in \mathcal{I}$ if and only if $(f(p), F(L)) \in \mathcal{I}$. If Σ is a primitive plane, it follows from Theorem 1.5.3 that a collineation may be denoted by a single function f where f is a one-to-one mapping from $\mathcal{P}$ onto $\mathcal{P}$ preserving collinearity. Then f is thought of not only as a mapping on points but also as a mapping on lines (sets of points).

Following are some examples of collineations on projective planes; Section 3.3 will focus on affine collineations.

EXAMPLE 3.1.2   Consider the plane $\pi_F = (\mathcal{P}, \mathcal{L}, \mathcal{I})$ and let g: $[x, y, z]$ $\rightarrow [ax + by + cz, dx + ey + fz, gx + hy + iz]$ where $a, b, c, d, e, f, g, h, i \in F$ and $g[x, y, z] \neq [0, 0, 0]$ for any point $p = [x, y, z]$.

To show that g is a collineation, it must first be shown that g is a well-defined, one-to-one, onto mapping. (The fact that g is used in two different ways here should cause no confusion.) Let G be the following function defined on F_3, the three-dimensional vector space over F:

$$G \colon (x, y, z) \rightarrow (ax + by + cz, dx + ey + fz, gx + hy + iz)$$

Since G maps only $(0, 0, 0)$ into $(0, 0, 0)$, it follows from elementary results of linear algebra that G is a nonsingular linear transformation and thus is a one-to-one mapping from F_3 onto F_3. Suppose that (x', y', z') $\in [x, y, z]$; then there exists $r \in F$, $r \neq 0$, such that $x' = xr$, $y' = yr$, $z' = zr$. Observe that

$$G((x', y', z')) =$$
$$(ax' + by' + cz', dx' + ey' + fz', gx' + hy' + iz') =$$
$$(axr + byr + czr, dxr + eyr + fzr, gxr + hyr + izr) =$$
$$((ax + by + cz)r, (dx + ey + fz)r, (gx + hy + iz)r)$$

It follows that $G((x', y', z')) \in [G((x, y, z))]$. Thus $G[x, y, z] =$ $G\{(xr, yr, zr): r \in F, r \neq 0\} = \{G((xr, yr, zr))\} = [G((x, y, z))] = [ax + by + cz, dx + ey + fz, gx + hy + iz]$. Since $g([x, y, z]) = G[x, y, z]$ for $[x, y, z]$ $\neq [0, 0, 0]$, we may conclude that g is a well-defined, one-to-one mapping from $\mathcal{P}$ onto $\mathcal{P}$.

The next step is to show that g preserves collinearity. Recall that G may be denoted in matrix form:

$$M = \begin{pmatrix} a & b & c \\ d & e & f \\ g & h & i \end{pmatrix}$$

and that $G((x,y,z))$ may be calculated by the matrix multiplication

$$\begin{pmatrix} a & b & c \\ d & e & f \\ g & h & i \end{pmatrix}\begin{pmatrix} x \\ y \\ z \end{pmatrix}$$

Suppose that $[x_i, y_i, z_i]$ are collinear, $i = 1,2,3$; then there exists $r,s,t \in F$ such that $rx_i + sy_i + tz_i = 0$; in matrix notation,

$$(r,s,t)\begin{pmatrix} x_i \\ y_i \\ z_i \end{pmatrix} = 0$$

Since G is nonsingular, M has an inverse; thus we may write

$$(r,s,t)(M^{-1}M)\begin{pmatrix} x_i \\ y_i \\ z_i \end{pmatrix} = ((r,s,t)M^{-1})\,M\begin{pmatrix} x_i \\ y_i \\ z_i \end{pmatrix} = 0$$

Let the triple $(r,s,t)M^{-1} = (r',s',t')$. Then $r'(ax + by + cz) + s'(dx + ey + fz) + t'(gx + hy + iz) = 0$, and thus the points $g([x_i,y_i,z_i])$ are collinear. Note that $G: \langle r,s,t \rangle \to \langle r',s',t' \rangle$ where $(r',s',t') = (r,s,t)M^{-1}$.

DEFINITION 3.1.3 *A collineation on* π_F *of the type* $f: [x,y,z] \to [ax + by + cz,\ dx + ey + fz,\ gx + hy + iz]$ *is called a* matrix-representable collineation.

Notation. This collineation is denoted by f_M where M denotes the matrix

$$\begin{pmatrix} a & b & c \\ d & e & f \\ g & h & i \end{pmatrix}$$

A matrix does not uniquely determine a collineation, as the following theorem shows.

THEOREM 3.1.4 *Let M and N be nonsingular matrices over a field F. Then $f_M = f_N$ if and only if there exists $r \in F$ such that $M = rN$.*

Proof. The proof is left as an exercise.

EXAMPLE 3.1.5

i. Consider the plane $\pi_{S_n} = (\mathscr{P}, \mathscr{L}, \in)$. Let $f\colon p_i \to p_{i+k}$, $F\colon L_j \to L_{j-k}$. Clearly, f and F are one-to-one functions of $\mathscr{P}$ onto $\mathscr{P}$ and $\mathscr{L}$ onto $\mathscr{L}$, respectively. Now $p_i \in L_j$ if and only if $i + j \in S_n$ if and only if $i + k + j - k \in S_n$ if and only if $p_{i+k} \in L_{j-k}$ if and only if $f(p_i) \in F(L_j)$. Thus (f, F) is a collineation on π_{S_n}.

ii. Consider the plane $\pi_H(9)$. Let

$$
\begin{array}{ll}
f\colon a_i \to a_i & d_i \to b_i \\
 b_i \to c_i & e_i \to g_i \\
 c_i \to d_i & f_i \to e_i \\
 & g_i \to f_i \qquad i = 0, \ldots, 12
\end{array}
$$

$$
\begin{array}{ll}
g\colon a_i \to a_i & d_i \to g_i \\
 b_i \to e_i & e_i \to b_i \\
 c_i \to f_i & f_i \to c_i \\
 & g_i \to d_i \qquad i = 0, \ldots, 12
\end{array}
$$

The student is left the task of verifying that both f and g are collineations on $\pi_H(9)$.

iii. Consider the plane $\pi_N(9)$. Let r, s be fixed elements of N. Let

$$
\begin{array}{l}
f\colon [x, y, 1] \to [x + r, y + s, 1] \text{ where } x, y \in N \\
 [1, x, 0] \to [1, x, 0] \text{ where } x \in N \\
 [0, 1, 0] \to [0, 1, 0]
\end{array}
$$

$$
\begin{array}{l}
F\colon \langle m, 1, k \rangle \to \langle m, 1, -rm - s + k \rangle \text{ where } m, k \in N \\
 \langle 1, 0, k \rangle \to \langle 1, 0, k - r \rangle \\
 \langle 0, 0, 1 \rangle \to \langle 0, 0, 1 \rangle
\end{array}
$$

Let r, s be fixed nonzero elements of N. Let

$$
\begin{array}{l}
g\colon [x, y, 1] \to [xr, ys, 1] \\
 [1, x, 0] \to [1, r^{-1}xs, 0] \\
 [0, 1, 0] \to [0, 1, 0]
\end{array}
$$

$$G: \langle m,1,k \rangle \rightarrow \langle r^{-1}ms,1,ks \rangle$$
$$\langle 1,0,k \rangle \rightarrow \langle 1,0,kr \rangle$$
$$\langle 0,0,1 \rangle \rightarrow \langle 0,0,1 \rangle$$

Both (f,F) and (g,G) are collineations on $\pi_N(9)$. This is shown here for (f,F); the proof for (g,G) is left as an exercise.

Case 1. Suppose that $[x,y,1] \in \langle m,1,k \rangle$; then $xm+y+k=0$. Thus $xm+rm+y+s-rm-s+k=0$ and $(x+r)m+y+s-rm-s+k=0$, by right distributivity. Therefore, $[x+r,y+s,1] \in \langle m,1,-rm-s+k \rangle$, and so $f([x,y,1]) \in F(\langle m,1,k \rangle)$.

Case 2. Suppose that $[x,y,1] \in \langle 1,0,k \rangle$; then $x+k=0$. Therefore, $x+r+k-r=0$, and so $[x+r,y+s,1] \in \langle 1,0,k-r \rangle$. Thus $f([x,y,1]) \in F(\langle 1,0,k \rangle)$.

Case 3. Suppose that $[1,x,0] \in \langle m,1,k \rangle$; then $m+x=0$. Thus $[1,x,0] \in \langle m,1,-rm-s+k \rangle$, and so $f([1,x,0]) \in F(\langle m,1,k \rangle)$.

Case 4. It is clear that $[1,x,0] \in \langle 0,0,1 \rangle$ and $f([1,x,0]) \in F(\langle 0,0,1 \rangle)$; that $[0,1,0] \in \langle 1,0,k \rangle$ and $f([0,1,0]) \in F(\langle 1,0,k \rangle)$; and that $[0,1,0] \in \langle 0,0,1 \rangle$ and $f([0,1,0]) \in F(\langle 0,0,1 \rangle)$.

EXERCISES

1. Find a collineation on Σ where Σ is:
    a. $\Sigma(10_3)$, the Desargues configuration.
    b. $\Sigma(9_3)$, the Pappus configuration.
    c. $\Sigma(8_3)$.
    d. $\Sigma_1(9_3)$ (see exercise 1.6.6).
    e. $\Sigma_2(9_3)$.

2. Prove Theorem 3.1.4.

3. The composition of two functions f and g is written $g \circ f$ and is read from right to left; that is, $g \circ f(x) = g(f(x))$. The multiplication of matrices M and N is written MN and is read from left to right in the same manner as ordinary multiplication of real numbers.
    a. Show that $f_N \circ f_M = f_{MN}$.
    b. Show that if $f_N \circ f_M = f_P$, then there exists a real number r such that $NM = rP$.

4. Verify that the following mappings are collineations on $\pi_H(9)$:
   a. f of Example 3.1.5(ii).
   b. g of Example 3.1.5(ii).
   c. f_j where f_j: $x_i \rightarrow x_{i+j}$. (Here $j = 0, \ldots, 12$; $x_i = a_i, \ldots, g_i$; and the addition $i + j$ is mod 13.)

5. Show that the pair (g, G) of Example 3.1.5(iii) is a collineation of $\pi_N(9)$.

6. Refer back to exercises 2.1.6 and 2.1.7. Let $P = (R, +, \cdot)$ be a Cartesian group and note the following definition:

   The mapping γ on P is an *automorphism* if and only if γ is an isomorphism from R onto itself.

   a. Show that γ maps 0 into 0 and 1 into 1.
   b. Suppose that F is a field, γ is an automorphism on F, and f_γ and F_γ are mappings defined on the points and lines, respectively, of π_F as follows:

   $$f_\gamma: [x, y, z] \rightarrow [\gamma(x), \gamma(y), \gamma(z)]$$
   $$F_\gamma: \langle a, b, c \rangle \rightarrow \langle \gamma(a), \gamma(b), \gamma(c) \rangle$$

   Show that (f_γ, F_γ) is a collineation on π_F.
   c. Suppose that P is a Cartesian group, γ is an automorphism on P, and f_γ and F_γ are mappings defined on the points and lines of $(\mathscr{P}, \mathscr{L}, \mathscr{I})$ (the plane constructed over P in exercise 2.1.6) as displayed in 6(b). Show that (f_γ, F_γ) is a collineation on $(\mathscr{P}, \mathscr{L}, \mathscr{I})$. Observe that the points and lines of π_F and $(\mathscr{P}, \mathscr{L}, \mathscr{I})$ are denoted similarly, so the mappings f_γ and F_γ make sense formally (that is, f_γ maps triples to triples as does F_γ, since both $F_\gamma(p)$ and $F_\gamma(L)$ are ordered triples). Care must be taken here (but not in 6(b)) to show that $f_\gamma(p)$ and $F_\gamma(L)$ are of the right form. In other words, points must be of the form $[x, y, 1]$, $[1, x, 0]$, or $[0, 1, 0]$; lines must be of the form $\langle m, 1, k \rangle$, $\langle 1, 0, k \rangle$, or $\langle 0, 0, 1 \rangle$ — see exercise 2.1.6.
   d. Give examples of automorphic collineations on $\pi(4)$; $\pi_F(9)$; $\pi_N(9)$; π_C, the plane defined over the field of complex numbers.

7. As Example 3.1.5(i) shows, the mapping f: $p_i \rightarrow p_{i+k}$ is a collineation on the plane π_{S_n}. The mapping f: $p_i \rightarrow p_{ki}$ is also a collineation on π_{S_n} if k is chosen properly.

   Let S_n be a perfect difference set and let $k \in S_n$. Then k is called a *multiplier* of S_n if and only if f: $p_i \rightarrow p_{ki}$ is a collineation on π_{S_n} where ki is multiplication mod $n^2 + n + 1$.

   a. Find examples of multipliers of S_n (see Table 7, p. 54) where $n = 2$; where $n = 3$; where $n = 4$.
   b. Show that if s and t are multipliers of S_n, then st mod $n^2 + n + 1$ is a multiplier of S_n.

   c. Let $n = 9$ and verify that 3 is a multiplier of S_9; then deduce that k is also a multiplier where $k = 9,27,61,81$.

   d. Is it true that if $n = p$ (where p is a prime number), then p is a multiplier for S_n?

3.2 CENTRAL COLLINEATIONS

The most important type of collineation in projective geometry is the central collineation. It arises naturally in the study of three-dimensional geometry as a "projective mapping" from a plane onto a plane. We may visualize such a mapping in Euclidean three-space as a transformation that maps a point p of a given plane onto point q of the image plane in such a way that the line pq always passes through a fixed point o lying outside both planes. Real projective geometry can be considered to be the study of the properties that are held invariant under compositions of projective mappings. Since we are limiting our study to two dimensions, our definition must necessarily be restricted to the concepts of point and line. However, the definition can retain the essential ingredient of the projective map—namely, the pointwise invariance of one line (the line of intersection of the given plane and the image plane). Therefore, this section begins with a brief study of fixed points and fixed lines.

DEFINITION 3.2.1 *Let (f,F) be a collineation on Σ. If $f(p) = p$, then p is a* fixed point *or an* invariant point *of f. If $F(L) = L$, then L is a* fixed line *or an* invariant line *of F. If every point on line L is held fixed, L is* pointwise invariant (fixed); *if every line through point p is held fixed, p is* linewise invariant (fixed). *If (f,F) holds every point and line fixed, then it is the* identity collineation, *or simply the* identity.

Several facts follow immediately from these definitions.

THEOREM 3.2.2 *Let (f,F) be a collineation on π.*

   i. *If p and q are distinct fixed points, then pq is a fixed line.*

   ii. *If L and M are distinct fixed lines, then $L \cap M$ is a fixed point.*

iii. *If L and M are pointwise invariant, then (f,F) is the identity.*

iv. *If p and q are linewise invariant, then (f,F) is the identity.*

Proof. The proofs of (i) and (ii) are immediate. To prove (iii), note that if $p \in L \cup M$, then $f(p) = p$. If $p \notin L \cup M$, then we may let N and N' be two lines that pass through p but not through $L \cap M$. Then N and N' each meet L and M in two fixed points, and therefore each is a fixed line. Thus, by (ii), $N \cap N' = p$ is a fixed point. We conclude that $f(p) = p$ for all points p, and so (f,F) is the identity. Statement (iv) is true because it is the dual of statement (iii).

DEFINITION 3.2.3   *A collineation on π is called a* central collineation *if and only if at least one line is held pointwise invariant.*

EXAMPLE 3.2.4

i. The identity collineation is a central collineation for any plane π.

ii. Consider the plane π_F. Let $f = f_M$ where

$$M = \begin{pmatrix} 1 & 0 & r \\ 0 & 1 & s \\ 0 & 0 & t \end{pmatrix}$$

Now

$$\begin{pmatrix} 1 & 0 & r \\ 0 & 1 & s \\ 0 & 0 & t \end{pmatrix} \begin{pmatrix} x \\ y \\ 0 \end{pmatrix} = \begin{pmatrix} x \\ y \\ 0 \end{pmatrix}$$

so $f([x,y,0]) = [x,y,0]$, and thus f leaves $\langle 0,0,1 \rangle$ pointwise invariant.

iii. Consider the plane π_{S_3}. Let $f: p_i \to p_{r(i)}$ where

$$
\begin{array}{llll}
r: 0 \to 0 & 4 \to 10 & 7 \to 11 & 10 \to 4 \\
1 \to 1 & 5 \to 6 & 8 \to 8 & 11 \to 7 \\
2 \to 2 & 6 \to 5 & 9 \to 3 & 12 \to 12 \\
3 \to 9 & & &
\end{array}
$$

A check of Table 5 (p. 52) shows that f leaves L_1 pointwise invariant. Let $g: p_i \to p_{s(i)}$ where

$$
\begin{array}{llll}
s: 0 \to 0 & 4 \to 11 & 7 \to 10 & 10 \to 3 \\
1 \to 5 & 5 \to 6 & 8 \to 8 & 11 \to 9 \\
2 \to 2 & 6 \to 1 & 9 \to 4 & 12 \to 12 \\
3 \to 7 & & &
\end{array}
$$

A check of this table also reveals that g also holds L_1 pointwise invariant.

iv. Consider the plane $\pi_N(9)$. The collineation (f,F) of Example 3.1.5(iii) is a central collineation holding $\langle 0,0,1 \rangle$ pointwise invariant; the collineation (g,G) of the same example is not necessarily a central collineation. If, for example, $r = 1$, then line $\langle 0,1,0 \rangle$ is held pointwise invariant; if $s = 1$, then $\langle 1,0,0 \rangle$ is held pointwise invariant. If, however, $r = a$ and $s = a$, then the collineation is not a central collineation. Proof of this statement is left as an exercise.

v. There are no central collineations other than the identity on the plane $\pi_H(9)$, as will be shown in Chapter 9. It is interesting to note that in $\pi_H(9)$, although no collineation leaves a line pointwise fixed, both collineations of Example 3.1.5(ii) hold a subplane $\pi(3)$ pointwise fixed. As will be seen later, this phenomenon cannot occur in planes π_F, π_{S_n}, or π_N.

Naturally, many collineations in planes π_F and π_{S_n}, as well as in planes $\pi_N(9)$ and $\pi_H(9)$, are not central collineations. The following two theorems help identify certain central collineations from among the matrix-representable collineations in π_F.

THEOREM 3.2.5 *Let f_M be a matrix-representable collineation on π_F. Then the line $\langle 0,0,1 \rangle$ is held pointwise invariant if and only if M is of the form*

$$\begin{pmatrix} 1 & 0 & r \\ 0 & 1 & s \\ 0 & 0 & t \end{pmatrix}$$

Proof. It was shown in Example 3.2.4(ii) that f_M holds $\langle 0,0,1 \rangle$ pointwise fixed. Now suppose that f_M holds $\langle 0,0,1 \rangle$ pointwise fixed where

$$M = \begin{pmatrix} a & b & c \\ d & e & f \\ g & h & i \end{pmatrix}$$

Thus $f: [x,y,0] \to [x,y,0]$, and so

$$\begin{pmatrix} a & b & c \\ d & e & f \\ g & h & i \end{pmatrix} \begin{pmatrix} x \\ y \\ 0 \end{pmatrix} = \begin{pmatrix} ax + by \\ dx + ey \\ gx + hy \end{pmatrix} = \begin{pmatrix} kx \\ ky \\ 0 \end{pmatrix}$$

for some $k \neq 0$, $k \in F$. Since x and y are arbitrary, $a = k$, $b = 0$, $d = 0$, $e = k$, and $g = h = 0$. Thus

$$M = \begin{pmatrix} k & 0 & c \\ 0 & k & f \\ 0 & 0 & i \end{pmatrix}$$

By Theorem 3.1.4, M is equivalent to

$$\begin{pmatrix} 1 & 0 & \dfrac{c}{r} \\ 0 & 1 & \dfrac{f}{r} \\ 0 & 0 & \dfrac{i}{r} \end{pmatrix}$$

which is of the desired form.

The following theorem may be similarly established:

THEOREM 3.2.6 *Let f_M be a matrix-representable collineation on π_F. Then the line $\langle 0,1,0 \rangle$ is held pointwise fixed if and only if M is of the form*

$$\begin{pmatrix} 1 & r & 0 \\ 0 & s & 0 \\ 0 & t & 1 \end{pmatrix}$$

and the line $\langle 1,0,0 \rangle$ is held pointwise fixed if and only if M is of the form

$$\begin{pmatrix} r & 0 & 0 \\ s & 1 & 0 \\ t & 0 & 1 \end{pmatrix}$$

In every central collineation of Example 3.2.4, there exists a point that is held linewise invariant. In (i), every point is held linewise invariant; in (ii), the point $[-r,-s,1-t]$ is held linewise invariant; in (iii), f leaves p_1 linewise fixed and g leaves p_8 linewise fixed; in (iv), f leaves the point $[1,r^{-1}s,0]$ linewise fixed. Statement (ii) is proved in Theorem 3.2.7; (iii) is left for the student to verify and (iv) is left as an exercise.

THEOREM 3.2.7 *If*

$$M = \begin{pmatrix} 1 & 0 & r \\ 0 & 1 & s \\ 0 & 0 & t \end{pmatrix}$$

then f_M leaves $[-r,-s,1-t]$ linewise fixed.

Proof. Recall that f_M: $\langle a,b,c \rangle \rightarrow \langle a',b',c' \rangle$ where $(a',b',c') = (a,b,c)M^{-1}$. It is easily verified that

$$M^{-1} = \frac{1}{t}\begin{pmatrix} t & 0 & -r \\ 0 & t & -s \\ 0 & 0 & 1 \end{pmatrix}$$

so

$$(a',b',c') = \frac{1}{t}(at,\ bt,\ -ar - bs + c)$$

(Why is it that $t \neq 0$?) Thus $\langle a,b,c \rangle$ is held fixed iff $-ar - bs + c = ct$ iff $-ar - bs + c(1 - t) = 0$. Therefore, all these lines pass through $[-r,-s,1-t]$.

Example 3.2.4 and Theorem 3.2.7 are simply special cases of the general theorem stating that every central collineation leaves a point linewise fixed. This theorem can be proved with the help of a lemma.

LEMMA 3.2.8 *If f is a central collineation that leaves L pointwise invariant and also leaves invariant two points p and q not on L, then f is the identity.*

Proof. The proof is left as an exercise.

THEOREM 3.2.9 *If f is a central collineation, there exists a point that is held linewise invariant.*

Proof. Let L be a pointwise invariant line. Clearly, if f is the identity, then all points are held linewise invariant. Suppose that f is not the identity. Then, by Theorem 3.2.7, there is at most one point not on L that is held fixed.

Case 1.   Suppose that $p \notin L$ and $f(p) = p$. Then any line M through p is held fixed because it contains two fixed points, p and $L \cap M$.

Case 2.   Suppose that there are no fixed points other than those on L. Let p be any point not on L. Because $f(p)$ is a point distinct from p, we may consider the line $pf(p)$. Let M denote the line $pf(p)$ and let $q = M \cap L$; then it is clear that f leaves M invariant because $M = pq$ and $f(M) = f(p)f(q) = f(p)q = M$. Thus p lies on an invariant line. By the same argument, if r is a point not on L or M, then there exists an invariant line N passing through r. Let $s = M \cap N$. Now s, being the intersection of two fixed lines, must be held fixed, and so s lies on L. But $M \cap L = q$, so $s = q$, and therefore $N \cap L = q$. Thus both p and r lie on invariant lines passing through q. Since r was arbitrarily chosen, we may say that every point not on L lies on an invariant line passing through q. Since L is also invariant, every line through q is invariant.

DEFINITION 3.2.10   *If f is a central collineation other than the identity, then the unique line held pointwise invariant is called the* axis *of f and the unique point held linewise invariant is called the* center *of f.*

DEFINITION 3.2.11   *A central collineation f other than the identity is called an* elation *if its center lies on its axis. If the center is not the axis, f is called a* homology. *If f is the identity, it is called both an elation and a homology. If f is a homology different from the identity such that f^2 (that is, f composed with itself) is the identity, then f is a* harmonic homology.

EXAMPLE 3.2.12   Refer back to Example 3.2.4.

   i. The identity is both an elation and a homology by definition.

  ii. By Theorem 3.2.7, f_M is an elation with axis $\langle 0,0,1 \rangle$ if and only if $[-r, -s, 1 - t] \in \langle 0,0,1 \rangle$, that is, if and only if $t = 1$. Thus matrix-representable elations with axis $\langle 0,0,1 \rangle$ may be written in the form

$$\begin{pmatrix} 1 & 0 & r \\ 0 & 1 & s \\ 0 & 0 & 1 \end{pmatrix}$$

Homologies with axis $\langle 0,0,1 \rangle$ and center $[0,0,1]$ may be represented by

$$\begin{pmatrix} 1 & 0 & 0 \\ 0 & 1 & 0 \\ 0 & 0 & t \end{pmatrix}$$

for some $t \in F$, $t \neq 0$. If F does not have characteristic 2,

$$\begin{pmatrix} 1 & 0 & 0 \\ 0 & 1 & 0 \\ 0 & 0 & -1 \end{pmatrix}$$

represents a harmonic homology.

iii. The collineation f of Example 3.2.4(iii) is a harmonic homology with center p_1 and axis L_1; g is an elation with center p_8 and axis L_1.

iv. The collineation (f,F) is an elation with center $[1,r^{-1}s,0]$ and axis $\langle 0,0,1 \rangle$. If $r = 1$, then the collineation (g,G) is a homology with center $[0,1,0]$ and axis $\langle 0,1,0 \rangle$. Furthermore, if $s = -1$, the homology is harmonic. If $s = 1$, the collineation (g,G) is a homology with center $[1,0,0]$ and axis $\langle 1,0,0 \rangle$. Also, if $r = -1$, the homology is harmonic. Verification of the latter statements is left to the student.

This section is concluded with a minimal characterization of uniqueness for elations and homologies.

THEOREM 3.2.13 *An elation is uniquely determined by its axis L and the image of one point q where $q \notin L$. A homology is uniquely determined by its center p, its axis L, and the image of one point q where $q \neq p$ and $q \notin L$.*

Proof. The proof is left as an exercise.

EXERCISES

1. Refer back to Example 3.1.5(iii).
   a. Show that the collineation (f,F) holds the point $[1,r^{-1}s,0]$ linewise fixed.

b. Show that the collineation (g, G) is not a central collineation if $r = s = a$.

2. Prove Lemma 3.2.8.

3. Determine if the following central collineations f in π_F are elations, homologies, or harmonic homologies:
   a. f: $[x, y, z] \rightarrow [z, y, x]$.
   b. f: $[x, y, z] \rightarrow [x, y, x - y + z]$.
   c. f: $[x, y, z] \rightarrow [x + z, y + z, 2z]$ where F does not have characteristic 2.

4. Prove Theorem 3.2.13.

5. Let f and g be central collineations with distinct centers p and q and axes L and M, respectively. Show that if $p \in M$ and $q \in L$, then $f \circ g = g \circ f$.

6. Observe that f_M holds the point $[x, y, z]$ fixed in π_F if and only if

$$M \begin{pmatrix} x \\ y \\ z \end{pmatrix} = r \begin{pmatrix} x \\ y \\ z \end{pmatrix}$$

for some real number r. This equation is equivalent to the equation

$$(M - rI) \begin{pmatrix} x \\ y \\ z \end{pmatrix} = \begin{pmatrix} 0 \\ 0 \\ 0 \end{pmatrix}$$

where I denotes the matrix

$$\begin{pmatrix} 1 & 0 & 0 \\ 0 & 1 & 0 \\ 0 & 0 & 1 \end{pmatrix}$$

and $M - rI$ is the difference between the two matrices M and rI. Recall that the latter equation implies that the determinant of the matrix $(M - rI)$ is 0.

   a. Find a collineation f_M on π_R that holds exactly three noncollinear points fixed; two points fixed; one point fixed.
   b. Show that every matrix-representable collineation on π_R holds at least one point fixed and one line fixed.
   c. State and prove a theorem giving necessary and sufficient conditions on a field F such that every matrix-representable collineation on π_F has at least one fixed point.
   d. (1) Give examples of fields F such that there exist collineations f_M that hold no points fixed.
      (2) Give examples of collineations f_M that hold no points fixed.

7. A collineation f is called an *involution* or *involutory collineation* if and only if f is not the identity but $f \circ f$ is the identity.

A collineation is called a *quasi-central collineation* if and only if the fixed points of f form a Baer subplane (refer back to exercise 2.7.4). Using these two definitions,

a. Find examples of the following mappings: an involutory homology (a harmonic homology); an involutory elation (an elation that is an involution); a quasi-central collineation.

b. Show that in a plane satisfying Fano's Axiom, every involutory central collineation is a homology.

c. Show that in a Fano plane, every involutory central collineation is an elation.

d. Show that in a Fano plane, every elation is an involutory central collineation.

e. Show that every involution is either a central collineation or a quasi-central collineation.

3.3   AFFINE COLLINEATIONS

There is a close relationship between collineations on an affine plane and collineations on a projective plane, as the first theorem in this section demonstrates.

THEOREM 3.3.1

i. *If f is a collineation on π that holds line L fixed, then $f|\mathscr{P}-L$ is a collineation on $\alpha(\pi_L^-)$.*

ii. *If g is a collineation on α, then there exists a unique collineation f on $\pi(\alpha^+)$ such that $f|\mathscr{P}=g$ where $\mathscr{P}$ denotes the points of α.*

Proof.   The proof of (i) is immediate. For the proof of (ii), define f on $\pi(\alpha^+)$ as follows:

$$f(p) = g(p) \text{ if } p \in \mathscr{P}$$
$$f([L]) = [g(L)]$$

Now we must prove that: (1) f is well defined, (2) f is one-to-one, (3) f is onto, (4) f preserves collinearity, and (5) if f' is a collineation on $\pi(\alpha^+)$ and $f'|\mathscr{P} = g$, then $f' = f$. Each of these statements can be proved as follows:

1. Suppose that $[M] = [L]$. Then $M \parallel L$ and so $g(M) \parallel g(L)$, since an

affine collineation preserves parallelism (because it preserves incidence). Thus $[g(M)] = [g(L)]$, $f([M]) = f([L])$, and f is well defined.

2. Clearly, $f|\mathscr{P}$ is one-to-one. Suppose that $f([L]) = f([M])$. Then $[g(L)] = [g(M)]$, so $g(L) \parallel g(M)$. Since g is a collineation, g^{-1} is also a collineation. Therefore, $L \parallel M$ and thus $[L] = [M]$. This implies that f is one-to-one.

3. Since $f(g^{-1}(p)) = p$ for $p \in \mathscr{P}$ and since $f([g^{-1}(L)]) = [L]$, we know that f is an onto map.

4. Since $f|\mathscr{P} = g$, we know that collinearity is preserved in α. Furthermore, f maps the ideal line onto itself. All that must be shown is that if $p \in L^+$, where L^+ is not the ideal line and p is an ideal point, then $f(p) \in f(L^+)$. A nonideal line L^+ of $\pi(\alpha^+)$ is of the form $L \cup [L]$, where L is a line of α. Thus $p = [L]$, and it follows that $f(p) = f([L]) = [g(L)]$. Since $f(L^+) = \{f(p): p \in L^+\} = \{g(p): p \in L\} \cup \{[g(L)]\}$, it follows that $f(p) \in f(L^+)$.

5. Consider the map $h = f - f'$. Since $f|\mathscr{P} = f'|\mathscr{P}$, we know that $h|\mathscr{P} = I$ (the identity on α). For the point $[L]$, we have $h([L]) = [I(L)] = [L]$. Thus h is the identity on $\pi(\alpha^+)$, and it follows that $f = f'$.

COROLLARY 3.3.2 *If g is a collineation on α and π is any principal extension of α, then there exists a unique collineation f on π such that $f|\mathscr{P} = g$, where $\mathscr{P}$ denotes the set of points of α.*

Proof. The proof follows from Theorem 3.3.1 and from the fact that principal extensions are unique (Corollary 1.7.12).

DEFINITION 3.3.3 *If g is a collineation on α and f is a collineation on $\pi(\alpha^+)$ such that $f|\mathscr{P} - L = g$, where $\mathscr{P}$ is the set of points of π, then g is the* restriction *of f and f is the* extension *of g.*

Examples of affine collineations can easily be derived from this definition.

EXAMPLE 3.3.4 Consider the plane α_F and let $g: (x,y) \rightarrow (ax+by+c, dx+ey+f)$ where $a,b,c,d,e,f \in F$ and the matrix

$$M = \begin{pmatrix} a & b & c \\ d & e & f \\ 0 & 0 & 1 \end{pmatrix}$$

is a nonsingular matrix. Using the notation $[x, y, 1]$ for points of α_F, denote $f([x, y, 1])$ by $[x', y', 1]$ where

$$\begin{pmatrix} x' \\ y' \\ 1 \end{pmatrix} = \begin{pmatrix} a & b & c \\ d & e & f \\ 0 & 0 & 1 \end{pmatrix} \begin{pmatrix} x \\ y \\ 1 \end{pmatrix}$$

Clearly, $g = f_M | \mathscr{P}$. It follows from Theorem 3.3.1 that g is a collineation on α_F.

DEFINITION 3.3.5 *A collineation g on α_F of the type $g: (x, y) \to (ax + by + c, dx + ey + f)$ is called a* matrix-representable affine collineation.

Notation. This collineation is denoted by g_M where

$$M = \begin{pmatrix} a & b & c \\ d & e & f \\ 0 & 0 & 1 \end{pmatrix}$$

There is clearly a close relationship between matrix-representable collineations on π_F and matrix-representable affine collineations on α_F. Theorem 3.3.6 follows almost immediately from Theorem 2.1.10 and from Definitions 3.1.3, 3.3.3, and 3.3.5.

THEOREM 3.3.6 *The collineation g on α_F is representable by the matrix M if and only if the extension of f on π_F is representable by the matrix M.*

EXAMPLE 3.3.7

i. Let $\alpha_1 = \alpha((\pi_{S_3})_{L_1}^-)$ and let $\alpha_2 = \alpha((\pi_{S_3})_{L_8}^-)$. Tables 14 and 15 display α_1 and α_2, respectively.

TABLE 14

						j						
	0	2	3	4	5	6	7	8	9	10	11	12
	1	11	10	9	9	7	6	5	4	3	3	1
i	3	1	11	10	11	10	7	6	5	4	5	4
	9	7	6	5	4	3	9	1	7	6	11	10

TABLE 15

	j											
	0	1	2	3	4	5	6	7	9	10	11	12
	0	12	11	10	9	9	7	7	4	3	2	2
i	3	0	12	11	10	11	10	9	7	4	3	4
	9	2	7	0	12	4	3	2	0	12	11	10

Define mappings f_1, f_2 on the points of α_1 and f_3, f_4 on the points of α_2 as follows:

$f_1(p_i) = p_{g_1(i)}$ where

$$g_1: \quad 1 \to 1 \qquad 5 \to 6 \qquad 9 \to 3$$
$$3 \to 9 \qquad 6 \to 5 \qquad 10 \to 4$$
$$4 \to 10 \qquad 7 \to 11 \qquad 11 \to 7$$

$f_2(p_i) = p_{g_2(i)}$ where

$$g_2: \quad 1 \to 5 \qquad 5 \to 6 \qquad 9 \to 4$$
$$3 \to 7 \qquad 6 \to 1 \qquad 10 \to 3$$
$$4 \to 11 \qquad 4 \to 10 \qquad 11 \to 9$$

$f_3(p_i) = p_{g_3(i)}$ where

$$g_3: \quad 0 \to 0 \qquad 5 \to 10 \qquad 10 \to 4$$
$$2 \to 2 \qquad 7 \to 11 \qquad 11 \to 7$$
$$3 \to 9 \qquad 9 \to 3 \qquad 12 \to 12$$

$f_4(p_i) = p_{g_4(i)}$ where

$$g_4: \quad 0 \to 0 \qquad 4 \to 11 \qquad 10 \to 3$$
$$2 \to 2 \qquad 7 \to 10 \qquad 11 \to 9$$
$$3 \to 7 \qquad 9 \to 4 \qquad 12 \to 12$$

A check of Example 3.2.4(iii) shows that $f_1 = f \,|\, \mathscr{P} - L_1; f_2 = g \,|\, \mathscr{P} - L_1; f_3 = f \,|\, \mathscr{P} - L_8; f_4 = g \,|\, \mathscr{P} - L_8$. Thus all these mappings are collineations.

ii. Consider the plane $\alpha_N(9)$. Let $f': [x,y,1] \to [x+r, y+s, 1]$; $g': [x,y,1] \to [xr, ys, 1]$. Recalling Example 3.1.5(iii), we see that $f' = f \,|\, \mathscr{P} - \langle 0,0,1 \rangle$ and that $g' = g \,|\, \mathscr{P} - \langle 0,0,1 \rangle$; therefore, f' and g' are collineations on $\alpha_N(9)$. We may represent f' and g' as:

$$f': (x,y) \to (x+r, y+s)$$
$$g': (x,y) \to (xr, ys)$$

Definition 3.3.8 lists five different types of affine collineations, all of which are restrictions of central collineations.

DEFINITION 3.3.8 *Let f be an affine collineation on α such that its extension f^+ on $\pi(\alpha^+)$ is a central collineation. Then:*

1. *f is a* dilatation *(also called dilation) if and only if f^+ has axis L_∞.*

2. *f is a* translation *if and only if f^+ is an elation with axis L_∞.*

3. *f is a* point reflection *about p if and only if f^+ is a harmonic homology with axis L_∞ and center p.*

4. *f is a* line reflection *(or reflection) with axis L if and only if f^+ is a harmonic homology with axis L and center on L_∞.*

5. *f is a* shear *with axis L if and only if f^+ is an elation with axis L and center on L_∞.*

EXAMPLE 3.3.9

i. Consider the plane α_F. By Theorems 3.2.5 and 3.3.6, the matrix-representable dilatations g_M can be represented by

$$M = \begin{pmatrix} t & 0 & r \\ 0 & t & s \\ 0 & 0 & 1 \end{pmatrix}$$

Thus g_M: $(x,y) \rightarrow (ts + r, ty + s)$. By Example 3.2.12(ii), the matrix-representable translations g_M can be represented by

$$M = \begin{pmatrix} 1 & 0 & r \\ 0 & 1 & s \\ 0 & 0 & 1 \end{pmatrix}$$

Thus g_M: $(x,y) \rightarrow (x+r, y+s)$. If F does not have characteristic 2, then we can represent a point reflection g_M about $p = (0,0)$ by

$$\begin{pmatrix} -1 & 0 & 0 \\ 0 & -1 & 0 \\ 0 & 0 & 1 \end{pmatrix}$$

Thus g_M: $(x,y) \rightarrow (-x,-y)$. Again assuming that the characteristic of F is not 2, Theorem 3.2.6 and Example 3.2.12(ii) indicate that a harmonic homology with axis $\langle 1,0,0 \rangle$ and center $[1,0,0]$ is represented by

$$\begin{pmatrix} -1 & 0 & 0 \\ 0 & 1 & 0 \\ 0 & 0 & 1 \end{pmatrix}$$

Therefore, a line reflection g_M with axis $x = 0$ (the y-axis) is represented by

$$\begin{pmatrix} -1 & 0 & 0 \\ 0 & 1 & 0 \\ 0 & 0 & 1 \end{pmatrix}$$

Thus g_M: $(x,y) \rightarrow (-x,y)$. Theorems 3.2.6 and 3.2.7 also indicate that an elation g_M with axis $\langle 1,0,0 \rangle$ and center $[0,1,0]$ is represented by

$$\begin{pmatrix} 1 & 0 & 0 \\ s & 1 & 0 \\ 0 & 0 & 1 \end{pmatrix}$$

Therefore, a shear g_M with axis $\langle 1,0,0 \rangle$ is represented by

$$\begin{pmatrix} 1 & 0 & 0 \\ s & 1 & 0 \\ 0 & 0 & 1 \end{pmatrix}$$

and so g_M: $(x,y) \rightarrow (x, sx + y)$.

ii. Reviewing Example 3.3.7(i), we can easily check that f_1 is a point reflection about p_1 on α_1; that f_2 is a translation on α_1; that f_3 is a line reflection with axis L_1 on α_2; and that f_4 is a shear with axis L_1 on α_2.

iii. Reviewing Example 3.3.7(ii), we see that f' is a translation and that g' is a dilatation (with center $(0,0)$).

As can be seen from these examples, the names of the five affine collineations are suggestive, especially in α_R. A dilatation (for example, f: $(x,y) \rightarrow (xr, ys)$) dilates, that is, expands or contracts; a translation (f: $(x,y) \rightarrow (x + r, y + s)$) translates; a point reflection (f: $(x,y) \rightarrow (-x,-y)$) and a line reflection (f: $(x,y) \rightarrow (-x,y)$) reflect symmetrically about a point and a line, respectively; a shear (f: $(x,y) \rightarrow (x, sx + y)$) depicts the action of two equal forces operating on the two half-planes formed by the axis in directions opposite and parallel to the axis.

As might be expected, not all affine collineations fall into one of these five classes. Two that do not are given in Example 3.3.10; both are restrictions of noncentral collineations.

EXAMPLE 3.3.10

i. Let

$$M = \begin{pmatrix} 1 & 1 & 0 \\ 1 & 0 & 0 \\ 0 & 0 & 1 \end{pmatrix}$$

and consider g_M in α_F; g_M: $(x,y) \to (x + y, x)$. Clearly, g_M is a restriction of f_M: $[x,y,z] \to [x + y, x, z]$, and this is not a central collineation, as can be seen by noting that the fixed points $[x,y,z]$ must satisfy the equations $x = rx + ry$; $y = rx$, $z = rz$ for some $r \neq 0$. If $z = rz$, then $r = 1$ or $z = 0$. If $r = 1$, then $x = x + y$ and $y = x$, which together imply that $x = y = 0$. If $z = 0$, then $x = rx + r^2 x = r(1 + r)x$; thus $r^2 + r - 1 = 0$ or $x = y = 0$. We may conclude that there are at most three fixed points: $[0,0,1]$, $[r_1, r_1^2, 0]$, $[r_2, r_2^2, 0]$ (where r_1, r_2 are the possible roots of $r^2 + r - 1 = 0$), depending on the field F. In any case (including $\pi(2)$), no line is held pointwise fixed. Thus g_M is not one of our five types of affine collineations.

ii. Consider the plane α_{R_4} of Example 2.3.12. Let f: $p_i \to p_{i+1}$; f: $q \to q$. In this example there is not readily available a collineation of $\pi(4)$ for which f is a restriction. We must therefore work within the plane α_{R_4} to discover the nature of f. A check of Example 2.3.12 shows that no line is held pointwise fixed and no point is held linewise fixed. Thus if the extension of f to $\pi(\alpha^+)$ is a central collineation, it must be an elation with axis L_∞. But f holds q fixed and thus f^+ cannot be such an elation. We may conclude that f^+ is not a central collineation and so f is not one of our five types of affine collineations.

Example 3.3.10(ii) suggests that it would be convenient to characterize our special affine collineations within the context of the affine plane rather than in terms of its projective extension. The following preliminary observations regarding the nature of affine collineations are easily established.

THEOREM 3.3.11   *Let f be an affine collineation.*

i. *If f is a translation, then no points are held fixed unless f is the identity.*

ii. *If f leaves no points fixed, then f may or may not be a translation.*

iii. *If f holds both a point linewise invariant and a line pointwise invariant, then f is the identity.*

iv. *If f leaves a point linewise invariant, then f is a dilatation.*

v. *If f leaves a line pointwise invariant, then it may be a line reflection, a shear, or neither.*

Proof. The proof is left as an exercise.

Several different alternate definitions can be given for all five of our affine collineations. A standard characterization of dilatation and translation in affine terms is given in Theorem 3.3.12; the student is encouraged to think of characterizations for point reflection, line reflection, and shear.

THEOREM 3.3.12 *An affine collineation f on α is a dilatation if and only if, for any two points p and q, pq $\parallel$ f(p)f(q); f is a translation if and only if f is a dilatation that leaves no points fixed or is the identity collineation on α.*

Proof. Suppose that f is a dilatation. Then its extension f^+ is a central collineation on $\pi(\alpha^+)$ leaving L_∞ fixed. Let r be the ideal point on pq and let r' be the ideal point on $f(p)f(q)$, that is, $r = [pq], r' = [f(p)f(q)]$. Since f maps line pq into line $f(p)f(q)$, we have $f^+ [pq] = [f(p)f(q)] = [pq]$. Thus $r = r'$, and it follows that $pq \parallel f(p)f(q)$. This argument may be reversed to prove the converse.

The second part of the theorem follows easily from the first part.

This section is concluded with a minimal characterization of uniqueness for our affine collineations. The proof of Theorem 3.3.13 follows immediately from the three definitions in this section and from Theorem 3.2.13. The details are left as an exercise.

THEOREM 3.3.13

i. *A dilatation is uniquely determined by the images of two of its points. If the dilatation is a translation, it is determined by the image of one of its points.*

ii. *A line reflection is uniquely determined by its axis and the image of one point not on the axis.*

iii. *A point reflection is uniquely determined by its center and the image of one point different from the center.*

iv. *A shear is uniquely determined by its axis and the image of one of its points not on the axis.*

EXERCISES

1. Prove Theorem 3.3.11.

2. Prove Theorem 3.3.13.

3. Identify the following collineations f in α_R:
   a. f: $(x,y) \rightarrow (-x + 1, y + 1)$.
   b. f: $(x,y) \rightarrow (x - 2y + 4, -y + 4)$.
   c. f: $(x,y) \rightarrow (y,x)$.

4. Write a matrix representation for the following collineations on α_F (F does not have characteristic 2):
   a. f is a dilation with center $(1,1)$ mapping $(0,0)$ onto the point $(-1,-1)$.
   b. f is a translation mapping $(1,2)$ onto $(2,1)$.
   c. f is a shear with axis $y = 1$ mapping $(0,0)$ onto $(0,2)$.

5. Recall the affine plane of exercise 1.3.8 (see also exercise 2.1.6(f)).
   a. Find a translation on this plane.
   b. Find a line reflection on this plane.

6. Let $\Sigma_0 = (\mathscr{P}, \mathscr{L}, \mathscr{I})$ be a primitive plane and let f be a collineation on Σ_0. Show that there exists one and only one collineation g on Σ_0^+ such that $g|\mathscr{P} = f$.

3.4   PROJECTIVITIES

In the first three sections, the concept of a collineation, a mapping from a plane onto itself, was introduced. Also important to our study is another kind of transformation that maps points of one line onto points of another line (or the same line). In the projective plane this transformation is called a projectivity; for affine planes, the term "affine projectivity" is used.

DEFINITION 3.4.1   *Let L and L' be two lines of π and let p be a point that is not on either line. Then the mapping f from L to L' such*

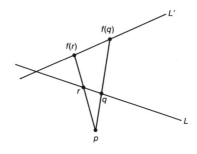

FIGURE 23

that $f(q) = pq \cap L'$ for all $q \in L$ is called a perspectivity *(see Figure 23). The point p is called the* center of the perspectivity. *The mapping f is denoted by* $L \overset{p}{\wedge} L'$. *If $L = L'$, then f is the identity on L.*

DEFINITION 3.4.2 *A composition of a finite number of perspectivities is called a* projectivity.

EXAMPLE 3.4.3 The first four examples are derived from π_{S_3} as displayed in Table 5 (p. 52). Let $f: p_i \to p_{t(i)}$ where:

i. $\qquad\qquad t: 2 \to 2 \qquad 7 \to 11$
$\qquad\qquad\qquad\quad 6 \to 5 \qquad 9 \to 3$

Here $f: L_7 \to L_{11}$ and $f = L_7 \overset{p_1}{\wedge} L_{11}$.

ii. $\qquad\qquad t: 1 \to 1 \qquad 6 \to 5$
$\qquad\qquad\qquad\quad 5 \to 6 \qquad 8 \to 8$

Here $f: L_8 \to L_8$ and $f = hg$ where $g = L_8 \overset{p_7}{\wedge} L_{12}$ and $h = L_{12} \overset{p_3}{\wedge} L_8$. This relationship is written simply as $f = L_8 \overset{p_7}{\wedge} L_{12} \overset{p_3}{\wedge} L_8$.

iii. $\qquad\qquad t: 1 \to 5 \qquad 6 \to 1$
$\qquad\qquad\qquad\quad 5 \to 6 \qquad 8 \to 8$

Here $f: L_8 \to L_8$ where $f = L_8 \overset{p_2}{\wedge} L_5 \overset{p_0}{\wedge} L_8$.

iv. $\qquad\qquad t: 0 \to 9 \qquad 3 \to 1$
$\qquad\qquad\qquad\quad 1 \to 0 \qquad 9 \to 3$

Here $f: L_0 \to L_0$ where $f = L_0 \overset{p_5}{\wedge} L_1 \overset{p_{10}}{\wedge} L_2 \overset{p_4}{\wedge} L_0$.

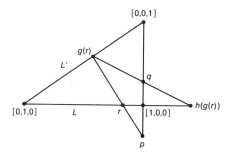

FIGURE 24

Note that (ii), (iii), and (iv) show that we can permute two, three, or four points, respectively, on L_8. The last two examples come from π_F.

v.
$$f: [x,y,0] \to [x,y,-x]$$

Here f: $\langle 0,0,1 \rangle \to \langle 1,0,1 \rangle$ and $f = L \overset{p}{\overline{\wedge}} L'$ where $L = \langle 0,0,1 \rangle$, $L' = \langle 1,0,1 \rangle$, and $p = [0,0,1]$. To verify this, let $q = [x,y,0]$ and check that $pq = \langle -y,x,0 \rangle$ and that $pq \cap L' = [x,y,-x]$.

vi.
$$f: [x,y,0] \to [-x,y,0]$$

Here f: $\langle 0,0,1 \rangle \to \langle 0,0,1 \rangle$ and $f = h \circ g$ where $g = L \overset{p}{\overline{\wedge}} L'$, $h = L' \overset{q}{\overline{\wedge}} L$, and $L = \langle 0,0,1 \rangle$; $L' = \langle 1,0,0 \rangle$, $p = [1,0,1]$, and $q = [-1,0,1]$ (see Figure 24). (We assume here that F does not have characteristic 2.) To verify this, let $r = [x,y,0]$ and check that $pr = \langle -y,x,y \rangle$ and that $pr \cap L' = [0,-y,x]$; thus $g(r) = [0,-y,x]$. Also, $qg(r) = \langle y,x,y \rangle$ and $qg(r) \cap L = [-x,y,0]$, so $h(g(r)) = [-x,y,0]$. Since $f(r) = [-x,y,0]$, we have shown that $f = h \circ g$.

In general it appears difficult to ascertain whether a given map is a projectivity. Evidently we must search for points and lines that will act as centers and images, respectively, of perspectivities whose composition will yield the map. The following theorem helps us considerably in this respect.

THEOREM 3.4.4 *Let f be a collineation of π. If $f|M$ is a projectivity for some line M, then $f|L$ is a projectivity for every line L in π.*

Proof. Let f be a collineation on π and suppose that $f|M$ is a projectivity. Suppose also that L is an arbitrary line, that p is a point not on

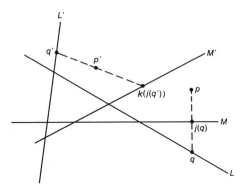

FIGURE 25

L or M, and that $L' = f(L)$, $M' = f(M)$, $p' = f(p)$, $g = f|M$, $h = f|L$, and $j = L \stackrel{p}{\overline{\wedge}} M$ (see Figure 25). The composition of mappings uses the notation introduced in exercise 3.1.3.

First it must be shown that the mapping $k = L' \stackrel{p'}{\overline{\wedge}} M'$ where $k = g \circ j \circ h^{-1}$. Since $g: M \to M'$, $h: L \to L'$, and $j: L \to M$, it is obvious that $k: L' \to M'$. Let q' be an arbitrary point on L' and let $f^{-1}(q') = q$. Clearly, $q \in L$, and since j is a perspectivity with center p, it follows that q, $j(q)$, and p are collinear. Since f is a collineation, q', $f(j(q))$, and p' are collinear. Recall that $f|L = h$; it follows that $q = h^{-1}(q')$, and so $j(q) = j(h^{-1}(q'))$. Similarly, recall that $g = f|M$; it follows that $f(j(q)) = g(j(q))$, and so $f(j(q)) = g \circ j \circ h^{-1}(q')$. Therefore, the points q', $g \circ j \circ h^{-1}(q')$, and p' are collinear. Since q' was arbitrary, $k = L' \stackrel{p'}{\overline{\wedge}} M'$.

Since $k = g \circ j \circ h^{-1}$, it follows that $h = k^{-1} \circ g \circ j$. But k, g, and j are all projectivities, so we may conclude that h is a projectivity.

DEFINITION 3.4.5 *A collineation that induces a projectivity on every line is called a* projective collineation.

THEOREM 3.4.6 *A central collineation is a projective collineation.*

Proof. Since the axis L of a central collineation f is held pointwise invariant, $f|L$ is a projectivity because the identity on L is a projectivity. By Theorem 3.4.4, f is therefore a projective collineation.

COROLLARY 3.4.7 *The product of central collineations is a projective collineation.*

Notice that the projectivities of (i), (ii), and (iii) of Example 3.4.3 are restrictions of the central collineations of Example 3.2.4. The projectivities of Example 3.4.3(i) and (ii) are $f|L_7$ and $f|L_8$, respectively, where f is the collineation of Example 3.2.4(ii). The projectivity of Example 3.4.3(iii) is $f|L_8$, where f is the collineation of Example 3.2.4(iii). The projectivity of Example 3.4.3(iv) is not the restriction of one collineation but the restriction of a product of collineations. Proof of this assertion is left to the student. The projectivity of Example 3.4.3(v) is $f|L$, where $L = \langle 0,0,1 \rangle$ and f is the central collineation represented by

$$\begin{pmatrix} 1 & 0 & 0 \\ 0 & 1 & 0 \\ -1 & 0 & 1 \end{pmatrix}$$

Here f is an elation with center $[0,0,1]$ and axis $\langle 1,0,0 \rangle$. The projectivity of Example 3.4.3(vi) is $f|L$, where $L = \langle 0,0,1 \rangle$ and f is the central collineation represented by

$$\begin{pmatrix} -1 & 0 & 0 \\ 0 & 1 & 0 \\ 0 & 0 & 1 \end{pmatrix}$$

Here f is a harmonic homology with center $[1,0,0]$ and axis $\langle 1,0,0 \rangle$.

The converse of Corollary 3.4.7 is not true, as the following example shows.

EXAMPLE 3.4.8   Let π be the plane $\pi_N(9)$ and let the collineation (h,H) be defined as follows:

$$\begin{aligned} h: \quad [x,y,1] \quad &\rightarrow [x + y, -x + y, 1] \\ [1,x,0] \quad &\rightarrow [1,x,0] \text{ where } x \neq 0,1,-1 \\ [0,1,0] \quad &\rightarrow [1,1,0] \\ [1,1,0] \quad &\rightarrow [1,0,0] \\ [1,0,0] \quad &\rightarrow [1,-1,0] \\ [1,-1,0] \quad &\rightarrow [0,1,0] \end{aligned}$$

$$\begin{aligned} H: \quad \langle m,1,k \rangle \quad &\rightarrow \langle m,1,km + k \rangle \text{ where } m \neq 0,1,-1 \\ \langle 1,0,k \rangle \quad &\rightarrow \langle -1,1,k \rangle \\ \langle 1,1,k \rangle \quad &\rightarrow \langle 1,0,k \rangle \\ \langle -1,1,k \rangle \quad &\rightarrow \langle 0,1,k \rangle \\ \langle 0,1,k \rangle \quad &\rightarrow \langle 1,1,k \rangle \\ \langle 0,0,1 \rangle \quad &\rightarrow \langle 0,0,1 \rangle \end{aligned}$$

The student will find it a challenging exercise to show that (h,H) is a collineation on $\pi_N(9)$. It is easily checked that (h,H) does not hold a line pointwise invariant and thus is not a central collineation. Proof that it is not a product of central collineations is deferred until Chapter 12. The following argument shows that it is a projective collineation.

Let $L = \langle 1,0,0 \rangle$ and notice that L is made up of points q where $q = [0,y,1]$ for $y \in N$ or $q = [0,1,0]$. Observe that $h([0,y,1]) = [y,y,1]$ and $h([0,1,0]) = [1,1,0]$ and let $p = [1,0,0]$. It is then easily checked that p, $[0,y,1]$, and $[y,y,1]$ lie on the line $\langle 0,1,-y \rangle$ and that p, $[0,1,0]$, and $[1,1,0]$ lie on the line $\langle 0,0,1 \rangle$. Thus p, q, and $h(q)$ are collinear for all $q \in L$, and so $h|L = L \overset{p}{\barwedge} M$ where $M = \langle -1,1,0 \rangle$. It follows from Theorem 3.4.4 that (h,H) is a projective collineation.

Two questions should be raised here. First, is every projectivity extendable to a collineation? In other words, if $g = L \overset{p}{\barwedge} M$, does there exist a collineation f such that $f|L = g$? Second, does every collineation induce a projectivity on some line (and thus on every line)? In other words, is every collineation a projective collineation? The answer to both questions is negative. To understand the reason for such an answer to the first question, consider Σ_o^+, the plane of Example 2.5.8(iii). Let L be any line containing a point of $\Sigma(7_3)$ and let L' be any line not containing a point of $\Sigma(7_3)$; for example, let $L = pr$ and $L' = pq$ as in Theorem 2.7.9. Let s be any point not on L or L' and let $g = L \overset{s}{\barwedge} L'$. If there were a collineation f such that $f|L = g$, it would follow that $\alpha(\pi_L{}^-) \sim \alpha(\pi_{L'}{}^-)$, which would contradict Theorem 2.7.9. Thus g is not extendable. A negative answer to the second question can be explained if we consider the plane $\pi(4)$. Using the representation π_F where $F = GF(4)$, our collineation f is constructed as follows: f: $[x,y,z] \rightarrow [\gamma(x), \gamma(y), \gamma(z)]$ where γ: $GF(4) \rightarrow GF(4)$ and γ: $0,1,a,b \rightarrow 0,1,b,a$. Since γ is an automorphism of F (that is, a one-to-one mapping from F onto F preserving addition and multiplication — see exercise 3.1.6), the mapping f is one-to-one and onto and preserves collinearity. This last fact follows because $[x_i,y_i,z_i]$ are collinear if and only if there exists $r,s,t \in F$ such that $rx_i + sy_i + tz_i = 0$. This is true if and only if $\gamma(rx_i) + \gamma(sy_i) + \gamma(tz_i) = 0$ if and only if $\gamma(r)\gamma(x_i) + \gamma(s)\gamma(y_i) + \gamma(t)\gamma(z_i) = 0$ if and only if $[\gamma(x_i),\gamma(y_i),\gamma(z_i)]$ are collinear. Clearly,

$$f|\langle 0,0,1 \rangle: [1,0,0] \rightarrow [1,0,0]$$
$$[0,1,0] \rightarrow [0,1,0]$$

$$[1,1,0] \to [1,1,0]$$
$$[a,1,0] \to [b,1,0]$$
$$[b,1,0] \to [a,1,0]$$

This mapping is not a projectivity. (Formal proof of this fact is delayed until Chapter 4 — see exercise 4.2.7(b)).

A class of projectivities in π_F from L onto L may be represented by 2×2 matrices over F in much the same way that a class of collineations on π_F was represented by 3×3 matrices in Example 3.1.2: Let the line $L = \langle 0,0,1 \rangle$ and observe again that points of the form $[x,y,0]$ are the residents of L. Clearly, we may represent these points by $[x,y]$ where it is understood that $[x,y] = [x',y']$ if and only if there exists $r \in F$, $r \neq 0$, such that $x = rx'$ and $y = ry'$. In other words, points of $\langle 0,0,1 \rangle$ may be considered to be members of $(F_2 - (0,0))/\sim$ where F_2 represents the two-dimensional vector space over F and $\sim$ represents the previously defined equivalence relation. In a manner analogous to that of Example 3.1.2 and Definition 3.1.3, we may define the matrix-representable mapping $f_M: [x,y] \to [x',y']$ where

$$M = \begin{pmatrix} a & b \\ c & d \end{pmatrix}$$

is a nonsingular matrix on F_2 and

$$\begin{pmatrix} x' \\ y' \end{pmatrix} = \begin{pmatrix} a & b \\ c & d \end{pmatrix} \begin{pmatrix} x \\ y \end{pmatrix}$$

Thus $f_M: [x,y] \to [ax + by, cx + dy]$. A theorem similar to Theorem 3.1.4 states that $f_M = f_N$ if and only if there exists $r \in F, r \neq 0$, such that $M = rN$. Theorem 3.4.9 shows that these matrix-representative mappings on $\langle 0,0,1 \rangle$ are projectivities.

THEOREM 3.4.9   *If $f_M: [x,y] \to [ax + by, cx + dy]$, then f_M is a projectivity.*

Proof.   If $d \neq 0$, we may assume that $d = 1$ since there is an equivalent matrix N such that $f_M = f_N$ and

$$N = \begin{pmatrix} a' & b' \\ c' & 1 \end{pmatrix}$$

Similarly, if $d = 0$, we may assume that $c = 1$. Assume that $d = 1$ and let

$$M = \begin{pmatrix} a & b \\ c & 1 \end{pmatrix} \quad \text{and} \quad N = \begin{pmatrix} a & b & 0 \\ c & 1 & 0 \\ 0 & 0 & 1 \end{pmatrix}$$

Note that $g_N | \langle 0,0,1 \rangle = f_M$. It is easily checked that if $a \neq 0$, then

$$N = \begin{pmatrix} a & 0 & 0 \\ 0 & 1 & 0 \\ 0 & 0 & 1 \end{pmatrix} \begin{pmatrix} 1 & 0 & 0 \\ c & 1 & 0 \\ 0 & 0 & 1 \end{pmatrix} \begin{pmatrix} 1 & 0 & 0 \\ 0 & r & 0 \\ 0 & 0 & 1 \end{pmatrix} \begin{pmatrix} 1 & s & 0 \\ 0 & 1 & 0 \\ 0 & 0 & 1 \end{pmatrix}$$

where

$$r = \frac{a - bc}{a} \quad \text{and} \quad s = \frac{b}{a}$$

(Multiplication is from left to right.) Note that $a - bc \neq 0$ because N is nonsingular. If $a = 0$, then

$$N = \begin{pmatrix} 0 & b & 0 \\ c & 1 & 0 \\ 0 & 0 & 1 \end{pmatrix} = \begin{pmatrix} 0 & 1 & 0 \\ 1 & 0 & 0 \\ 0 & 0 & 1 \end{pmatrix} \begin{pmatrix} 1 & 0 & 0 \\ 0 & b & 0 \\ 0 & 0 & 1 \end{pmatrix} \begin{pmatrix} c & 0 & 0 \\ 0 & 1 & 0 \\ 0 & 0 & 1 \end{pmatrix} \begin{pmatrix} 1 & t & 0 \\ 0 & 1 & 0 \\ 0 & 0 & 1 \end{pmatrix}$$

where $t = 1/c$. Note that $b \neq 0$, $c \neq 0$ because N is nonsingular.

If $d = 0$, then

$$M = \begin{pmatrix} a & b \\ 1 & 0 \end{pmatrix}$$

and

$$N = \begin{pmatrix} a & b & 0 \\ 1 & 0 & 0 \\ 0 & 0 & 1 \end{pmatrix} = \begin{pmatrix} 0 & 1 & 0 \\ 1 & 0 & 0 \\ 0 & 0 & 1 \end{pmatrix} \begin{pmatrix} 1 & 0 & 0 \\ 0 & b & 0 \\ 0 & 0 & 1 \end{pmatrix} \begin{pmatrix} 1 & 0 & 0 \\ u & 1 & 0 \\ 0 & 0 & 1 \end{pmatrix}$$

where $u = a/b$. Note that $b \neq 0$. In each instance we have factored N into a product of elementary matrices N_i, and in each instance g_{N_i} is a central collineation holding $\langle 0,0,1 \rangle$ fixed. Since each central collineation g_{N_i} induces a projectivity $g_{N_i} | \langle 0,0,1 \rangle$, we may conclude that f_M is the product of projectivities and thus is itself a projectivity.

This section is concluded with a comment about the similarities between collineations (two-dimensional mappings) and projectivities mapping a line onto itself (one-dimensional mappings). The "projective mapping" mentioned in Section 3.2 is clearly the three-dimensional generalization of the projectivity in a plane. Similarly, the central collineation is a two-dimensional generalization of the one-dimensional projectivity that leaves exactly one point fixed. Therefore, finite compositions of central collineations are analogues of finite compositions of one-dimensional projectivities that leave one point fixed. In π_F, it is true that these finite compositions of central collineations generate the set of all projective collineations and that finite compositions of one-dimensional mappings generate the set of all one-dimensional projectivities. Thus the projective collineation is the natural generalization of the one-dimensional projectivity in π_F. Recall, however, that there are collineations in π_F (for example, $\pi(4)$) that are not projective collineations. Therefore, in a sense, the projective plane π_F has a wider variety of isomorphisms on itself than does the projective line. The comparison of sets of transformations will be placed in its proper setting and discussed thoroughly for planes π_F in Chapter 4.

EXERCISES

1. Give an example of a projective collineation on π_F that is not a central collineation.

2. If $f = L \overset{p}{\barwedge} M \overset{q}{\barwedge} L$ where $L \neq M$ and $p \neq q$, then the point $r = L \cap M$ is called the *axial invariant point* of f and $s = pq \cap L$ is called the *central invariant point* of f.
   a. Suppose that f is the product of two perspectivities such that $f: L \to L$. Show that if f is not the identity, then f leaves only two points fixed, the axial invariant point and the central invariant point.
   b. Let p,q,r,s be distinct collinear points on L in π. Construct a projectivity $f: L \to L$ that is the product of two perspectivities such that p is the axial invariant point, q is the central invariant point, and $f: r \to s$.
   c. Let p,p',q,q',r be collinear points on L such that p,q,r and p',q',r are triples of distinct points. Construct a projectivity f that is the product of two perspectivities such that $f: p,q \to p',q'$ and r is: the axial invariant point; the central invariant point.

3. Let p,p',q,q' be collinear points such that p,p',q and p,p',q' are triples of distinct points. Show that there exists a projectivity f such that $f: p,p',q,q' \to p',p,q',q$ (that is, f exchanges p with p' and q with q').

4. Show that (h,H) of Example 3.4.8 is a collineation.

5. Let

$$f: \langle 1,0,0 \rangle \xrightarrow{[1,1,0]} \langle -1,0,1 \rangle \xrightarrow{[1,0,0]} \langle 1,0,0 \rangle$$

be a projectivity on π_F. Let n be the positive integer such that $f^n([0,0,1]) = [0,0,1]$ and f^k does not fix $[0,0,1]$ for $k < n$. (f^n denotes the mapping f composed with itself n times.) Show that p is the characteristic of F if and only if $p = n$. (Compare this exercise with exercise 2.5.2.)

6. Refer back to exercise 3.2.6.
   a. In π_R find a projectivity f_M: $\langle 0,0,1 \rangle \rightarrow \langle 0,0,1 \rangle$ that holds exactly two points fixed; one point fixed; no points fixed.
   b. Show that in π_C (the plane over the complex numbers), every projectivity f_M that maps $\langle 0,0,1 \rangle$ onto itself holds at least one point fixed.
   c. State and prove a theorem giving necessary and sufficient conditions on a field F such that every projectivity f_M that maps $\langle 0,0,1 \rangle$ onto itself has at least one fixed point.
   d. Show that for finite fields F there exist projectivities f_M that map $\langle 0,0,1 \rangle$ onto itself and leave no points fixed. Hint: Consider all equations of the form $x^2 + bx + c = 0$ where $b,c \in F$ and consider all equations of the form $(x - r)(x - s) = 0$ where $r,s \in F$ and r and s are not necessarily distinct.

CHAPTER 4

GROUPS OF
TRANSFORMATIONS

In this chapter the concept of "group" is used as a tool for the study of projective planes. A group consists of a set, G, and a binary operation, $\cdot$, such that four axioms are satisfied:

1. G is closed under $\cdot$.
2. The operation $\cdot$ is associative.
3. There is an identity element.
4. Every member of G has an inverse.

Only the most elementary results of group theory will be used here.

Letting S be any set of elements, it is easily checked that the set of all one-to-one mappings from S onto S is a group under the operation of composition. This group is called the *permutation group* of S. If $S = \mathscr{P}$ and the set of points of a primitive plane $\Sigma = (\mathscr{P}, \mathscr{L}, \mathscr{I})$, then the set of collineations on Σ is a subgroup of the permutation group of $\mathscr{P}$. This group and its subgroups will be the subjects of study in this chapter. Group theory provides a setting within which the study of collineations becomes a most effective tool for the analysis of projective and affine planes.

4.1   PRELIMINARY RESULTS

As mentioned in the preceding paragraph, the set of collineations on a primitive plane is a group under the operation of composition. The following notation is used for projective and affine collineation groups.

Notation.   The group of collineations of a projective plane π is denoted by $C(\pi)$; the group of collineations of an affine plane α is denoted by $C(\alpha)$. As stated in exercise 3.1.3, the composition $g(f(p))$ of the collineations f and g is denoted by $g \circ f$; in other words, the operation of composition is read from right to left. The term "product" is used interchangeably with the term "composition."

There are several subgroups that will be useful to our study. Recall that a subset of a group is a subgroup if and only if it is closed under the group operation. Except for elations with a common axis or common center, it is easily seen that the subsets of $C(\pi)$ listed following the proof of Theorem 4.1.1 are subgroups. Theorem 4.1.1 assures us that the set of elations with a common center or common axis is also a group.

THEOREM 4.1.1

   i. *If f and g are elations with axis L, then $g \circ f$ is an elation with axis L.*

   ii. *If f and g are elations with center p, then $g \circ f$ is an elation with center p.*

Proof.   The proof is left as an exercise.

Notation.   The following notation is used for projective planes π:

   1. The group of projective collineations on π is denoted by $PC(\pi)$.
   2. The group of elations with axis L is denoted by $El(L)$.
   3. The group of elations with center p is denoted by $El(p)$.
   4. The group of central collineations with center p and axis L is denoted by $CC(p,L)$. If $p \in L$, the notation used is $El(p,L)$; if $p \notin L$, the notation used is $Hom(p,L)$.
   5. The group of central collineations with axis L is denoted by $CC(L)$.

6. The group of central collineations with center p is denoted by CC(p).
7. The group of finite compositions of central collineations on π is denoted by CC(π).
8. The group of matrix-representable collineations on π_F is denoted by PGL(2,F).

The following notation is used for affine planes α:

1. The group of dilatations is denoted by Dil(α).
2. The group of translations is denoted by Trans(α).
3. The group of dilatations with center p is denoted by Dil(p).
4. The group of translations with direction L is denoted by Trans(L) (that is, $pf(p) \parallel L$ for all $p \in \alpha$).
5. The group of matrix-representable collineations on α_F is denoted by Af(2,F).

The following notation is used for one-dimensional projective planes:

1. The group of all projectivities from L onto L is denoted by Proj(L).
2. The group of all matrix-representable projectivities on L is denoted by PGL(1,F).

This system of notation is, for the most part, self-explanatory. The notation PGL(n,F) is read as "the projective general linear group of dimension n over the field F." This group is equal to the quotient group GL($n + 1$, F)/N, where GL($n + 1$, F) is the general linear group (also called the full linear group) over F_{n+1} (the $n + 1$-dimensional vector space over F) and N is the normal subgroup whose elements are scalar multiples of the identity transformation. Recall that the general linear group over F_{n+1} is the group of all nonsingular linear transformations on F_{n+1}. The notation Af(2,F) is read as "the affine group of dimension 2 over F."

Theorems 4.1.2 and 4.1.3 follow immediately from Definitions 3.2.3, 3.2.11, 3.3.8, and Corollary 3.4.7.

THEOREM 4.1.2 *Let α be an affine plane and let $\pi(\alpha^+)$ be its projective extension. Then the following groups are isomorphic:*

$CC(L_\infty) \sim \text{Dil}(\alpha)$

$El(L_\infty) \sim \text{Trans}(\alpha)$

$\text{Hom}(p, L_\infty) \sim \text{Dil}(p)$

$El(p, L_\infty) \sim \text{Trans}(L)$ where $p = [L]$

(This notation was defined in Construction 1.7.7.)

THEOREM 4.1.3   *The following group inclusions hold:*

$$\left. \begin{array}{l} \text{Hom}(p,L) \\ El(p,L) \subseteq El(L) \end{array} \right\} \subseteq CC(L) \subseteq CC(\pi) \subseteq PC(\pi) \subseteq C(\pi)$$

$$\left. \begin{array}{l} \text{Trans}(L) \subseteq \text{Trans}(\alpha) \\ \text{Dil}(p) \subseteq \text{Dil}(\alpha) \end{array} \right\} \subseteq C(\alpha)$$

Every inclusion can be made proper, or it can be made an equality. For planes $\pi(4)$ and $\alpha(4)$, all inclusions are proper except $CC(\pi)$ and $PC(\pi)$. Example 3.4.8 is a projective collineation that cannot be generated by finite combinations of central collineations, and thus $CC(\pi_N(9)) \subset PC(\pi_N 9)$. (This will be proved in Chapter 12.) Table 16 (p. 139) describes a projective plane π that admits only one collineation, the identity. Therefore, the groups of Theorem 4.1.3(i) are trivial (that is, they contain only the identity), and thus equality holds throughout. Any affine restriction α of the plane π also admits only the identity collineation, so for α, equality holds for all groups of Theorem 4.1.3(ii).

In general, little can be said about the structure of the groups and subgroups of Theorem 4.1.3. A few basic facts can be established, however. $CC(\pi)$ is a normal subgroup of $C(\pi)$ (recall that N is a normal subgroup of G if and only if $xnx^{-1} \in N$ for all $n \in N$ and $x \in G$); $CC(p, L)$ and $CC(q, M)$ are often isomorphic; and planes and their duals share the same collineation group structure. These statements are expressed more formally in the following three theorems.

THEOREM 4.1.4   *If $f \in C(\pi)$ and $g \in CC(p,L)$, then $f \circ g \circ f^{-1} \in$* $CC(f(p), f(L))$.

Proof. Suppose that $g \in CC(p,L)$ and let q be a fixed point of g. Then $f \circ g \circ f^{-1}(f(q)) = f(q)$, and so $f(q)$ is a fixed point of $f \circ g \circ f^{-1}$. Conversely, if $f(r)$ is held fixed by $f \circ g \circ f^{-1}$, then $f \circ g \circ f^{-1}(f(r))$ $= f(g(r)) = f(r)$, and so r is a fixed point of g. We may conclude that $f \circ g \circ f^{-1} \in CC(f(p), f(L))$.

This theorem suggests that we may relate the group structure of $CC(p,L)$ and $CC(q,M)$ if there exists a collineation $f: p \to q, L \to M$. In fact, if such a collineation exists, the two groups are isomorphic.

THEOREM 4.1.5 *Let* $f \in C(\pi)$ *such that* $f: p \to q, L \to M$; *then* $CC(p,L)$ *is isomorphic to* $CC(q,M)$.

Proof. Let $\beta: g \to f \circ g \circ f^{-1}$ for $g \in CC(p,L)$. With the help of Theorem 4.1.4 we can easily establish that β is a one-to-one mapping from $CC(p,L)$ onto $CC(q,M)$. Since $\beta(h \circ g) = f \circ h \circ g \circ f^{-1} = f \circ h \circ f^{-1} \circ f \circ g \circ f^{-1} = \beta(h) \circ \beta(g)$, β preserves the group operation and is therefore an isomorphism.

The converse is trivially false. For example, $CC(p,L) = CC(q,M) = \{I\}$ for all such central collineation groups in $\pi_H(9)$; yet, as will be explained in Section 4.3, there are points in $\pi_H(9)$—for example, a_0 and b_0—that cannot be related by a collineation.

The concepts of collineation and central collineation are self-dual; this means that if (f,F) is a (central) collineation of π, then (F,f) is a (central) collineation of π^d. Furthermore, if $(f,F) \in El(p,L)$ in π, then $(F,f) \in El(L,p)$ in π^d; and if $(f,F) \in Hom(p,L)$ in π, then $(F,f) \in Hom(L,p)$ in π^d. Verification of these statements and proof of the following theorem are left as an exercise.

THEOREM 4.1.6 *The following group isomorphisms are found in* π *and* π^d:

$$C(\pi) \sim C(\pi^d)$$
$$CC(\pi) \sim CC(\pi^d)$$
$$El(p,L) \text{ in } \pi \sim El(L,p) \text{ in } \pi^d$$
$$Hom(p,L) \text{ in } \pi \sim Hom(L,p) \text{ in } \pi^d$$

EXERCISES

1. a. Show that if $(f,F) \in El(p,L)$ in π, then $(F,f) \in El(L,p)$ in π^d.
   b. Show that if $(f,F) \in Hom(p,L)$ in π, then $(F,f) \in Hom(L,p)$ in π^d.
   c. Prove Theorem 4.1.6.

2. Show that the set of all collineations that leaves a set of points $\mathscr{P}_0$ in π fixed (not necessarily pointwise) is a group.

3. Prove Theorem 4.1.1.

4. Show that the subset of all homologies with axis L is not necessarily a subgroup of $CC(L)$.

5. Refer back to exercises 2.1.6, 2.1.7, and 3.1.6.
   a. Show that the set of automorphisms on a Cartesian group forms a group with respect to the operation of composition.
   b. Show that the set of automorphic collineations is a subgroup of $C(\pi)$.

6. Using Example 3.2.12 and Theorem 4.1.4, find a matrix M such that f_M is the following central collineation on π_F:
   a. A harmonic homology with center $[1,1,1]$ and axis $\langle 1,-1,1 \rangle$.
   b. An elation with center $[1,0,0]$ and axis $\langle 0,1,1 \rangle$ that maps $[0,0,1] \rightarrow [1,0,1]$.
   c. A homology with center $[-1,0,1]$ and axis $\langle -1,0,1 \rangle$ that maps $[1,0,0] \rightarrow [0,0,1]$.

4.2   TRANSFORMATION GROUPS FOR π_F

Considerably more can be said about the nature of the groups and subgroups of projective planes π_F (and therefore of affine planes α_F) than of general planes π. The most important group in π_F is $PGL(2,F)$. It will be analyzed first so that its place can be found in the inclusion chain of subgroups. The first theorem in this section states that there is a collineation $f \in PGL(2,F)$ associating any pair of four-points in π_F. This theorem is preceded by two lemmas.

LEMMA 4.2.1   *The points* $p = [x_1,x_2,x_3]$, $q = [y_1,y_2,y_3]$, $r = [z_1,z_2,z_3]$ *are collinear if and only if the matrix*

$$M = \begin{pmatrix} x_1 & y_1 & z_1 \\ x_2 & y_2 & z_2 \\ x_3 & y_3 & z_3 \end{pmatrix}$$

is singular.

Proof. If p, q, r are collinear, a common line $\langle a,b,c \rangle$ can be found containing them. Thus $ax_1 + bx_2 + cx_3 = ay_1 + by_2 + cy_3 = az_1 + bz_2 + cz_3 = 0$. It follows that $(a,b,c)M = (0,0,0)$, and thus M is singular. (Here (a,b,c) and $(0,0,0)$ denote 1×3 matrices.) The converse is true by a reverse argument.

LEMMA 4.2.2 *If p,q,r,s is a four-point where p, q, r have the coordinates of Lemma 4.2.1 and $s = [w_1, w_2, w_3]$, then there exists a unique triple (a,b,c) where $a,b,c \in F$ such that $w_i = ax_i + by_i + cz_i$, $i = 1,2,3$.*

Proof. Since p, q, r, are noncollinear, the matrix

$$M = \begin{pmatrix} x_1 & y_1 & z_1 \\ x_2 & y_2 & z_2 \\ x_3 & y_3 & z_3 \end{pmatrix}$$

is nonsingular and therefore has an inverse M^{-1}. Thus the system of equations

$$w_1 = kx_1 + my_1 + nz_1$$
$$w_2 = kx_2 + my_2 + nz_2$$
$$w_3 = kx_3 + my_3 + nz_3$$

has a unique solution $k = a$, $m = b$, $n = c$. In matrix notation,

$$M \begin{pmatrix} a \\ b \\ c \end{pmatrix} = \begin{pmatrix} w_1 \\ w_2 \\ w_3 \end{pmatrix}, \quad \text{and so} \quad \begin{pmatrix} a \\ b \\ c \end{pmatrix} = M^{-1} \begin{pmatrix} w_1 \\ w_2 \\ w_3 \end{pmatrix}$$

THEOREM 4.2.3 *If p,q,r,s and p',q',r',s' are two four-points in π_F, there exists a unique collineation $f \in PGL(2,F)$ such that $f: p,q,r,s \to p',q',r',s'$.*

Proof. First, to show existence, let p,q,r,s be an arbitrary four-point and let $u = [1,0,0]$, $v = [0,1,0]$, $o = [0,0,1]$, and $e = [1,1,1]$. By Lemma 4.2.2, we may choose coordinates $[p_1,p_2,p_3]$ for p, $[q_1,q_2,q_3]$ for q, $[r_1,r_2,r_3]$ for r, and $[s_1,s_2,s_3]$ for s such that $s_i = p_i + q_i + r_i$, $i = 1,2,3$. Thus the collineation f_M, where

$$M = \begin{pmatrix} p_1 & q_1 & r_1 \\ p_2 & q_2 & r_2 \\ p_3 & q_3 & r_3 \end{pmatrix}$$

maps $u,v,o,e \to p,q,r,s$. Let p,q,r,s and p',q',r',s' be two four-points and let $h = g_N (f_M)^{-1}$ where $f_M: u,v,o,e \to p,q,r,s$ and $g_N: u,v,o,e \to p',q',r',s'$. Thus $h: p,q,r,s \to p',q',r',s'$.

Next, to show uniqueness, suppose that a collineation f_M holds u,v,o,e fixed where

$$M = \begin{pmatrix} a & b & c \\ d & e & f \\ g & h & i \end{pmatrix}$$

Then we can infer that: f_M: $u \to u$ implies that $a \neq 0$, $d = g = 0$; f_M: $v \to v$ implies that $e \neq 0$, $b = h = 0$; f_M: $o \to o$ implies that $i \neq 0$, $c = f = 0$. It follows that g_M: $e = [1,1,1] \to [a,e,i]$. (Note that e is being used here in two different senses.) Therefore, $a = e = i$, and so

$$M = \begin{pmatrix} a & 0 & 0 \\ 0 & a & 0 \\ 0 & 0 & a \end{pmatrix}$$

Thus f_M is the identity.

Suppose that f and g are matrix-representable collineations mapping $p,q,r,s \to p',q',r',s'$. Let h_M: $u,v,o,e \to p,q,r,s$; then $h_M^{-1} \circ g^{-1} \circ f \circ h_M$: $u,v,o,e \to u,v,o,e$. Thus $h_M^{-1} \circ g^{-1} \circ f \circ h_M$ is the identity, and so $f = g$. We may conclude that the mapping f is unique.

This theorem is central to our study of the plane π_F; it facilitates completion of the proofs of Theorems 2.8.6 through 2.8.9. It is used here as a tool for studying other subgroups.

THEOREM 4.2.4 If $f \in CC(\pi_F)$, then $f \in PGL(2,F)$, that is, $CC(\pi_F) \subseteq PGL(2,F)$.

Proof. Suppose that $f \in El(\langle 0,0,1 \rangle)$. By Theorem 3.2.13, f is uniquely determined by the image of $[0,0,1]$. Let $f([0,0,1]) = [a,b,1]$; then $f = f_M$ where

$$M = \begin{pmatrix} 1 & 0 & a \\ 0 & 1 & b \\ 0 & 0 & 1 \end{pmatrix}$$

If $f \in Hom([0,0,1],\langle 0,0,1 \rangle)$, then by Theorem 3.2.13, f is uniquely determined by the image of $[0,1,1]$. Since f holds every line through $[0,0,1]$ fixed (in particular, the line $\langle 1,0,0 \rangle$), we may write $f([0,0,1])$ as $[0,1,r]$. Clearly, $f = f_N$ where

$$N = \begin{pmatrix} 1 & 0 & 0 \\ 0 & 1 & 0 \\ 0 & 0 & r \end{pmatrix}$$

Thus any central collineation in $El(\langle 0,0,1 \rangle)$ or $Hom([0,0,1],\langle 0,0,1 \rangle)$ is in $PGL(2,F)$.

Suppose that $g \in CC(p,L)$ for an arbitrary p and L. By Theorem

4.2.3, there exists $f_M \in \mathrm{PGL}(2,F)$ such that $f_M: L \rightarrow \langle 0,0,1 \rangle$; and if $p \notin L$, then $p \rightarrow [0,0,1]$. By Theorem 4.1.4, $f_M \circ g \circ f_M^{-1} \in \mathrm{El}(\langle 0,0,1 \rangle)$ if $p \in L$, and $f_M \circ g \circ f_M^{-1} \in \mathrm{Hom}([0,0,1],\langle 0,0,1 \rangle)$ if $p \notin L$. In either instance some collineation $h_N \in \mathrm{PGL}(2,F)$ exists such that $f_M \circ g \circ f_M^{-1} = h_N$. Thus $g = f_M^{-1} \circ h_N \circ f_M \in \mathrm{PGL}(2,F)$.

The two lemmas that follow help to prove that $\mathrm{PC}(\pi_F) \subseteq \mathrm{PGL}(2,F)$.

LEMMA 4.2.5 *Let* $g: L \overset{p}{\barwedge} L'$ *where* $L = \langle 0,1,0 \rangle$, $L' = \langle 0,1,1 \rangle$, *and* $p = [0,1,0]$. *Then* $f_M | \langle 0,0,1 \rangle = g$ *where*

$$M = \begin{pmatrix} 1 & 0 & 0 \\ 0 & 1 & -1 \\ 0 & 0 & 1 \end{pmatrix}$$

Proof. Note that $f_M \in \mathrm{El}([0,1,0],\langle 0,0,1 \rangle)$ such that $f_M: [0,0,1] \rightarrow [0,-1,1]$. Thus $f_M: [1,0,0],[0,0,1] \rightarrow [1,0,0],[0,-1,1]$, and so $f_M: \langle 0,1,0 \rangle \rightarrow \langle 0,1,1 \rangle$. Therefore, if $q \in \langle 0,1,0 \rangle$, then $f_M(q) \in \langle 0,1,1 \rangle$. It follows that $p,q,f_M(q)$ are collinear since f_M leaves fixed every line passing through p, and so $f_M | L = g$.

LEMMA 4.2.6 *If* $g: L \overset{p}{\barwedge} L'$ *is an arbitrary perspectivity in* π_F, *then there exists* $f \in \mathrm{PGL}(2,F)$ *such that* $f | L = g$.

Proof. It follows from Theorem 4.2.3 that there exists $h \in \mathrm{PGL}(2,F)$ such that $h: p \rightarrow [0,1,0]$, $L \rightarrow \langle 0,1,0 \rangle$, $L' \rightarrow \langle 0,1,1 \rangle$. (Proof of this is left as an exercise). Letting $f = h^{-1} \circ f_M \circ h$ where f_M is defined as in Lemma 4.2.5, it is easily seen that $f | L = g$.

THEOREM 4.2.7 *If* $f \in \mathrm{PC}(\pi_F)$, *then* $f \in \mathrm{PGL}(2,F)$, *that is*, $\mathrm{PC}(\pi_F) \subseteq \mathrm{PGL}(2,F)$.

Proof. Let $f \in \mathrm{PC}(\pi_F)$ and let $g = f | L$. Since g is a projectivity, $g = g_n \circ \ldots \circ g_i \circ \ldots \circ g_1$ where $g_i = L_i \overset{p_i}{\barwedge} L_{i+1}$. By Lemma 4.2.6, there exists $f_i \in \mathrm{PGL}(2,F)$ such that $f_i | L_i = g_i$. Let $h = f_n \circ \ldots \circ f_1$; then $h \in \mathrm{PGL}(2,F)$ and $h | L = f | L$. Thus $f \circ h^{-1}$ holds L pointwise fixed, and therefore $f \circ h^{-1}$ is a central collineation with axis L; call this collineation j. By Theorem 4.2.4, $j \in \mathrm{PGL}(2,F)$, so $f = j \circ h \in \mathrm{PGL}(2,F)$.

The following theorem allows us to establish that $CC(\pi_F) = PGL(2,F) = PC(\pi_F)$.

THEOREM 4.2.8 *Every matrix-representable collineation is representable by the product of at most three central collineations.*

Proof. Let

$$M = \begin{pmatrix} a & b & c \\ d & e & f \\ g & h & i \end{pmatrix}$$

be a nonsingular matrix and let

$$M_x = \begin{pmatrix} x_1 & 0 & 0 \\ x_2 & 1 & 0 \\ x_3 & 0 & 1 \end{pmatrix} \quad M_y = \begin{pmatrix} 1 & y_1 & 0 \\ 0 & y_2 & 0 \\ 0 & y_3 & 1 \end{pmatrix} \quad M_z = \begin{pmatrix} 1 & 0 & z_1 \\ 0 & 1 & z_2 \\ 0 & 0 & z_3 \end{pmatrix}$$

A simple computation shows that

$$M_x M_y M_z = \begin{pmatrix} x_1 & x_1 y_1 & x_1 z_1 + x_1 y_1 z_2 \\ x_2 & x_2 y_1 + y_2 & x_2 z_1 + (x_2 y_1 + y_2) z_2 \\ x_3 & x_3 y_1 + y_3 & x_3 z_1 + (x_3 y_1 + y_3) z_2 + z_3 \end{pmatrix}$$

and that $M = M_x M_y M_z$ iff

$$x_1 = a$$
$$x_2 = d$$
$$x_3 = g$$

$$y_1 = \frac{b}{a}$$

$$y_2 = e - \frac{db}{a}$$

$$y_3 = h - \frac{gb}{a}$$

$$z_1 = \frac{ce - fb}{ae - bd}$$

$$z_2 = \frac{af - cd}{ae - bd}$$

$$z_3 = i - (gz_1 + hz_2)$$

Thus M may be represented by $M_x M_y M_z$ iff $a \neq 0$ and $ae - bd \neq 0$. Similarly, it may be shown that

$$M = M_x\, M_z\, M_y \;\; \text{iff } a \neq 0 \text{ and } af - cd \neq 0$$
$$M = M_y\, M_x\, M_z \;\; \text{iff } b \neq 0 \text{ and } ae - bd \neq 0$$
$$M = M_y\, M_z\, M_x \;\; \text{iff } b \neq 0 \text{ and } bf - ec \neq 0$$
$$M = M_z\, M_x\, M_y \;\; \text{iff } c \neq 0 \text{ and } af - cd \neq 0$$
$$M = M_z\, M_y\, M_x \;\; \text{iff } c \neq 0 \text{ and } bf - ec \neq 0$$

It is not difficult to check that since M is nonsingular, one of these six possibilities must hold. Clearly, $a = b = c = 0$ is impossible; $a = b = 0$ and $af - cd = bf - ec = 0$ implies either that $a = b = c = 0$ or that $a = b = d = e = 0$, and both results are impossible. Further checking is left to the student.

Now $f_{M_x}, f_{M_y}, f_{M_z}$ are central collineations with axes $\langle 1,0,0 \rangle$, $\langle 0,1,0 \rangle$, and $\langle 0,0,1 \rangle$, respectively. From the preceding discussion, it follows that f_M may be represented by at most three central collineations.

COROLLARY 4.2.9 *In π_F, $\mathrm{CC}(\pi_F) = \mathrm{PGL}(2,F) = \mathrm{PC}(\pi_F)$.*

Theorems 4.2.3 and 4.1.4 imply that all subgroups of the form $\mathrm{El}(p,L)$ are isomorphic and that all subgroups of the form $\mathrm{Hom}(p,L)$ are isomorphic. The following theorem shows that these subgroups are of a convenient form.

THEOREM 4.2.10 *The group $\mathrm{El}(p,L)$ is isomorphic to $(F,+)$, that is, the field F with the operation of addition; the group $\mathrm{Hom}(p,L)$ is isomorphic to $(F - \{0\},\cdot)$, that is, the nonzero elements of F with the operation of multiplication.*

Proof. Since all groups $\mathrm{El}(p,L)$ are isomorphic, we may let $p = [1,0,0]$ and $L = \langle 0,0,1 \rangle$. Define f_a as f_M where

$$M = \begin{pmatrix} 1 & 0 & a \\ 0 & 1 & 0 \\ 0 & 0 & 1 \end{pmatrix}$$

and define a mapping from $\mathrm{El}(p,L)$ to F by $\beta \colon f_a \to a$. Clearly, β is a one-to-one, onto mapping. Also β preserves the operation by the following argument: $f_a \colon [0,0,1] \to [a,0,1]$ and $f_b \colon [a,0,1] \to [a+b,0,1]$, so

$f_b \circ f_a ([0,0,1]) = [a + b,0,1] = f_{a+b}([0,0,1])$, and therefore $f_b \circ f_a = f_{a+b}$. Thus $\beta(f_b \circ f_a) = \beta(f_{a+b}) = a + b = \beta(f_a) + \beta(f_b)$, and we may conclude that β is an isomorphism.

Since all groups $\text{Hom}(p,L)$ are isomorphic, let us choose $p = [1,0,0]$ and $L = \langle 1,0,0 \rangle$. Define g_a as g_M where

$$M = \begin{pmatrix} a^{-1} & 0 & 0 \\ 0 & 1 & 0 \\ 0 & 0 & 1 \end{pmatrix}$$

and define a mapping γ from $\text{Hom}(p,L)$ to $(F - \{0\},\cdot)$ by $\gamma \colon g_a \to a$. Clearly, γ is a one-to-one, onto mapping. Also γ preserves the operation by the following argument: $g_a \colon [1,0,1] \to [a^{-1},0,1]$ and $g_b \colon [a^{-1},0,1] \to [b^{-1}a^{-1},0,1] = [(ab)^{-1},0,1]$, so $g_b \circ g_a ([1,0,1]) = [(ab)^{-1},0,1] = g_{ab}([1,0,1])$, and therefore $g_b \circ g_a = g_{ab}$. Thus $\gamma(g_b \circ g_a) = \gamma(g_{ab}) = ab = \gamma(g_a) \cdot \gamma(g_b)$, and so γ is an isomorphism from $\text{Hom}(p,L)$ to $(F - \{0\},\cdot)$.

Notice that the plane π_F is much more amenable to analysis than any other type of plane, mainly because of the presence of an underlying algebraic structure that permits the use of algebraic tools in the analysis. Theorem 4.2.10 suggests that, in the absence of an underlying algebraic structure, the groups $\text{El}(p,L)$ and $\text{Hom}(p,L)$ may act as a substitute. This exciting suggestion will be clarified in Section 4.4; a powerful approach to the subject of projective and affine planes based on this idea will be developed in Chapters 8–14.

This section is concluded with a theorem concerning the subgroups of $C(\alpha)$ and two theorems dealing with the groups $\text{Proj}(L)$ in π_F. The first theorem follows directly from Theorem 4.1.2, Corollary 4.2.9, and Theorem 4.2.10; the second is simply a restatement of Theorem 3.4.9 and may be proved in a manner analogous to the proof of Theorem 4.2.3; proof of the third is left as an exercise.

THEOREM 4.2.11 *The following subgroups of $C(\alpha_F)$ are isomorphic:*

$\text{Dil}(\alpha) \sim \text{Af}(2,F)$
$\text{Trans}(L) \sim (F,+)$ *for any line L*
$\text{Dil}(p) \sim (F - \{0\},\cdot)$ *for any point p*

THEOREM 4.2.12 *If (p,q,r) and (p',q',r') are ordered triples of* *distinct points on $\langle 0,0,1 \rangle$ in π_F, then there exists a unique projectivity $f \in$ PGL$(1,F)$ such that $f: p,q,r \rightarrow p',q',r'$.*

THEOREM 4.2.13 *In π_F the group* Proj(L) *is isomorphic to* PGL$(1,F)$ *for any line L in π_F. If $L = \langle 0,0,1 \rangle$,* Proj$(L)$ = PGL$(1,F)$.

EXERCISES

1. Refer back to exercise 2.1.5.
   a. Show that π_F satisfies Fano's Axiom if and only if F does not have characteristic 2.
   b. Show that π_F is a Fano plane if and only if F has characteristic 2.

2. Let p,q,L,L',M,M' be a set of two points and four lines in π_F such that $p \in L$ if and only if $q \in M$, and $p \in L'$ if and only if $q \in M'$. Show that there exists $f \in$ PGL$(2,F)$ such that $f: p,L,L' \rightarrow q,M,M'$.

3. Let M be a nonsingular 3×3 matrix over the field F. Show that the collineation f_M is a central collineation on π_F if and only if there exist two nonequivalent vectors v and w (that is, $v \neq rw$ for any $r \in F$) and a scalar $s \in F$ such that $Mv = sv$ and $Mw = sw$. (Mv denotes matrix multiplication of M by the column vector v and sv denotes scalar multiplication of v by s.)

4. Prove Theorem 4.2.13.

5. In π_F let $u = [1,0,0]$, $v = [0,1,0]$, $o = [0,0,1]$, and $S = ou - \{u\}$. Thus the elements of S are of the form $[x,0,1]$, $x \in F$. Also let f_x and g_x be represented by the matrices

$$\begin{pmatrix} 1 & 0 & x \\ 0 & 1 & 0 \\ 0 & 0 & 1 \end{pmatrix}$$

and

$$\begin{pmatrix} x & 0 & 0 \\ 0 & 1 & 0 \\ 0 & 0 & 1 \end{pmatrix}$$

respectively, where $x \in F$ and, for g_x only, $x \neq 0$. Notice that $f_x \in$ El(u,uv) such that $f_x: o \rightarrow [x,0,1]$, and $g_x \in$ Hom(u,ov) such that $g_x: [1,0,1] \rightarrow [x,0,1]$. Define $p \oplus q$ and $p \odot q$ for $p,q \in S$ as follows: Let $p = [x,0,1]$ and $q = [y,0,1]$. Then $p \oplus q = f_y \circ f_x(o), p \odot q = g_y \circ g_x([1,0,1])$ if $x,y \neq 0$, and $p \odot o = o \odot p = o$.

   a. Show that the mapping $\gamma: g_x \rightarrow x$ is an isomorphism from Hom(u,ov) onto $(F - \{0\},\cdot)$. (Compare this mapping with the mapping γ in the proof of Theorem 4.2.10).

b. Show that the mapping α: $x \to [x,0,1]$ is an isomorphism from the field F onto $(S,\oplus,\odot)$.

6. Utilizing the notation of exercise 5, define the operation $\otimes$ on the group $\mathrm{El}(u,uv)$ as follows: $f_x \otimes f_y = g_y \circ f_x \circ g_y^{-1}$. Show that the field F is isomorphic to $(\mathrm{El}(u,uv),\circ,\otimes)$.

7. Refer back to exercise 4.1.5 and adopt the notation $\mathrm{Aut}(\pi_F)$ for the set of automorphic collineations on π_F.

a. Show that the only collineation f on π_F that is both an automorphic collineation and a matrix-representable collineation is the identity collineation, that is, $\mathrm{Aut}(\pi_F) \cap \mathrm{PGL}(2,F) = \{I\}$.

b. Show that if $f \in \mathrm{Aut}(\pi_F)$, then $f|L$ is not a projectivity for any line L in π_F.

c. Let f be a collineation on π_F that leaves fixed the four-point u,v,o,e where $e = [1,1,1]$. (The notation here corresponds to that of exercise 5.) Then f leaves the set $S = ou - \{u\}$ fixed and induces a mapping h: $x \to x'$ on the field F where x' is the member of F such that $f([x,0,1]) = [x',0,1]$. Show that h is an automorphism on F. Hint: Let f_x and g_x correspond to the collineations of exercise 5 and show that $f_{x'} = f \circ f_x \circ f^{-1}$ and $g_{x'} = f \circ g_x \circ f^{-1}$. Then show that $f([x + y,0,1]) = f([x,0,1]) \oplus f([y,0,1])$ and $f([xy,0,1]) = f([x,0,1]) \odot f([y,0,1])$ where $\oplus$ and $\odot$ are defined as in exercise 5.

d. Show that every collineation on π_F is the product of an automorphic collineation and a matrix-representable collineation.

e. Show that F admits only the identity automorphism if and only if $\mathrm{CC}(\pi_F) = \mathrm{PGL}(2,F) = \mathrm{PC}(\pi_F) = \mathrm{C}(\pi_F)$.

4.3   EXAMPLES OF COLLINEATION GROUPS

This section gives some examples of collineation groups of various projective planes. The affine analogies will be included in the exercises at the end of this section.

EXAMPLE 4.3.1   Consider the plane π_F.

i. The central collineation groups of π_F were established in Theorem 4.2.10. The group $\mathrm{El}(p,L)$ is isomorphic to $(F,+)$; the group $\mathrm{Hom}(p,L)$ is isomorphic to $(F - \{0\},\cdot)$.

ii. The groups $\mathrm{CC}(\pi_F)$, $\mathrm{PC}(\pi_F)$, and $\mathrm{PGL}(2,F)$ are identical by Corollary 4.2.9. It follows from Theorem 4.2.3 that for finite planes π_F, the order of this group can be found by counting the distinct four-points of π_F. Thus for $\pi(2)$, $CC(\pi_F)$ has $7 \cdot 6 \cdot 4$ $= 168$ members, and for $\pi(3)$, $CC(\pi_F)$ has $13 \cdot 12 \cdot 9 \cdot 4 = 5616$ members. In general, $CC(\pi_F)$ has $n(n - 1)(n - (m + 1))(n - 3m)$ $= m^3(m^3 - 1)(m^2 - 1)$ members where $n = m^2 + m + 1$, $m = p^k$, and $F = GF(p^k)$.

iii. The collineation group $C(\pi_F)$ is not always equal to $CC(\pi_F)$. For example, if $F = GF(p^k)$ and $k > 1$, then $CC(\pi_F) \subset C(\pi_F)$. This statement also holds true for the field of complex numbers C. In fact, if there exists an automorphism on F other than the identity, then $CC(\pi_F) \subset C(\pi_F)$. (Refer to exercise 4.2.7 for more details.) On the other hand, if $F = GF(p)$ or $F = Q$ or $F = R$, then $CC(\pi_F)$ $= C(\pi_F)$. This discussion will be continued in more detail in Chapter 7. The order of $C(\pi_F)$ for finite planes $F = GF(p^k)$ is $k(p^{3k})(p^{3k} - 1)(p^{2k} - 1)$. In particular, $\pi(4)$ has 120,960 collineations and $\pi_F(9)$ has 84,913,920 collineations.

In preparation for an examination of collineation groups in $\pi_N(9)$, two theorems will be proved.

THEOREM 4.3.2   *The group* $El([1,0,0],\langle 0,0,1 \rangle)$ *in* $\pi_N(9)$ *is isomorphic to* $(N,+)$.

Proof.   Let $r \in N$ and let

$$f_r: \ [x,y,1] \ \rightarrow [x + r,y,1]$$
$$[1,x,0] \ \rightarrow [1,x,0]$$
$$[0,1,0] \ \rightarrow [0,1,0]$$

$$F_r: \ \langle m,1,k \rangle \rightarrow \langle m,1,-rm + k \rangle$$
$$\langle 1,0,k \rangle \rightarrow \langle 1,0,k - r \rangle$$
$$\langle 0,0,1 \rangle \rightarrow \langle 0,0,1 \rangle$$

This was shown to be a collineation in Example 3.1.5(iii). Clearly, it is a member of $El([1,0,0],\langle 0,1,0 \rangle)$. Let the mapping β from $El([1,0,0],$ $\langle 0,1,0 \rangle)$ to $(N,+)$ be defined by $f_r \rightarrow r$. It is easily verified that β is one-to-one and onto and that $\beta(f_s \circ f_r) = \beta(f_{r+s}) = r + s = \beta(f_r) + \beta(f_s)$ and that therefore the group operation is preserved. Thus β is an isomorphism.

THEOREM 4.3.3   *The group* $Hom([1,0,0],\langle 1,0,0 \rangle)$ *in* $\pi_N(9)$ *is isomorphic to* $(N - \{0\},\cdot)$.

Proof.   Let $r \in N$ such that $r \neq 0$ and let

$$g_r\colon [x,y,1] \;\rightarrow\; [xr,y,1]$$
$$[1,x,0] \;\rightarrow\; [1,r^{-1}x,0]$$
$$[0,1,0] \;\rightarrow\; [0,1,0]$$

$$G_r\colon \langle m,1,k \rangle \rightarrow \langle r^{-1}m,1,k \rangle$$
$$\langle 1,0,k \rangle \rightarrow \langle 1,0,kr \rangle$$
$$\langle 0,0,1 \rangle \rightarrow \langle 0,0,1 \rangle$$

It is easily checked that $(g_r,G_r) \in \text{Hom}([1,0,0],\langle 1,0,0 \rangle)$. Let γ: $g_r \rightarrow r$; then $\gamma(g_s \circ g_r) = \gamma(g_{rs}) = rs = \gamma(g_r) \cdot \gamma(g_s)$, and so γ is an isomorphism.

EXAMPLE 4.3.4   Consider the plane $\pi_N(9)$.

i. The central collineation groups of $\pi_N(9)$ differ depending on the location of the center and the axis in the plane. For example, $\text{El}(p,L) \sim (N,+)$ if $L = \langle 0,0,1 \rangle$. This can be verified by observing that the point $[1,0,0]$ may be mapped onto any point of $\langle 0,0,1 \rangle$ by a composition of the collineation (h,H) of Corollary 3.4.7 with a collineation of the form (g_r,G_r). (Note that both h and g_r fix the line $\langle 0,0,1 \rangle$.) If $L \neq \langle 0,0,1 \rangle$, $\text{El}(p,L)$ contains only the identity. This will be proved in Chapter 12.

The homology groups are more varied than the elation groups. The group $\text{Hom}(p,L) \sim (N - \{0\},\cdot)$ if p and L satisfy one of the following conditions: $p = [1,0,0]$ and L passes through $[0,1,0]$; $p = [0,1,0]$ and L passes through $[1,0,0]$; $p = [1,x,0]$ and L passes through $[1,-x,0]$. (It is assumed that $L \neq \langle 0,0,1 \rangle$ because $p \notin L$.) Proof of this is left as an exercise.

The group $\text{Hom}(p,L)$ is isomorphic to the group of order 2 if $L = \langle 0,0,1 \rangle$. The nonidentity element of $\text{Hom}([0,0,1],\langle 0,0,1 \rangle)$ is (j,J) where:

$$j\colon [x,y,1] \;\rightarrow\; [-x,-y,1]$$
$$[1,x,0] \;\rightarrow\; [1,x,0]$$
$$[0,1,0] \;\rightarrow\; [0,1,0]$$

$$J\colon \langle m,1,k \rangle \rightarrow \langle m,1,-k \rangle$$
$$\langle 1,0,k \rangle \rightarrow \langle 1,0,-k \rangle$$
$$\langle 0,0,1 \rangle \rightarrow \langle 0,0,1 \rangle$$

The proof that $\text{Hom}([0,0,1],\langle 0,0,1\rangle)$ contains no other homology is given in Chapter 12. The group $\text{El}(\langle 0,0,1\rangle)$ provides us with enough collineations to map $[0,0,1]$ onto any arbitrary point p not on $\langle 0,0,1\rangle$, and therefore $\text{Hom}(p,\langle 0,0,1\rangle) \sim \text{Hom}([0,0,1], \langle 0,0,1\rangle)$. Finally, $\text{Hom}(p,L)$ contains only the identity for all other admissible choices of p and L.

ii. Since the collineation (h,H) is a projective collineation that is not generated by central collineations, we have the proper inclusion $CC(\pi_N(9)) \subset PC(\pi_N(9))$.

iii. The collineation group $C(\pi_N(9))$ is generated by $\text{El}(\langle 0,0,1\rangle)$, $\text{Hom}([0,0,1],\langle 0,0,1\rangle)$, $\text{Hom}([0,1,0],\langle 0,1,0\rangle)$, (h,H), (r_α,R_α), and (r_β,R_β), where the two latter collineations are defined as follows:

$$r_\alpha: [x,y,1] \rightarrow [\alpha(x),\alpha(y),1]$$
$$[1,x,0] \rightarrow [1,\alpha(x),0]$$
$$[0,1,0] \rightarrow [0,1,0]$$

$$R_\alpha: \langle m,1,k\rangle \rightarrow \langle \alpha(m),1, \alpha(k)\rangle$$
$$\langle 1,0,k\rangle \rightarrow \langle 1,0,\alpha(k)\rangle$$
$$\langle 0,0,1\rangle \rightarrow \langle 0,0,1\rangle$$

where α is the following mapping from N onto N:

$$\alpha: 0 \rightarrow 0 \quad a \rightarrow b \quad d \rightarrow f$$
$$1 \rightarrow 1 \quad b \rightarrow c \quad e \rightarrow d$$
$$2 \rightarrow 2 \quad c \rightarrow a \quad f \rightarrow e$$

Similarly, (r_β,R_β) is defined as follows:

$$r_\beta: [x,y,z] \rightarrow [\beta(x),\beta(y),\beta(z)]$$
$$R_\beta: \langle m,n,k\rangle \rightarrow \langle \beta(m),\beta(n),\beta(k)\rangle$$

where β is the mapping:

$$\beta: 0 \rightarrow 0 \quad a \rightarrow d \quad d \rightarrow a$$
$$1 \rightarrow 1 \quad b \rightarrow e \quad e \rightarrow b$$
$$2 \rightarrow 2 \quad c \rightarrow f \quad f \rightarrow c$$

It is easily verified that α and β are automorphisms on N. Therefore, $[x,y,z] \in \langle m,n,k\rangle$ if and only if $[\delta(x),\delta(y),\delta(z)] \in \langle \delta(m), \delta(n),\delta(k)\rangle$, and it follows that r_δ is a collineation where $\delta = \alpha$ or β.

The order of $C(\pi_N(9))$ is $162 \cdot 1{,}920 = 311{,}040$. This figure is obtained by first observing that there are exactly 162 central collineations with axis $\langle 0,0,1 \rangle$ and then by accepting, for now, the fact that there are 1,920 collineations (including the identity) that permute points on $\langle 0,0,1 \rangle$.

EXAMPLE 4.3.5   Consider the plane $\pi_N(9)^d$. By Theorem 4.1.5, the group structures of $C(\pi_N(9))$, $Hom(p,L)$, and $El(p,L)$ in $\pi_N(9)$ are the same as those for $C(\pi_N(9)^d)$, $Hom(L,p)$, and $El(L,p)$, respectively, in $\pi_N(9)^d$.

EXAMPLE 4.3.6   Consider the plane $\pi_H(9)$.

i. There are no central collineations in $\pi_H(9)$, so $El(p,L)$, $Hom(p,L)$, and $CC(\pi_H(9))$ contain only the identity. The collineation group $PC(\pi_H(9))$ has not been examined here.

ii. The collineation group $C(\pi_H(9))$ is generated by (f,F) and (g,G) of Example 3.1.5(ii) along with collineations of the form h^+, where h denotes a collineation on $\pi_H(9)|A$, $A = \{a_0, \ldots, a_{12}\}$, and $h^+|A = h$. Clearly, h^+ is an extension of h. Such extensions exist and are unique for every collineation h on $\pi_H(9)|A$. This fact will not be proved here, but further analysis in Chapter 9 will lend it credence. The order of $C(\pi_H(9))$ is $6 \cdot 5,616 = 33,696$. This figure is obtained as follows: (f,F) and (g,G) generate a normal subgroup of order 6; there are 5,616 collineations on $\pi_H(9)|A$ because it is the plane $\pi(3)$.

EXAMPLE 4.3.7   Consider free extension planes Σ_0^+. The collineation groups of free extension planes have not been easy to determine (see Sandler, 1963). It can be said that there are relatively few collineations on a plane Σ_0^+ and no central collineations other than the identity. Clearly, only those permutations that preserve confined configurations have a chance of being collineations. Recalling Theorem 2.7.8, we may conclude that only permutations that preserve the confined configurations of Σ_0 may be collineations.

i. Let $\Sigma_0 = \Sigma(7_3)$ plus an additional point. This is the plane of Example 2.5.8(iii). The collineation group $C(\Sigma_0^+)$ is made up of all those collineations that preserve the configuration $\Sigma(7_3)$. As we know from Example 4.3.1(ii), there are 168 such collineations.

ii. Consider the configuration Σ_o shown in Table 16. It is easily checked that Σ_o is a confined configuration. If f is a collineation on Σ_o, clearly $f: L_1 \to L_1$ and $L_2 \to L_2$ simply because they contain five and four points, respectively, and the other lines have three. Similarly, $f: p_{10} \to p_{10}$ because it is the only point with four lines through it. More detailed checking of the incidence matrix shows, in fact, that $f: p_i \to p_i$ for all i, $i = 1, \ldots, 12$. Thus Σ_o^+ is a projective plane with only one collineation, the identity.

TABLE 16

	L_1	L_2	L_3	L_4	L_5	L_6	L_7	L_8	L_9	L_{10}	L_{11}	L_{12}	L_{13}
p_1	1	0	0	1	0	0	0	1	0	0	0	0	0
p_2	1	0	0	0	1	0	0	0	1	0	0	0	0
p_3	1	0	0	0	0	1	0	0	0	0	1	0	0
p_4	1	0	0	0	0	0	1	0	0	0	0	1	0
p_5	1	0	0	0	0	0	0	0	0	1	0	0	1
p_6	0	1	0	1	0	0	0	0	1	0	0	0	1
p_7	0	1	0	0	1	0	0	1	0	0	1	0	0
p_8	0	1	0	0	0	1	0	0	0	0	0	1	0
p_9	0	1	0	0	0	0	1	0	0	1	0	0	0
p_{10}	0	0	1	1	1	1	1	0	0	0	0	0	0
p_{11}	0	0	1	0	0	0	0	1	1	1	0	0	0
p_{12}	0	0	1	0	0	0	0	0	0	0	1	1	1

The examples of this section show that the collineation groups of projective planes exhibit a wide range of behavior. Naturally, two planes with nonisomorphic collineation groups are themselves nonisomorphic; however, some nonisomorphic planes (such as $\pi_N(9)$ and $\pi_N(9)^d$) have isomorphic collineation groups. Thus the key to our study of projective planes is obviously something more than the mere group structure of $C(\pi)$ or $CC(\pi)$. In finite cases, however, the order of the group $C(\pi)$ provides some information. In general, the more collineations a plane has, the more "homogeneous" that plane is. For example, if there are enough collineations so that every two points can be related by such a map, then, in one sense, the plane is the "same" at these two points. If every two four-points can be related by a collineation, as is true of π_F because of Theorem 4.2.3, then the plane is the same around every four-point. Using this crude cardinality criterion for homogeneity of planes, we may grade the planes of order 9 as follows: $\pi_H(9)$ is the least homogeneous, planes $\pi_N(9)$ and $\pi_N(9)^d$

(having the same homogeneity index) are more homogeneous, and $\pi_F(9)$ is the most homogeneous. This idea is discussed further in the next section.

EXERCISES

1. Find the cardinality of $\mathrm{Af}(2,F)$ if $F = \mathrm{GF}(p^k)$.

2. Refer back to Example 4.3.4(i). Show that $\mathrm{Hom}(p,L) \sim (N - \{0\},\cdot)$ where:
   a. $p = [1,0,0]$, L passes through $[0,1,0]$, and $L \neq \langle 0,0,1\rangle$.
   b. $p = [0,1,0]$, L passes through $[1,0,0]$, and $L \neq \langle 0,0,1\rangle$.
   c. $p = [1,x,0]$, L passes through $[1,-x,0]$, and $L \neq \langle 0,0,1\rangle$.

3. Let $M = (R,+,\circ)$ as in exercise 2.1.6 and show that in π_M:
   a. $\mathrm{El}([0,1,0],\langle 0,0,1\rangle) \sim (R,+)$.
   b. $\mathrm{Hom}([1,0,0],\langle 1,0,0\rangle) \sim (R - \{0\},\circ)$.

4. Refer back to exercises 2.1.6, 2.4.4, and 2.8.6.
   a. Show that for each of the planes π_A, π_B, and π_C, $\mathrm{El}([1,0,0],\langle 0,0,1\rangle)$ $\sim (R,+)$.
   b. Show that for each of the planes in 4(a), the mapping (g_r, G_r) displayed in Theorem 4.3.3 is not a homology for $r \neq \pm 1$.
   c. Show that a point p and a line L do not exist such that:
      (1) $\mathrm{Hom}(p,L) \sim (R - \{0\},\circ)$ in π_A.
      (2) $\mathrm{Hom}(p,L) \sim (R - \{0\},\odot)$ in π_B.
      (3) $\mathrm{Hom}(p,L) \sim (R - \{0\},\circledcirc)$ in π_C.
   d. For each of these planes find a point p and a line L such that $\mathrm{Hom}(p,L)$ $\sim (R - \{0\},\cdot)$. (Such a point and line exist because π_A, π_B, π_C, and π_N are isomorphic.)

5. Refer back to Example 4.3.6(ii). Construct the projectivity h^+ on $\pi_H(9)$ where h is the projectivity on $\pi_H(9)|A$ defined as follows:
   a. h: $a_i \rightarrow a_{i+j}$ where $i = 0, \ldots, 12$; j is fixed and $0 \leq j \leq 12$; and $i + j$ is addition mod 13.
   b. h: $a_i \rightarrow a_{3i}$ where $i = 0, \ldots, 12$; and $3i$ is multiplication mod 13. Hint: Refer back to exercise 2.8.6(a).

4.4 TRANSITIVITY OF TRANSFORMATION GROUPS

Continuing the discussion begun at the conclusion of Section 4.3, this section will analyze the relationship between points and collineations.

As previously indicated, the collineation groups themselves are of minor interest here; the interplay of the group with the plane is the important factor.

DEFINITION 4.4.1 *Let X be a set, let G be its permutation group, and let H be a subgroup of G. For $x \in X$, the* orbit of x with respect to H *is defined as the set of all $f(x)$ where $f \in H$.*

Notation. The orbit of x with respect to H is denoted by O_x^H. Thus $O_x^H = \{f(x): f \in H\}$.

DEFINITION 4.4.2 *The group H is* transitive on X *if and only if $O_x^H = X$; that is, for $x, y \in X$, there exists $f \in H$ such that $f(x) = y$.*

EXAMPLE 4.4.3

  i. Let $\pi = (\mathscr{P}, \mathscr{L}, \mathscr{I})$ be the cyclic plane π_{S_n}. It is easily seen that $C(\pi)$ is transitive on $\mathscr{P}$. In fact, the subgroup $H = \{f_j: j = 0, \ldots, n^2 + n\}$ is transitive on $\mathscr{P}$ where $f_j: p_i \to p_{i+j}, i = 0, \ldots, n^2 + n$, and $i + j$ is modulo $n^2 + n + 1$.

  ii. If π is a plane of the form π_F, then $PGL(2,F)$ is transitive on $\mathscr{P}$ by Theorem 4.2.3.

  iii. If $\pi = \pi_N(9)$, then $C(\pi)$ is not transitive on $\mathscr{P}$. From the discussion in Section 4.3, it follows that $O_p^{C(\pi)} = \mathscr{P} - \langle 0,0,1 \rangle$ for any $p \notin \langle 0,0,1 \rangle$ and $O_p^{C(\pi)} = \langle 0,0,1 \rangle$ for $p \in \langle 0,0,1 \rangle$.

  iv. If $\pi = \pi_N(9)^d$, then $O_p^{C(\pi)} = \mathscr{P} - \{\langle 0,0,1 \rangle\}$ for $p \neq \langle 0,0,1 \rangle$. Thus $C(\pi)$ is not transitive on $\mathscr{P}$.

  v. If $\alpha = \alpha_N(9)$, then $C(\alpha)$ is transitive on $\mathscr{P}$. In fact, it follows from Section 4.3 that $\mathrm{Trans}(\alpha)$ is transitive on $\mathscr{P}$.

  vi. If $\pi = \pi_H(9)$, then $O_p^{C(\pi)} = \mathscr{P} - A$ for $p \notin A$ and $O_p^{C(\pi)} = A$ for $p \in A$. Here $A = \{a_0, \ldots, a_{12}\}$.

  vii. If $\pi = \Sigma_o^+$ and Σ_o is the configuration of Table 16, then $O_p^{C(\pi)} = \{p\}$ for any $p \in \mathscr{P}$.

DEFINITION 4.4.4 *The group H is* k-ply transitive *on X for a positive integer k if and only if, for any two k-tuples $(x_1, x_2, \ldots, x_k)$ and $(y_1, y_2, \ldots, y_k)$, there exists $f \in H$ such that $f: x_i \to y_i, i = 1, \ldots, k$.*

This definition must be altered somewhat for collineation groups. Clearly, no collineation group is triply transitive on $\mathscr{P}$ because no collineation maps p,q,r into p',q',r' where p,q,r are collinear and p',q',r' are not. In fact, this reasoning can be generalized to show that no collineation group is k-ply transitive for $k > 2$. Therefore, the following definitions are adopted.

DEFINITION 4.4.5   *A collineation group H on $\mathscr{P}$ is transitive on triangles if and only if, given two triples of noncollinear points p_i and q_i, $i = 1,2,3$, respectively, there exists a collineation f such that $f: p_i \to q_i$, $i = 1,2,3$. The group H is* transitive on four-points *if and only if, given two four-points p_i and q_i, $i = 1, \ldots, 4$, there exists a collineation f such that $f: p_i \to q_i$, $i = 1, \ldots, 4$.*

The word "triangle" in the term "transitive on triangles" is a possible source of confusion. The definition of "triangle" is not based on the concept of order, whereas the definition of "transitivity on triangles" is. Since both definitions are standard, they will be used here, although it is tempting to adopt the less familiar definition of a triangle as an ordered triple of noncollinear points.

EXAMPLE 4.4.6

i. By Theorem 4.2.3, PGL$(2,F)$ is transitive on four-points in π_F.

ii. The group Af$(2,F)$ is transitive on triangles in α_F. Proof of this fact is left as an exercise.

iii. The group C$(\alpha_N(9))$ is doubly transitive. Proof of this fact is also left as an exercise.

iv. In a projective plane π_F, the group Proj(L) is triply transitive on L for any L in π_F. This conclusion follows from Theorems 4.2.12 and 4.2.13.

The following theorem extends the result of Example 4.4.6(iv) to any projective plane.

THEOREM 4.4.7   *If (p,q,r) and (p',q',r') are two triples of distinct collinear points, then there exists a projectivity f such that $f: p,q,r \to p',q',r'$.*

Proof.   Let line $pq = L$, $p'q' = L'$.

Case 1. $p = p'$ and $L \neq L'$. Let $s = qq' \cap rr'$; then clearly $f = L \overset{s}{\overline{\wedge}} L'$ is a projectivity mapping $p,q,r \to p',q',r'$, as shown in Figure 26.

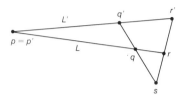

FIGURE 26

Case 2. $p \neq p'$ and $p' \notin L$. Let s be a point on line pp' distinct from p and p'. Such a point exists because every line has at least three points. Construct a line M through p' distinct from L' and pp'. Such a line exists because every point has at least three lines through it. Let $f = L \overset{s}{\overline{\wedge}} M$. This perspectivity is defined because s is on neither L nor M since $s \in L'$, $L \cap L' = p$, and $M \cap L' = p'$. Define q'' and r'' to be images of q and r under f. Clearly, p' is the image of p under f. By Case 1, there exists a perspectivity g: $p' \to p'$, $q'' \to q'$, $r'' \to r'$. Thus $g \circ f$: $p \to p'$, $q \to q'$, $r \to r'$ (see Figure 27).

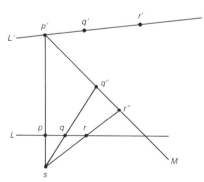

FIGURE 27

Case 3.   Suppose that neither Case 1 nor Case 2 holds. Let p'' be on neither L nor L'. Such a point exists by Pj3. Let M be a line through p'' containing neither p' nor p. Such a line exists because every point has at least three lines through it. Let q'' and r'' be two other distinct points on M. By Case 2, there exists a projectivity $f: p,q,r \to p'',q'',r''$ and a projectivity $g: p',q',r' \to p'',q'',r''$. Thus $g^{-1} \circ f$ maps $p,q,r \to p',q',r'$ (see Figure 28).

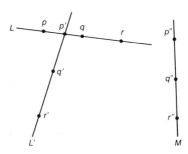

FIGURE 28

The study of the existence of a transformation relating k-tuples of points is naturally linked with study of the uniqueness of such a transformation. This fact leads to formulation of the following definition.

DEFINITION 4.4.8 *A group H of transformations on X is sharply k-transitive if and only if it is k-transitive and the function relating x_i to y_i, $i = 1, \ldots, k$, is unique; H is sharply transitive on four-points (or triangles) if and only if the function relating p_i to q_i, $i = 1, \ldots, 4$ ($i = 1,2,3$), is unique.*

EXAMPLE 4.4.9

   i. Theorem 4.2.3 shows that the group $PGL(2,F)$ is sharply transitive on four-points in $\mathscr{P}$. Since $PGL(2,F) = CC(\pi_F) = PC(\pi_F)$, all the groups are sharply transitive on four-points.

  ii. The group $AF(2,F)$, and therefore the group $\mathrm{Dil}(\alpha_F)$, is sharply transitive on triangles. Proof of this statement is left as an exercise.

 iii. Theorem 4.2.12 shows that $PGL(1,F)$ is sharply 3-transitive on $\langle 0,0,1 \rangle$. Theorem 4.2.13 implies that $\mathrm{Proj}(L)$ is sharply 3-transitive on L for any L in π_F.

  iv. From Example 4.3.4, it follows that $\mathrm{Dil}(\alpha_N(9))$ is sharply 2-transitive.

v. The group H of Example 4.4.3(i) is sharply transitive on the points of π_{S_n}. This is easily seen from the nature of H.

The importance of transitive transformation groups to projective and affine planes is not yet apparent. The following theorem is the first formal application in this text of collineation groups to projective planes.

THEOREM 4.4.10   If $C(\pi)$ is transitive on four-points, then $C(\pi)$ has a characteristic (see Definition 2.8.4 for the definition of "characteristic").

Proof.   The proof is left as an exercise.

The converse is false for planes of characteristic 0. Consider the free extension plane generated by a four-point (as in Example 2.5.8(ii)). The converse of Theorem 4.4.10 is true for finite planes of characteristic 2 (Gleason, 1956) but remains an open question for finite planes of other characteristics and for infinite planes of all characteristics other than 0.

It should be observed that, with the exception of the example of a transitive group $C(\pi)$ that is not transitive on four-points given in Section 4.6, the only examples of collineation groups $C(\pi)$ given here are either transitive on four-points (recall that all known finite cyclic planes are planes of the form π_F) or not transitive at all. It is suspected, but as yet unproved, that a finite plane with a transitive collineation group must necessarily be a plane of the form π_F. Chapter 14 will show that a finite plane π with a collineation group transitive on four-points must be a plane of the form π_F. The best result in this area (Ostrom and Wagner, 1959) establishes that a finite plane π with a doubly transitive collineation group must be a plane π_F.

One important aspect of transitivity of transformation groups involves the existence and uniqueness of projective collineations and projectivities. In fact, the two so-called fundamental theorems of projective planes are formulated in these terms:

THE FUNDAMENTAL THEOREM OF PROJECTIVE GEOMETRY (FT-I)
There exists one and only one projectivity mapping an ordered triple of distinct collinear points p, q, r onto an ordered triple of distinct collinear points p', q', r'.

THE FUNDAMENTAL THEOREM OF PROJECTIVE COLLINEATIONS
(FT-II)   *There exists one and only one projective collineation mapping a four-point p,q,r,s onto a four-point p', q', r', s'.*

THEOREM 4.4.11   *The plane π_F satisfies* FT-II.

Proof.   This conclusion follows because $PC(\pi_F)$ is sharply transitive
on four-points.

THEOREM 4.4.12   *A plane π satisfies* FT-I *if and only if, for some
fixed line L, the group* Proj(L) *is sharply 3-transitive.*

Proof.   Clearly, if π satisfies FT-I, then Proj(L) is sharply 3-transitive
for all L in π.
     Suppose that Proj(L) is sharply 3-transitive and let p,q,r and
p',q',r' be two arbitrary triples of collinear points. By Theorem
4.4.7, there exist projectivities f,g,h such that $f: p,q,r \rightarrow p',q',r'$;
$g: p,q,r \rightarrow a,b,c$; $h: p',q',r' \rightarrow a,b,c$ where a,b,c are three arbitrary
points on L. Suppose that f' also maps $p,q,r \rightarrow p',q',r'$. Then $h \circ f \circ g^{-1}$
and $h \circ f' \circ g^{-1}$ are projectivities fixing a,b,c, and therefore $h \circ f \circ g^{-1}$
$= h \circ f' \circ g^{-1}$. We may conclude that $f = f'$ and thus that the mapping
from p,q,r to p',q',r' is unique. Therefore, π satisfies FT-I.

COROLLARY 4.4.13   *The plane π_F satisfies* FT-I.

Proof.   The proof follows directly from Theorems 4.2.12 and 4.4.12.

EXAMPLE 4.4.14

   i. The plane $\pi_N(9)$ does not satisfy FT-II because $C(\pi_N(9))$ is not
      even transitive, as evidenced by Example 4.4.3(iii). Neither does
      $\pi_N(9)$ satisfy FT-I. To show this, let $f = \langle 0,1,0 \rangle \overset{p}{\barwedge} \langle 0,0,1 \rangle \overset{q}{\barwedge}$
      $\langle 0,1,0 \rangle$ where $p = [0,1,1]$ and $q = [e,c,1]$; also let $g = \langle 0,1,0 \rangle \overset{r}{\barwedge}$
      $\langle c,1,f \rangle \overset{s}{\barwedge} \langle 0,1,0 \rangle$ where $r = [b,1,1]$ and $s = [1,e,1]$. It may be
      verified that

$$f: [0,0,1] \rightarrow [0,1,0] \rightarrow [-1,0,1]$$
$$[-1,0,1] \rightarrow [1,1,0] \rightarrow [0,0,1]$$
$$[a,0,1] \rightarrow [1,a,0] \rightarrow [a,0,1]$$

$$g: \quad [0,0,1] \quad \rightarrow [e,a,1] \rightarrow [-1,0,1]$$
$$[-1,0,1] \rightarrow [1,e,0] \rightarrow [0,0,1]$$
$$[a,0,1] \quad \rightarrow [a,0,1] \rightarrow [a,0,1]$$

so f and g agree on the three points $[0,0,1]$, $[-1,0,1]$, and $[a,0,1]$. However, $f: [1,0,0] \rightarrow [1,0,0] \rightarrow [1,0,0]$ and $g: [1,0,0] \rightarrow [1,-1,1] \rightarrow [1,0,1]$, so $f \neq g$.

Figures 29 and 30, the constructions associated with these projectivities, explain the choice of f and g. Clearly, f was constructed in a $\pi(3)$ subplane; g was constructed in a $\pi(2)$ subplane. This should cause the student to wonder if the validity of FT-I is related to the existence of a characteristic in π. (Theorem 4.4.10 tells us that the existence of a characteristic is necessary for the validity of FT-II.)

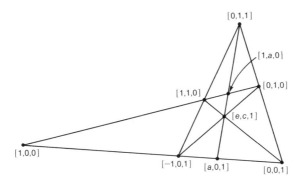

FIGURE 29

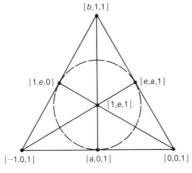

FIGURE 30

ii. The plane $\pi_H(9)$ does not satisfy FT-II because $C(\pi_H(9))$ is not even transitive, as indicated in Example 4.4.3(vi). Proof of the fact that the plane does not satisfy FT-I either is left as an exercise.

By the preceding theorems and examples, it appears that projective planes satisfy FT-II if and only if they satisfy FT-I. It would seem that FT-II is stronger than FT-I because the condition of transitivity on four-points on $PC(\pi)$ is not always satisfied on π, whereas the triple transitivity restriction of FT-I is always satisfied because of Theorem 4.4.7. Nevertheless, Chapter 7 will show that planes satisfying FT-I always satisfy FT-II; Chapter 14 will show that the two theorems are actually equivalent and that the fundamental theorems are satisfied only on planes π_F. It seems that the more we learn about projective planes, the more special the planes π_F become.

The most important application of transitivity of transformation groups to projective and affine planes is found in the study of central collineation groups. This application is discussed in the next section.

EXERCISES

1. a. Show that if G is an Abelian group transitive on a set X, then G is sharply transitive on X.
   b. Show that the converse of 1(a) is not true.

2. Show that if $C(\pi)$ is sharply transitive on four-points, then $C(\pi^d)$ is also sharply transitive on four-points.

3. a. Show that the group $Af(2,F)$ is sharply transitive on triangles in α_F.
   b. Show that $C(\alpha_N(9))$ is doubly transitive but not sharply 2-transitive.

4. Show that if π satisfies FT-I and if $PC(\pi)$ is transitive on four-points, then π also satisfies FT-II.

5. Show that the following theorems are equivalent to FT-I:

   For any two distinct lines L and L', if f is a projectivity mapping $L \to L'$ such that f leaves $L \cap L'$ fixed, then f is a perspectivity.

   There exist two distinct lines L and L' such that if f is a projectivity mapping $L \to L'$ and f leaves $L \cap L'$ fixed, then f is a perspectivity.

   There exist two ordered triples of three distinct collinear points a,b,c and a',b',c', such that exactly one projectivity maps $a,b,c \to a',b',c'$.

6. Show that $\pi_H(9)$ does not satisfy FT-I.

7. Prove Theorem 4.4.10.

4.5 TRANSITIVITY IN CENTRAL COLLINEATION GROUPS

The general definition of transitivity must be altered to derive the definition of transitivity for $CC(p,L)$, $CC(L)$, and $CC(p)$ for the same reasons that the definition of k-ple transitivity on $C(\pi)$ was altered in Definition 4.4.5. The natural definitions for this restricted transitivity may be stated as follows.

DEFINITION 4.5.1

1. *The group* $CC(p,L)$ *is transitive on* $\mathscr{P}$ *if and only if, for any two points* q,r *such that* $q,r \notin L$, $q,r \neq p$, *and* p,q,r *are collinear, there exists* $f \in CC(p,L)$ *such that* $f: q \rightarrow r$.
2. *The group* $CC(L)$ *is transitive on* $\mathscr{P}$ *if and only if, for any two points* $q,r \in L$, *there exists* $f \in CC(L)$ *such that* $f: q \rightarrow r$.
3. *The group* $CC(p)$ *is transitive on* $\mathscr{P}$ *if and only if, for any two points* q,r *distinct from and collinear with* p, *there exists* $f \in CC(p)$ *such that* $f: q \rightarrow r$.

Analogous definitions can be formulated for $\text{Dil}(\alpha)$, $\text{Trans}(\alpha)$, $\text{Dil}(p)$, and $\text{Trans}(L)$. The phrase "sharply transitive" will be used in the same sense as in Definition 4.4.8.

Theorems 4.5.2 and 4.5.3 are almost immediate and are left for the student to verify.

THEOREM 4.5.2

i. *If* $\text{El}(p,L)$ *is transitive for all* $p \in L$, *then* $CC(L)$ *is transitive.*

ii. *If* $\text{El}(p,L)$ *is transitive for all* L *through* p, *then* $CC(p)$ *is transitive.*

THEOREM 4.5.3 *If* $CC(p,L)$ *is transitive, then it is sharply transitive.*

EXAMPLE 4.5.4

  i. If $\pi = \pi_F$, then the group $CC(p,L)$ is transitive for all p and L. (Proof of this fact is left as an exercise.)

  ii. Let $\pi = \pi_N(9)$. From the discussion in Section 4.3, it is easily seen that $El(p,\langle 0,0,1 \rangle)$ is transitive for every $p \in \langle 0,0,1 \rangle$; that $Hom([0,0,1],\langle 0,0,1 \rangle)$ is not transitive; but that $Hom([0,1,0], \langle 0,1,0 \rangle)$, $Hom([1,0,0],\langle 1,0,0 \rangle)$, and $Hom([1,x,0],\langle x,1,0 \rangle)$, $x \neq 0$, are all transitive groups.

  iii. Let $\pi = \pi_H(9)$. Again recalling Section 4.3, we see that none of the central collineation groups are transitive.

The following theorem allows us to recognize transitivity in $CC(p,L)$ more easily.

THEOREM 4.5.5 *The group $H = CC(p,L)$ is transitive if and only if, given q and M such that $q,p \in M$, $q \notin L$, $q \neq p$, then $O_q^H = M - \{p,L \cap M\}$.*

Proof. Clearly, if $CC(p,L)$ is transitive, then $O_q^H = M - \{p,L \cap M\}$. Suppose that $O_q^H = M - \{p,L \cap M\}$ and let r, s, and N be such that $M \neq N$; $p,r,s \in N$; $r,s \neq p$; and $r,s \notin L$. Also let $qr \cap L = t$; let $ts \cap M = u$; and let f be the unique member of $CC(p,L)$: $f(q) = u$ (see Figure 31). Thus f: $qt \to ut$ and so f: $qt \cap N \to ut \cap N$, that is, f: $r \to s$. Since r and s were arbitrary points on $N - \{p,L \cap N\}$, it follows that $CC(p,L)$ is transitive.

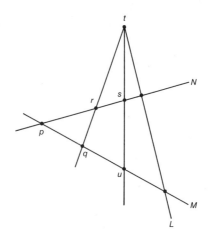

FIGURE 31

We are now ready to apply our study of collineation groups to projective and affine planes. Recall that for the projective plane π_F, the group $\text{El}(p,L)$ is isomorphic to $(F,+)$, and $\text{Hom}(p,L)$ is isomorphic to $(F - \{0\},\cdot)$. Specifically, Theorem 4.2.10 showed that, for planes π_F, $\text{El}(p,L) \sim (F,+)$ where $p = [1,0,0]$ and $L = \langle 0,0,1 \rangle$, and that $\text{Hom}(q,M) \sim (F - \{0\},\cdot)$ where $q = [1,0,0]$ and $M = \langle 1,0,0 \rangle$. Thus we could equally well have coordinatized π_F by elements of $\text{El}(p,L)$ and $\text{Hom}(q,M)$. Similarly, Theorems 4.3.2 and 4.3.3 showed that $\text{El}(p,L) \sim (N,+)$ and $\text{Hom}(q,M) \sim (N - \{0\},\cdot)$ for the same choice of points on $\pi_N(9)$. This suggests the possibility that for other planes π, these two central collineation groups can be used to coordinatize π, as is done in the following paragraphs.

Let u,v,o,e be a fixed four-point, let $i = ve \cap ou$, and let $S = ou - \{u\}$. Suppose that $O_o^G = S$ where $G = \text{El}(u,uv)$ (or, equivalently, suppose that $\text{El}(u,uv)$ is transitive). Suppose that $O_i^H = S - \{o\}$ where $H = \text{Hom}(u,ov)$ (equivalently, suppose that $\text{Hom}(u,ov)$ is transitive). Denote the unique elation mapping o to p by f_p and denote the unique homology mapping i to $p(p \neq o)$ by g_p. Then the following definition can be stated.

Definition 4.5.6 For $p,q \in S$, define $p \oplus q$ as $f_q \circ f_p(o)$; define $p \odot q$ for $p,q \neq o$ as $g_q \circ g_p(i)$; and define $p \odot o = o \odot p = o$.

Theorem 4.5.7 The following structures are isomorphic: $(S,\oplus) \sim \text{El}(u,uv)$ and $(S - \{o\},\odot) \sim \text{Hom}(u,ov)$.

Proof. Let β: $\text{El}(u,uv) \to S$ be defined by $\beta(f_p) = p$ and let γ: $\text{Hom}(u,ov) \to S - \{o\}$ be defined by $\gamma(g_p) = p$. Clearly, β and γ are isomorphisms.

Example 4.5.8

i. To relate this construction to π_F, let $u = [1,0,0]$, $v = [0,1,0]$, $o = [0,0,1]$, $e = [1,1,1]$. Thus $i = [1,0,1]$, $uv = \langle 0,0,1 \rangle$, $ov = \langle 1,0,0 \rangle$. It is easily verified (see exercise 4.2.5(b)) that $S = \{[x,0,1]: x \in F\}$, that $[x,0,1] \oplus [y,0,1] = [x + y,0,1]$, and that $[x,0,1] \odot [y,0,1] = [xy,0,1]$. Thus, for π_F, $(S,\oplus,\odot) \sim (F,+,\cdot)$.

ii. Let π be the plane $\pi_N(9)$ and let $u = [1,0,0]$, $v = [0,1,0]$, $o = [0,0,1]$, $e = [1,1,1]$; then $i = [1,0,1]$, $uv = \langle 0,0,1 \rangle$, $ov = \langle 1,0,0 \rangle$.

As in (i), it is easily seen that $S = \{[x,0,1]: x \in N\}$, $[x,0,1]$ $\oplus [y,0,1] = [x+y,0,1]$, and $[x,0,1] \odot [y,0,1] = [xy,0,1]$. Therefore, for $\pi_N(9)$, $(S,\oplus,\odot) \sim (N,+,\cdot)$.

In general, the structure $(S,\oplus,\odot)$ has the properties described in Theorems 4.5.9 and 4.5.10.

THEOREM 4.5.9 *The structures $(S,\oplus)$ and $(S - \{o\},\odot)$ are groups.*

Proof. It is easily seen that S is closed under both $\oplus$ and $\odot$ and that both operations are associative because the composition of mappings is associative. It is also obvious that the additive identity is o, the multiplicative identity is i, the additive inverse of p is $f_p^{-1}(o)$, and the multiplicative inverse of p for $p \neq o$ is $g_p^{-1}(i)$.

We may relate the operations of addition and multiplication with a right-hand distributive law.

THEOREM 4.5.10 *For $(S,\oplus,\odot)$, $(q \oplus r) \odot p = (q \odot p) \oplus (r \odot p)$.*

Proof. The first step is to show that $f_{p\odot q} = g_q \circ f_p \circ g_q^{-1}$ for $q \neq o$. Clearly, both mappings are in $El(u,uv)$. Also, $f_{p\odot q}(o) = p \odot q$ and $g_q \circ f_p \circ g_q^{-1}(o) = g_q \circ f_p(o) = g_q(p) = g_q \circ g_p(i) = g_{p\odot q}(i) = p \odot q$. Therefore, $f_{p\odot q}(o) = g_q \circ f_p \circ g_q^{-1}(o)$, and so they must be the same elation.
We now have the following equalities:

$$(q \oplus r) \odot p = f_{(q\oplus r)\odot p}(o) = g_p \circ f_{q\oplus r} \circ g_p^{-1}(o)$$
$$= g_p \circ f_r \circ f_q \circ g_p^{-1}(o) = g_p \circ f_r \circ g_p^{-1} \circ g_p \circ f_q \circ g_p^{-1}(o)$$
$$= f_{r\odot p} \circ f_{q\odot p}(o) = (q \odot p) \oplus (r \odot p)$$

All we need is commutativity of addition and $(S,\oplus,\odot)$ becomes a right nearfield. The following theorem provides a condition for commutativity.

THEOREM 4.5.11 *If there exist two points p and q on L such that $El(p,L)$ and $El(q,L)$ each have at least one elation other than the identity, then $El(L)$ is Abelian.*

Proof. Let $f \in El(p,L)$ and let $g \in El(q,L)$ such that neither f nor g is the identity and $p \neq q$. Let r be any point not on L. Both of the triples $p,r,f(r)$ and $q,r,g(r)$ are collinear since f and g are elations. Thus $f(p)$, $g(r)$, $g \circ f(r)$ and $f(q)$, $f(r)$, and $f \circ g(r)$ are collinear triples because f and g are collineations. Note that p and q are held fixed under g and f. Therefore, we have $p,g(r)$, $g \circ f(r)$ on one line, call it M, and $q,f(r)$, $f \circ g(r)$ on another line, call it N. Clearly, $M \neq N$ because $pq = L$ and all the other points are not on L. Since f and g are elations, $p,g(r)$, $f \circ g(r)$, and $q,f(r)$, $g \circ f(r)$ are collinear triples. Thus M contains $p,g(r)$, $g \circ f(r)$, and $f \circ g(r)$, and N contains $q,f(r),f \circ g(r)$, and $g \circ f(r)$. It follows that $f \circ g(r) = g \circ f(r)$. Since r was arbitrary, we have shown that $f \circ g = g \circ f$ for all elations f and g whose centers are different.

Suppose that $f,h \in \text{El}(p,L)$. Let $g \in \text{El}(q,L)$ as previously; then $h \circ g = g \circ h$. Clearly, $h \circ g \in \text{El}(s,L)$ for some $s \neq p,q$; therefore, $f \circ (h \circ g) = (h \circ g) \circ f$. But $h \circ g = g \circ h$, so $(f \circ h) \circ g = f \circ (h \circ g) = (g \circ h) \circ f = g \circ (h \circ f)$. Furthermore, g commutes with $h \circ f$ since $g \in \text{El}(q,L)$ and since $h \circ f \in \text{El}(p,L)$, so $g \circ (f \circ h) = (f \circ h) \circ g = g \circ (h \circ f)$. Therefore, $f \circ h = h \circ f$, and the proof is complete.

This proof has shown that for a plane π having three noncollinear points o,u,v such that the groups $\text{El}(u,uv)$ and $\text{Hom}(u,ov)$ are transitive, there is an associated algebraic structure $(S,\oplus,\odot)$. For planes π_F, $(S,\oplus,\odot)$ is isomorphic to $(F,+,\cdot)$, and for the plane $\pi_N(9)$, $(S,\oplus,\odot)$ is isomorphic to $(N,+,\cdot)$. The following important theorem has not yet been proved here: *Given the plane π and the associated system* $(S,\oplus,\odot)$, *if $(S,\oplus,\odot)$ is a field F, then $\pi \sim \pi_F$. If $(S,\oplus,\odot)$ is a right nearfield N, then $\pi \sim \pi_N$.* (π_N is constructed as in Construction 2.4.3.) If this theorem were true, we would have a method for representing certain planes by algebraic means. This would be a major breakthrough for us; we have had notable success in dealing with the planes π_F and $\pi_N(9)$ and very little success in analyzing other planes. Fortunately, the theorem is true; it will be proved for fields in Chapter 8 and for nearfields in Chapter 12. In general, if $\text{El}(u,uv)$ and $\text{Hom}(u,ov)$ are transitive groups in $C(\pi)$, then π is representable by an algebraic system $(S,\oplus,\odot)$. The representation of $\pi = (\mathscr{P},\mathscr{L},\mathscr{I})$ is similar to that in Construction 2.4.3:

$$\mathscr{P} = \{[p,q,i]: p,q \in S\} \cup \{[i,p,o]: p \in S\} \cup \{[o,i,o]\}$$
$$\mathscr{L} = \{\langle r,i,t \rangle: r,t \in S\} \cup \{\langle i,o,t \rangle: t \in S\} \cup \{\langle o,o,i \rangle\}$$

and $\mathscr{I}$ is defined as follows: $[p,q,k] \in \langle r,s,t \rangle$ if and only if $(p \odot r) \oplus (q \odot s) \oplus (k \odot t) = o$.

As we already are aware, there are projective planes that do not satisfy the transitivity requirements for $\text{El}(p,L)$ and $\text{Hom}(q,M)$; $\pi_H(9)$ is one example, another is Σ_o^+ (where Σ_o is a nontrivial configuration). Thus systems different from $(S,\oplus,\odot)$ must be developed in order to represent general projective planes. This is done in Part 3. The key concept in this development is that of the transitive collineation group.

This section concludes with some definitions and elementary theorems concerning transitive collineation groups. Most of the proofs are straightforward and are left as exercises.

DEFINITION 4.5.12

1. *The plane π is (p,L) transitive if and only if the group $\text{CC}(p,L)$ is transitive.*

2. *π is (M,L) transitive if and only if π is (p,L) transitive for all $p \in M$.*

3. *π is (p,q) transitive if and only if π is (p,L) transitive for all L through q.*

EXAMPLE 4.5.13

i. The plane π_F is (p,L) transitive for all p and L.

ii. The plane π_N is (L,L) transitive where $L = \langle 0,0,1 \rangle$. It is (p,q) transitive where $p = [0,1,0]$ and $q = [1,0,0]$, where $p = [1,0,0]$ and $q = [0,1,0]$, and where $p = [1,x,0]$, $q = [1,-x,0]$, and $x \in N$, $x \neq 0$.

iii. The plane π_H is not (p,L) transitive for any p or L.

THEOREM 4.5.14   *If π is (p,L) transitive and there exists $f \in \text{C}(\pi)$ such that $f: p \to q$ and $L \to M$, then π is (q,M) transitive.*

Proof.   The proof is left as an exercise.

THEOREM 4.5.15

i. *If π is (p,L) and (q,L) transitive, then π is (pq,L) transitive.*

ii. *If π is (p,L) and (p,M) transitive, then π is $(p, L \cap M)$ transitive.*

Proof.   Only (i) is proved here; (ii) follows by duality. Suppose that π is (p,L) and (q,L) transitive; then we have three cases:

Case 1. $p \in L, q \notin L$. Since π is (p,L) transitive, there exists, for any $r \in pq, r \notin L$, an $f_r \in El(p,L)$ such that $f_r(q) = r$. Thus by Theorem 4.5.14, π is (r,L) transitive for all $r \in pq, r \notin L$. Also, π is (p,L) transitive and $p \in L$, so our conclusion follows.

Case 2. $p \notin L, q \notin L$. Since π is (p,L) transitive, there exists, for any $r \in pq$ such that $r \neq p$ and $r \notin L$, a homology $f_r \in$ Hom(p,L) such that $f_r(q) = r$. Thus π is (r,L) transitive for all $r \notin L, r \in pq$. Let $t = pq \cap L$. The final step in the proof is to show that π is (t,L) transitive. We shall assume here that each line has at least four points. (We may make this assumption because the theorem is easily seen to be true for $\pi(2)$.) Let a and b be arbitrary points not on pq or L such that t, a, and b are collinear. Let $pa \cap qb = r$. We have two subcases.

Subcase A. $r \notin L$. Let $f \in$ Hom(p,L) such that $f(a) = r$ and let $g \in$ Hom(q,L) such that $g(r) = b$. Then $g \circ f \in El(t,L)$ and $g \circ f$: $a \rightarrow b$.

Subcase B. $r \in L$. Let $c \in pq$ such that $c \neq t, p, q$ and let $pa \cap cb = s$. Clearly, $s \notin L$ and $s \notin pq$. Let $f \in$ Hom(p,L) such that $f(a) = s$ and let $g \in$ Hom(q,L) such that $g(s) = b$. Then $g \circ f \in El(t,L)$ and $g \circ f$: $a \rightarrow b$. Thus π is (t,L) transitive.

Case 3. $p,q \in L$. This follows easily.

THEOREM 4.5.16

   i. *If π is (L,L) and (M,M) transitive, then π is (N,N) transitive for all N through $L \cap M$.*

   ii. *If π is (p,p) and (q,q) transitive, then π is (r,r) transitive for all $r \in pq$.*

Proof.   The proof is left as an exercise.

THEOREM 4.5.17

   i. *If π is (L,L), (M,M), and (N,N) transitive for three nonconcurrent lines L,M,N, then π is (X,X) transitive for all lines X in π.*

   ii. *If π is (p,p), (q,q), and (r,r) transitive for three noncollinear points p,q,r, then π is (x,x) transitive for all points x in π.*

Proof.   The proof is left as an exercise.

THEOREM 4.5.18 *If π is (p,L) transitive for all $p \notin L$, then π is (p,L) transitive for all p.*

Proof. The proof is left as an exercise.

THEOREM 4.5.19 *If π is (p,L) transitive for all points p and all lines L, then $PC(\pi)$ is transitive on four-points. Furthermore, $PC(\pi) = CC(\pi)$ and every member of $PC(\pi)$ is the product of at most five central collineations.*

Proof. Let p,q,r,s and p',q',r',s' be two arbitrary four-points in π. Let L be a line not containing p or p' and let $f \in El(L)$ such that f: $p \to p'$. Let M be a line that contains p' but does not pass through $f(q)$ or q'. Let $g \in El(M)$ such that g: $f(q) \to q'$. Let $N = p'q'$. We know that $g \circ f(r)$ and r' are not on N because p,q,r and p',q',r' are triples of noncollinear points and f and g are collineations, so there exists $h \in El(N)$ such that h: $g \circ f(r) \to r'$. Thus $h \circ g \circ f$: $p,q,r \to p',q',r'$.

Let $s'' = h \circ g \circ f(s)$. Then neither s' nor s'' is incident with any line of the triangle $p'q'r'$ because of the initial condition that p,q,r,s and p',q',r',s' be four-points and the fact that f, g, and h are collineations. Let $t = p's' \cap q's''$. Then there exists $j \in Hom(p',q'r')$ such that j: $s'' \to t$ and $k \in Hom(q',p'r')$ such that k: $t \to s'$. Thus $k \circ j \circ h \circ g \circ f$: $p,q,r,s \to p',q',r',s'$.

EXERCISES

1. Show that if π is (p,L) transitive, then π^d is (L,p) transitive.

2. Prove Theorem 4.5.14.

3. Prove the statement in Example 4.5.4(i).

4. Prove Theorem 4.5.16.

5. Prove Theorem 4.5.17.

6. Prove Theorem 4.5.18.

7. Show that π is (p,p) transitive for all points p if and only if π is (L,L) transitive for all lines L.

8. Consider the Moulton plane π_M of exercise 2.1.6(f) and let $u = [1,0,0]$, $v = [0,1,0]$, and $o = [0,0,1]$.
   a. Show that π_M is (v,uv) transitive.
   b. Show that π_M is (u,ov) transitive.

9. Refer back to exercise 2.1.6. Let $(R,+,\cdot)$ be a Cartesian group, let $\pi = (\mathscr{P},\mathscr{L},\mathscr{I})$ be the plane constructed over the Cartesian group, and let $u = [1,0,0]$, $v = [0,1,0]$, and $o = [0,0,1]$.
   a. Suppose that π is (u,uv) transitive. Show that $(R,+,\cdot)$ satisfies right distributivity if and only if $(R,+) \sim \mathrm{El}(u,uv)$.
   b. Suppose that π is (u,ov) transitive. Show that $(R - \{0\},\cdot)$ satisfies the associative law if and only if $(R - \{0\},\cdot) \sim \mathrm{Hom}(u,ov)$.

4.6 CYCLIC TRANSFORMATION GROUPS AND CYCLIC PLANES

This section includes another application of collineation groups to projective planes. Perhaps this application, more than any other, approximates what the uninitiated student expected of an interplay between groups and planes. Clearly, every finite cyclic plane has a cyclic transitive collineation subgroup. The converse is also true: If a finite plane π has a cyclic transitive collineation subgroup, then π is a cyclic plane. These observations are stated formally in Theorems 4.6.1 and 4.6.2.

THEOREM 4.6.1 *The plane π_{S_n} has a cyclic transitive group $G \subseteq C(\pi_{S_n})$.*

Proof. Simply let $G = \{f^k : k = 0, \ldots, n^2 + n\}$ where $f : p_i \to p_{i+1}$, $f^0 = I$, and $f^k = f \circ \ldots \circ f$ (k times).

THEOREM 4.6.2 *If π is a finite plane with a cyclic transitive collineation group $G \subseteq C(\pi)$, then π is a cyclic plane.*

Proof. Let f be the collineation that generates the transitive subgroup and number the points of π with subscripts $0, \ldots, n^2 + n$ where $p_k = f^k(p_0)$ and p_0 is arbitrary. Choose any line L of π and denote its points by $p_{a_0}, \ldots, p_{a_n}$. Let $S = \{a_0, \ldots, a_n\}$.

First it must be shown that S is a perfect difference set of order n. Let $N = n^2 + n + 1$. Notice that there are exactly $N - 1$ numbers of the form $d \bmod N$ where $d \neq 0$, and there are at most $N - 1$ numbers of the form $a_i - a_j \bmod N$ (since S contains $n + 1$ numbers). There will be exactly $N - 1$ differences mod N iff, for any $d \neq 0$, there exists $a_i, a_j \in S$ such that $a_i - a_j = d \bmod N$. It follows that if $a_i - a_j = d$ mod N has a solution for each $d \neq 0$, then the solution is unique and thus S is a perfect difference set. Let $d \neq 0$ be given. The line $f^d(L)$ intersects L in at least one point, p_{a_i}. Since $p_{a_i} \in f^d(L)$, $p_{a_i} = p_b$ where $b = a_j + d \bmod N$ for some $a_j \in S$. Thus $a_i - a_j = d \bmod N$.

The proof is concluded by showing that an isomorphism exists between π and the plane generated by S, call it π_S. Denote the points of π_S by $q_0, \ldots, q_{N-1}$ and denote the lines of π_S by $M_0, \ldots, M_{N-1}$. At the beginning of the proof the points of π were labeled with subscripts and were denoted by $p_0, \ldots, p_{N-1}$. Now label the lines of π as follows: $f^k(L) = L_k$, $k = 0, \ldots, N - 1$. This labeling exhausts the lines of π because each line is uniquely represented by $f^k(L)$. To see that this is true, suppose that $L \cap f^k(L)$ contains two points, p_{a_i} and p_{a_j}. Then $a_i, a_j, a_r, a_s \in S$ where $a_r + k = a_i \bmod N$ and $a_s + k = a_j$ mod N. But this implies that two distinct differences, $a_i - a_j$ and $a_r - a_s$, are congruent mod N, which contradicts the uniqueness of differences in S. Define h from the points of π to the points of π_S as follows: $h: p_i \rightarrow q_i$, $i = 0, \ldots, N - 1$. It is easy to check that h preserves incidences because $h: L_i \rightarrow M_i$, $i = 0, \ldots, N - 1$, and because both $p_i \in L_j$ and $q_i \in M_j$ occur exactly when $i+j \in S$.

This close relationship between cyclic groups and cyclic planes does not extend to subgroups and subplanes. As indicated in Example 2.7.4(ii), if $\pi = \pi_{S_9}$, the cyclic subgroup $H = \{f^{7i}: i = 0, \ldots, 12\}$ of G corresponds to a cyclic subplane $\pi | \{p_{7i}: i = 0, \ldots, 12\}$, but the cyclic subgroup $J = \{f^{13i}: i = 0, \ldots, 6\}$ of G does not correspond to any subplane of π. Also, some cyclic subplanes of cyclic planes do not correspond to cyclic subgroups. The example $\pi_{S_2} \subseteq \pi_{S_8}$ serves in this regard since 7 does not divide 73.

A generalization of cyclic planes to infinite planes has been examined by Hall (1947) but little has been done since then. This section is concluded with a brief examination of that generalization.

DEFINITION 4.6.3 *Let $(S,+)$ be a group under addition. The set T is a difference set of S if and only if $T \subset S$ and each nonzero element of S is generated by a unique difference of elements of T.*

THEOREM 4.6.4 *If $T_r = \{a_0, \ldots, a_r\}$ is a set of integers such that no two differences $a_i - a_j$, $i \neq j$, are equal and d is any integer not equal to any of these differences, then there exist integers a_{r+1}, a_{r+2} such that $a_{r+2} - a_{r+1} = d$ and such that no two differences of the set $T_{r+2} = \{a_0, \ldots, a_{r+2}\}$ are equal.*

Proof. Let $S = \{a_i - a_j: i \neq j\}$; let $M = \max\{|s|+|a|: s \in S$ and $a \in T_r\}$; let $x > M$ and let $y = x + |d|$. For all $s \in S$ and $a \in T_r$, we have $a + s < x$, $a - s < x$, $a + s - |d| < x$, and $a - s - |d| < x$. Thus there does not exist $s \in S$ and $a \in T_r$ such that any of the following equalities hold: $s = x - a$, $s = a - x$, $s = y - a$, $s = a - y$. Furthermore, either $x - y$ or $y - x = d$. Letting a_{r+1} and a_{r+2} be the appropriate choice of x or y, we have our set T_{r+2}.

COROLLARY 4.6.5 *There exists an infinite difference set of $(J,+)$, the group of integers under addition.*

EXAMPLE 4.6.6 This example does not give an infinite difference set in closed form; it exhibits only the first few elements.

Let $T_4 = \{-8,-6,0,1,4\}$. The differences here are $\pm 1, \pm 2, \pm 3, \pm 4, \pm 6, \pm 7, \pm 8, \pm 9, \pm 10, \pm 12$. Using the constructive method developed in Theorem 4.6.4, we have $M = 20$; thus we may take $x = 21$ and $y = 26$ to generate ± 5. Therefore, we let $T_6 = \{-8,-6,0,1,4,21,26.\}$.

CONSTRUCTION 4.6.7 Let T be a difference set of the integers J and construct $\mathscr{P}$, $\mathscr{L}$, and $\mathscr{I}$ as follows: $\mathscr{P} = \{p_i: i \in J\}$; $\mathscr{L} = \{L_i: i \in J\}$; $(p_i, L_j) \in \mathscr{I}$ if and only if $i + j \in T$.

THEOREM 4.6.8 *The triple $(\mathscr{P}, \mathscr{L}, \mathscr{I})$ is a projective plane.*

Proof. The proof is left as an exercise.

DEFINITION 4.6.9 *The plane of Construction 4.6.7 is called an infinite cyclic plane, denoted by π_T.*

THEOREM 4.6.10   *The plane π_T has a cyclic transitive subgroup of*
$C(\pi_T)$.

Proof.   The proof is left as an exercise.

THEOREM 4.6.11   *If π has an infinite cyclic transitive subgroup of*
$C(\pi)$, *then π is an infinite cyclic plane.*

Proof.   The proof is left as an exercise.

EXAMPLE 4.6.12   The difference set T generated in Example 4.6.6
defines an infinite cyclic plane that obviously has a transitive collineation group. Chapter 5 will show that $C(\pi_T)$ is not transitive on
four-points.

EXERCISES

1. Prove Theorem 4.6.8.

2. Prove Theorem 4.6.10.

3. Prove Theorem 4.6.11.

*4. Give an example of an infinite perfect difference set of J in closed form.

PART 2

DESARGUESIAN PLANES

This part introduces the two classical planes of
projective geometry: the Desarguesian plane
and the Pappian plane. The tools of Part 1
are used to analyze these
planes thoroughly.

CHAPTER 5

DESARGUESIAN PLANES

This chapter and the next will examine in detail two classical projective planes, the Desarguesian plane and the Pappian plane. Like the Fano plane, these two planes are defined essentially in terms of the embeddability of a confined configuration. For the Fano plane it is the Fano configuration, $\Sigma(7_3)$; for the Desarguesian plane it is the Desargues configuration, $\Sigma(10_3)$; and for the Pappian plane it is the Pappus configuration, $\Sigma(9_3)$. These configurations were displayed in Example 1.6.12.

5.1 THE DESARGUES THEOREM

The embeddability requirement for the Desargues configuration may be stated as follows:

DESARGUES THEOREM *In a projective plane π, if p, q, r and p', q', r' are triples of noncollinear points such that lines pp', qq', and rr' are concurrent, then the points $pq \cap p'q'$, $pr \cap p'r'$, and $qr \cap q'r'$ are collinear* (see Figure 32).

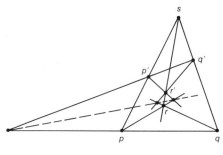

FIGURE 32

DEFINITION 5.1.1 *If a plane satisfies the Desargues Theorem, it is called a* Desarguesian plane.

For the sake of convenience, the following definitions are introduced so that the Desargues Theorem may be restated.

DEFINITION 5.1.2

1. *Two triangles pqr and p'q'r' are* centrally perspective *from point s if and only if the respective vertices are collinear with s, that is, pp', qq', and rr' pass through s.*

2. *The two triangles are* axially perspective *from line L if and only if the respective sides of the triangles meet on L, that is, pq ∩ p'q', pr ∩ p'r', and qr ∩ q'r' are on L.*

Thus the Desargues Theorem may be reformulated as follows: *Every two centrally perspective triangles in π are axially perspective.*

Before proceeding the student should note that the class of Desarguesian planes satisfies the principle of duality.

THEOREM 5.1.3 *If π is a Desarguesian plane, then every two axially perspective triangles are centrally perspective.*

Proof. The proof is left as an exercise.

COROLLARY 5.1.4 *The class of Desarguesian planes satisfies the principle of duality.*

First it should be pointed out that two triangles pqr and $p'q'r'$, which are both centrally perspective and axially perspective, need not form the Desargues configuration. For example, we may have $p = r'$ and $q' \in pq$ as in Figure 33. In fact, even if p,q,r,p',q',r' are distinct points and $pq \cap p'q'$, $pr \cap p'r'$, and $qr \cap q'r'$ are distinct points, the configuration need not be Desarguesian, as Figure 34 illustrates.

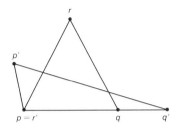

FIGURE 33

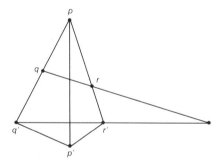

FIGURE 34

This text does not consider all the different configurations that two centrally and axially perspective triangles may assume. However, from now on it will be presumed that the quadruples p,p',q,q'; p,p',r,r'; and q,q',r,r' are four-points.

If this assumption is not satisfied, the fact that the Desargues Theorem is true is easily verified. Furthermore, it is easily checked that this assumption can never be satisfied in $\pi(2)$ and thus $\pi(2)$ is a Desarguesian plane.

THEOREM 5.1.5 *If π has at least four points on a line, then there exist two triangles pqr and $p'q'r'$ that are centrally perspective from a seventh point s.*

Proof. Let p,p',q,q' be a four-point. Let $s = pp' \cap qq'$ and let $a = pq \cap p'q'$. Furthermore, let L be a line through s distinct from sp, sq, and sr. Such a line exists because π has at least four points on one line; thus π has at least four points on every line; and dually, π has at least four lines through every point. Let r be any point on L other than $pq \cap L$ and let r' be a point on L other than $p'q' \cap L$. Then pqr and $p'q'r'$ are vertices of triangles satisfying the required conditions (see Figure 35).

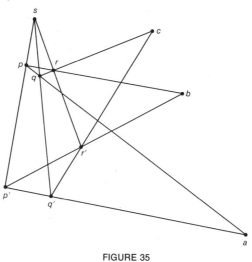

FIGURE 35

THEOREM 5.1.6   *If π has at least four points on a line and π is a Desarguesian plane, then π has a confined configuration of ten points and ten lines.*

Proof.   Using the notation of the preceding proof, let $b = pr \cap p'r'$ and $c = qr \cap q'r'$. It is then easily checked that p,q,r,p',q',r',s,a,b,c are ten distinct points. Furthermore, s,p,p'; s,q,q'; s,r,r'; p,q,a; p',q',a; p,r,b; p',r',b; q,r,c; q',r',c are all sets of collinear points on nine different lines. If π is Desarguesian, then a,b,c are also collinear and thus we have a confined configuration of ten points and ten lines.

Notice that the configuration described in Theorem 5.1.6 need not be the Desargues configuration $\Sigma(10_3)$. Indeed, in both $\pi(3)$ and $\pi(4)$ it cannot be $\Sigma(10_3)$.

EXAMPLE 5.1.7

i. $\pi(2)$ satisfies the Desargues Theorem since the hypothesis of the theorem cannot be met in $\pi(2)$.

ii. π_F satisfies the Desargues Theorem. This may be shown as follows. Let $p = [0,0,1]$, $q = [0,1,0]$, $p' = [1,0,1]$, and $q' = [1,1,0]$. Thus $s = [1,0,0]$. Let r and r' be arbitrary and subject only to the restriction that pqr and $p'q'r'$ are triangles centrally perspective from s and let $rr' \neq pp'$ or qq'. Thus $r = [x,y,1]$ and $r' = [z,y,1]$ where $x,y \neq 0$, $z \neq x$, and $z \neq y + 1$. It is verified in the partial proof of Theorem 2.8.6 that $pq \cap p'q'$, $pr \cap p'r'$, and $qr \cap q'r'$ all lie on the line $\langle z - y - x - 1, y, x \rangle$.

Suppose that a,b,c and a',b',c' are centrally perspective triangles from a seventh point d and that $aa' \neq bb' \neq cc' \neq aa'$. Thus a,a',b,b' is a four-point. Since $C(\pi_F)$ is transitive on four-points, there exists a collineation f such that f: $a,a',b,b' \rightarrow p,p',q,q'$. Letting $r = f(c)$ and $r' = f(c')$, it follows that rr' passes through $[1,0,0]$. Thus, by the preceding argument, $pq \cap p'q'$, $pr \cap p'r'$, and $qr \cap q'r'$ are collinear. Since f^{-1}: $p,p',q,q',r,r' \rightarrow a,a',b,b',c,c'$, and since f^{-1} is a collineation, $ab \cap a'b'$, $ac \cap a'c'$, and $bc \cap b'c'$ are collinear.

iii. $\pi_H(9)$ does not satisfy the Desargues Theorem. Consider the triangles a_0,b_1,a_5 and c_1,b_2,d_{11} in Figure 36. They are perspective from e_3. It is easily checked that $a_0b_1 \cap c_1b_2 = e_4$, $a_0a_5 \cap c_1d_{11} = a_6$, and $b_1a_5 \cap b_2d_{11} = a_4$ but that e_4, a_6, a_4 are not collinear because $a_4e_4 = L_{28}$ and $a_6 \notin L_{28}$.

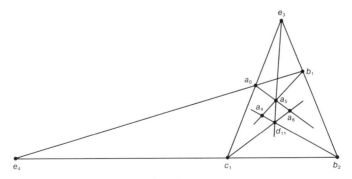

FIGURE 36

iv. $\pi_N(9)$ does not satisfy the Desargues Theorem. Proof of this fact is left as an exercise.

v. Let $\pi = \Sigma_o^+$ where Σ_o is not a projective plane. The plane π has at least four points on each line (actually, infinitely many points on each line) and yet, by Theorem 2.7.8, π has no confined configurations other than those of Σ_o. It follows from Theorem 5.1.6 that π does not satisfy the Desargues Theorem.

EXERCISES

1. a. Prove Theorem 5.1.3.
   b. Explain why the principle of duality for projective planes does not apply to your proof in 1(a).

2. Suppose that π is a Desarguesian plane. If p,q,r,s is a four-point and if $a = pq \cap rs$, $b = pr \cap qs$, $c = ps \cap qr$, $d = bc \cap rs$, $e = ac \cap pr$, and $f = ab \cap ps$, then d,e,f are collinear.

3. Suppose that π is a Desarguesian plane. Let $a,b,c \in L$, $a',b',c' \in L'$, and $d = L \cap L'$ where $d \neq a,b,c,a',b',c'$. Also let $p = ab' \cap a'b$, $q = ac' \cap a'c$, and $r = bc' \cap b'c$. Show that if d,p,q are collinear, then p,q,r are collinear.

4. A plane satisfies *Reidemeister's Condition* if and only if the following is true. Let p,q,r be distinct points and let L,M,N be distinct lines concurrent at r such that $pq \neq L,M,N$; $p \notin L,M,N$; and $q \notin L,M,N$ (see Figure 37). Also let $a_1,a_2 \in L$ such that $a_i \notin pq$, $a_i \neq r$ for $i = 1,2$, and let $b_i = M \cap a_ip$, $c_i = N \cap a_iq$, and $d_i = b_iq \cap c_ip$ for $i = 1,2$. Then r,d_1,d_2 are collinear. Prove that if π is Desarguesian, then π satisfies Reidemeister's Condition. The converse is also true but it is more difficult to show (see Klingenberg, 1955).

5. Show that the following planes are not Desarguesian:
   a. $\pi_N(9)$.
   b. π_M, the Moulton plane defined in exercise 2.1.6.
   c. π_T, the cyclic plane defined in Example 4.6.6.

6. a. Let p, q, and r be three noncollinear points in π and let s and L be an arbitrary point and line, respectively. Show that if $f \in CC(s,L)$, then p,q,r and $f(p),f(q),f(r)$ form two triangles that are centrally perspective from s and axially perspective from L.
   b. Show that if π has more than five points on a line and admits a homology other than the identity, then the Desargues configuration can be embedded in π.

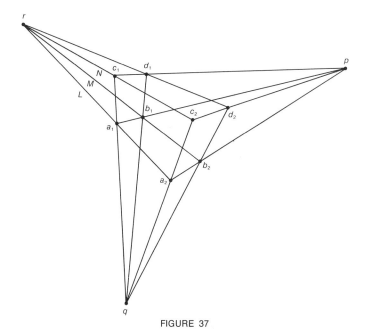

FIGURE 37

7. a. Show that the Desargues configuration cannot be embedded in a plane of order less than five.

   b. Show that the Desargues configuration can be embedded in a plane of order n if $n \geq 5$. Hint: Let L_1, L_2, and L_3 be distinct lines concurrent at point p and let r and s be two points such that $p \notin rs$ and $r, s \notin L_1, L_2, L_3$. Consider the set T of triangles that have one vertex each on L_1, L_2, and L_3 and that have one side containing r and a second side containing s. Compare the number of triangles in T with the number of points on the line rs that are available for intersection with the third sides of the respective triangles.

8. Let $\Sigma(10_3)$ be the Desargues configuration made up of points $\{p, a, b, c, a', b', c', d, e, f\}$ and lines $\{p, a, a'\}$, $\{p, b, b'\}$, $\{p, c, c'\}$, $\{a, b, d\}$, $\{a', b', d\}$, $\{a, b, e\}$, $\{a', b', e\}$, $\{a, c, f\}$, $\{a', c', f\}$, $\{d, e, f\}$ (see Figure 38). Relabel the points with the following doubletons: $p = \{1, 2\}$, $a = \{1, 3\}$, $b = \{1, 4\}$, $c = \{1, 5\}$, $a' = \{2, 3\}$, $b' = \{2, 4\}$, $c' = \{2, 5\}$, $d = \{3, 4\}$, $e = \{3, 5\}$, $f = \{4, 5\}$. Let line xy be denoted by $x \cup y$. The lines are then characterized as follows:

$$
\begin{array}{lll}
\{p, a, a'\} = \{1, 2, 3\} & \{b', c', f\} = \{2, 4, 5\} & \{a, c, e\} = \{1, 3, 5\} \\
\{p, c, c'\} = \{1, 2, 5\} & \{p, b, b'\} = \{1, 2, 4\} & \{b, c, f\} = \{1, 4, 5\} \\
\{a', b', d\} = \{2, 3, 4\} & \{a, b, d\} = \{1, 3, 4\} & \{d, e, f\} = \{3, 4, 5\} \\
\{a', c', e\} = \{2, 3, 5\} & &
\end{array}
$$

Thus $L \cap M$ is set theoretically accurate.

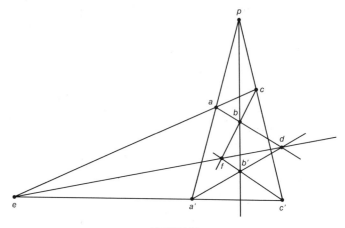

FIGURE 38

a. A *3-point* in $\Sigma(10_3)$ is a set of three points that form the vertices of a triangle in $\Sigma(10_3)$.
   (1) Characterize the 3-points of $\Sigma(10_3)$. Hint: For the 3-point x,y,z, consider $x \cup y \cup z$.
   (2) Define and characterize the 3-lines of $\Sigma(10_3)$.
   (3) Determine how many distinct triangles are embedded in $\Sigma(10_3)$.

b. (1) Characterize the pairs of centrally perspective triangles from a seventh point.
   (2) Characterize the pairs of axially perspective triangles from a seventh line.
   (3) How many centrally perspective (and axially perspective) triangles are there in $\Sigma(10_3)$?

c. A *4-point* in $\Sigma(10_3)$ is a set of four points that form the vertices of a complete quadrangle.
   (1) Give an example of a four-point that is not a 4-point in $\Sigma(10_3)$.
   (2) Characterize the 4-points of $\Sigma(10_3)$.
   (3) Define and characterize the 4-lines of $\Sigma(10_3)$.
   (4) How many complete quadrangles are embedded in $\Sigma(10_3)$?

d. A *permutation on a set* X is a one-to-one mapping of X onto itself. The set of permutations on a set X is a group under composition. If X has five members, this group is denoted by S_5. Show that $C(\Sigma(10_3))$, the group of collineations on $\Sigma(10_3)$, is:
   (1) isomorphic to S_5.
   (2) transitive on points.
   (3) transitive on triangles.
   (4) transitive on quadrangles.

5.2 SPECIALIZED DESARGUES THEOREMS AND TRANSITIVITY

This section analyzes Desarguesian planes with the aid of the tools that were developed in Chapter 4 and demonstrates that there is a close relationship between perspective triangles and central collineation groups.

DESARGUES (p,L) THEOREM *If abc and a'b'c' are two triangles centrally perspective from p such that ab $\cap$ a'b' and ac $\cap$ a'c' lie on L, then bc $\cap$ b'c' $\in$ L.*

FIRST DESARGUES (L) THEOREM *If abc and a'b'c' are two triangles centrally perspective from any arbitrary point p $\in$ L such that ab $\cap$ a'b' and ac $\cap$ a'c' lie on L, then bc $\cap$ b'c' $\in$ L.*

SECOND DESARGUES (L) THEOREM *If abc and a'b'c' are two triangles centrally perspective from any arbitrary point p $\notin$ L such that ab $\cap$ a'b' and ac $\cap$ a'c' lie on L, then bc $\cap$ b'c' $\in$ L.*

Notation. The Desargues (p,L) Theorem is often used for a specific point and line. For example, if the point is a and the line is bc, the appropriate theorem will be called the Desargues (a,bc) Theorem. This is shorthand for "the Desargues (p,L) Theorem where $p = a$ and $L = bc$." Similarly, "the First Desargues (bc) Theorem" is shorthand for "the First Desargues (L) Theorem where $L = bc$."

LITTLE DESARGUES THEOREM *The First Desargues (L) Theorem holds for all lines L in π.*

It is immediate from the preceding theorems that the following relationships hold: π satisfies the Desargues (p,L) Theorem for all $p \in L$ if and only if π satisfies the First Desargues (L) Theorem; π satisfies the Desargues (p,L) Theorem for all $p \notin L$ if and only if π satisfies the Second Desargues (L) Theorem; π satisfies the Desargues Theorem if and only if π satisfies the Desargues (p,L) Theorem for all p and L in π.

THEOREM 5.2.1   *The projective plane π satisfies the Desargues(p,L)*
Theorem if and only if π is (p,L) transitive.

Proof.   Suppose that π is (p,L) transitive and suppose that abc and
$a'b'c'$ are two triangles centrally perspective from p such that ab
$\cap\ a'b'$ and $ac\ \cap\ a'c'$ lie on L. Since $CC(p,L)$ is transitive, there exists
a central collineation f with center p and axis L such that $f\colon a\to a'$.
Let $ab\ \cap\ a'b' = d$. Then $f\colon d\to d$, and thus $f\colon ad\to a'd$. Let M denote
the line containing p,b,b'. Then $f\colon M\ \to\ M$, and so $f\colon M\ \cap\ ad\to$
$M\ \cap\ a'd$; in other words, $f\colon b\to b'$. Similarly, $f\colon c\to c'$; it follows
that $f\colon bc\to b'c'$. Since f fixes the line L pointwise, $f\colon bc\ \cap\ L\to b'c'$
$\cap\ L$ and $bc\ \cap\ L = b'c'\ \cap\ L$. Thus $bc\ \cap\ b'c'\in L$, and we may conclude that π satisfies the Desargues (p,L) Theorem.

Suppose that π satisfies the Desargues (p,L) Theorem. Let a and a'
be two points subject to the restrictions that p,a,a' are collinear,
$a,a'\neq p$, and $a,a'\notin L$. First define a function g on $\mathcal{P}-pa$ as follows:
$g(b) = b$ if $b\in L$; $g(b) = b'$ if $b\notin L$ where $b' = ta'\ \cap\ pb$ and $t = ab$
$\cap\ L$ (see Figure 39). The next step is to show that g is an affine collineation on $\alpha(\pi_{pa}^{-})$. Proof that g is a one-to-one, onto mapping is left
as an exercise. The following argument show that g preserves collinearity. Suppose that q,r,s are collinear in α. If $p\in qr$, then clearly
$g(q)g(r) = qr$ and $g(q)g(s) = qs$, and since $qr = qs$, it follows that
$g(q),g(r),g(s)$ are collinear. If $p\notin qr$, we have two cases:

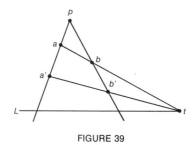

FIGURE 39

Case 1. $a\in qr$. It is easily seen that $g(q),g(r),a'$ are collinear,
and since $a\in sr$, we know that $g(s),g(r),a'$ are also collinear. Thus
$g(q),g(r),g(s)$ are collinear.

Case 2. $a\notin qr$. Consider triangles aqr and $a'g(q)g(r)$ (see
Figure 40). Clearly, they are centrally perspective from p, and so,
by the Desargues (p,L) Theorem, they are axially perspective

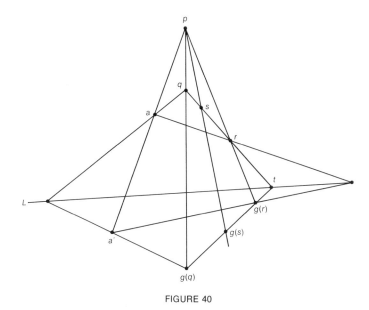

FIGURE 40

from L. Thus $qr \cap g(q)g(r) = t \in L$. A similar argument shows that $sr \cap g(s)g(r) = t \in L$. Therefore, $g(q),g(r),g(s)$ are collinear.

Let f be the unique projective extension of g. Clearly, $f \in CC(p,L)$. Also $f: a \rightarrow a'$ by the following argument. If b and b' are such that $ab \cap ab' = t$ and $b,b' \notin pa$, then $g: bt \rightarrow b't$. Thus $f: a = bt \cap pa \rightarrow b't \cap pa = a'$. This completes the proof that π is (p,L) transitive.

Section 4.5 reveals the close relationship between transitive central collineation groups and algebraic representations of π; Theorem 5.2.1, the most important theorem in Part 2, reveals the relationship between transitive central collineation groups and Desarguesian configurations. The result is that we see a close relationship between a purely geometric condition on π and an algebraic representation of π.

Several results follow immediately from Theorem 5.2.1.

COROLLARY 5.2.2

   i. *The following statements are equivalent:*

   π *satisfies the First Desargues (L) Theorem.*
   $El(p,L)$ *is transitive for all* $p \in L$.
   π *is (p,L) transitive for all* $p \in L$.

ii. *The following statements are equivalent:*

π *satisfies the Second Desargues (L) Theorem.*
Hom(p,L) *is transitive for all* $p \notin L$.
π *is* (p,L) *transitive for all* $p \notin L$.

iii. *The following statements are equivalent:*

π *satisfies the Little Desargues Theorem.*
El(p,L) *is transitive for all* p *and* L *such that* $p \in L$.
π *is* (p,L) *transitive for all* p *and* L *such that* $p \in L$.

iv. *The following statements are equivalent:*

π *satisfies the Desargues Theorem.*
CC(p,L) *is transitive for all* p *and* L.
π *is* (p,L) *transitive for all* p *and* L.

Parts (iii) and (iv), respectively, may be stated informally as follows:

π satisfies the Little Desargues Theorem if and only if all possible elations exist.

π is a Desarguesian plane if and only if all possible central collineations exist.

THEOREM 5.2.3 *If π satisfies the First Desargues (L) Theorem, then* El(L) *is Abelian.*

Proof. The proof is based on Theorems 4.5.10 and 5.2.1.

THEOREM 5.2.4 *The following statements are equivalent:*

π *satisfies the Desargues Theorem.*
Hom (p,L) *is transitive for all* p *and* L.
π *is* (p,L) *transitive for* p *and* L *such that* $p \in L$.

Proof. The proof follows from Corollary 5.2.2(iv) and Theorem 4.5.18.

THEOREM 5.2.5 *If π satisfies the Second Desargues (L) Theorem, then π satisfies the First Desargues (L) Theorem.*

Proof.   The proof follows from Theorems 4.5.18 and 5.2.1.

THEOREM 5.2.6   *If π is a Desarguesian plane, then* $PC(\pi) = CC(\pi)$ *and $CC(\pi)$ is transitive on four-points.*

Proof.   The proof follows from Theorems 4.5.18 and 5.2.1.

COROLLARY 5.2.7   *If π is a Desarguesian plane, then π has a characteristic.*

EXERCISES

1. Show that the mapping g defined in Theorem 5.2.1 is a one-to-one, onto mapping.

2. Show that the class of Little Desarguesian planes (planes satisfying the Little Desargues Theorem) satisfies the principle of duality.

3. Show that if π satisfies the Little Desargues Theorem, then $CC(\pi) = PC(\pi)$.

4. Show that if π is a Little Desarguesian plane that is (p, L) transitive for some $p \notin L$, then π is Desarguesian.

5. a. Show that if π is (p, L) transitive for some $p \in L$ and if $C(\pi)$ is transitive, then π is a Little Desarguesian plane.
   b. Show that if π is (p, L) transitive for some $p \notin L$ and if $C(\pi)$ is transitive, then π is a Desarguesian plane.
   c. Show that $C(\pi_M)$ is not transitive. (Refer back to exercise 2.1.6 for the definition of π_M.)
   d. Show that $C(\pi_T)$ is not (p, L) transitive for any p and L. (Refer back to Example 4.6.6 for the definition of π_T.)

6. Let abc and $a'b'c'$ be two triangles centrally perspective from p and axially perspective from L. Let $r = ab \cap a'b'$, $s = ac \cap a'c'$, and $t = bc \cap b'c'$ and let the lines of this configuration be the following sets: $\{p, a, a'\}$, $\{p, b, b'\}$, $\{p, c, c'\}$, $\{a, b, r\}$, $\{a', b', r\}$, $\{a, c, s\}$, $\{a', c', s\}$, $\{b, c, t\}$, $\{b', c', t\}$, and $\{p, r, s, t\}$. This configuration is called the *Little Desargues configuration*, denoted by Σ.
   a. Draw a figure depicting Σ.
   b. Determine whether Σ can be embedded in $\Sigma(21_5)$ (see Figure 14, p. 36).

c. State and prove a theorem about Σ that is analogous to the theorem about the Desargues configuration stated in exercise 5.1.6(b).

7. Show that if π is a Little Desarguesian plane, then $C(\pi)$ is transitive on triangles. (Actually, $C(\pi)$ is transitive on four-points but this has never been shown geometrically.)

5.3   AFFINE DESARGUES THEOREMS

There is no obvious analogue of the Desargues Theorem in the affine plane. Probably the most natural analogue would be a theorem that holds in α if and only if its projective counterpart holds in $\pi(\alpha^+)$. The statement of such a theorem appears to be cumbersome, however, because of the many different ways in which L_∞ can enter a Desargues configuration in π. For example, two triangles may lie in α and be centrally perspective from $p \in L_\infty$; or two triangles may lie in α and two associated sides may intersect on L_∞; or, worse still, a vertex of one of the triangles may lie on L_∞; or there may be combinations of these situations. The predicament is not as bad as it seems at first since the Desargues Theorem holds in $\pi(\alpha^+)$ under conditions weaker than those existing under the hypothesis that all possible homologies exist. This fact will be verified in Chapter 8. The following are two special cases of an affine theorem that will be left unstated for the present.

First Affine Desargues Theorem (Minor Desargues Theorem)
If abc and $a'b'c'$ are two triangles such that aa', bb', and cc' are parallel and $ab \parallel a'b'$ and $ac \parallel a'c'$, then $bc \parallel b'c'$ (see Figure 41).

Second Affine Desargues Theorem (Major Desargues Theorem) *If abc and $a'b'c'$ are centrally perspective from any given point p and if $ab \parallel a'b'$ and $ac \parallel a'c'$, then $bc \parallel b'c'$* (see Figure 42).

Theorem 5.3.1

i. *The affine plane α satisfies the First Affine Desargues Theorem if and only if $\pi(\alpha^+)$ satisfies the First Desargues (L_∞) Theorem.*

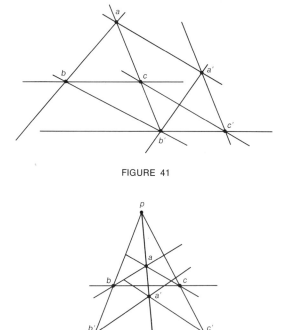

FIGURE 41

FIGURE 42

ii. *The affine plane α satisfies the Second Affine Desargues Theorem if and only if $\pi(\alpha^+)$ satisfies the Second Desargues (L_∞) Theorem.*

Proof. The proof follows directly from the statements of the theorems.

The following theorems are direct consequences of the definitions of Section 5.2, as well as Corollary 5.2.2.

THEOREM 5.3.2

    i. *If α satisfies the First Affine Desargues Theorem, then* Trans(L) *is transitive for all lines L in α.*

    ii. *If α satisfies the Second Affine Desargues Theorem, then* Dil(p) *is transitive for all points p in α.*

THEOREM 5.3.3 *If α satisfies the First Affine Desargues Theorem, then* Trans(α) *is Abelian.*

EXERCISES

1. Give an example of an affine plane α such that
   a. α does not satisfy the Minor Desargues Theorem.
   b. α satisfies the Minor Desargues Theorem but not the Major Desargues Theorem.
   c. α satisfies the Major Desargues Theorem.

2. Show that the following statement is false: If $\alpha(\pi_L^-)$ satisfies the Minor Desargues Theorem for some line L in π, then $\alpha(\pi_M^-)$ satisfies the Minor Desargues Theorem for all lines M in π.

*3. Show that if $\alpha(\pi_L^-)$ satisfies the Major Desargues Theorem for some L in π, then $\alpha(\pi_M^-)$ satisfies the Major Desargues Theorem for all M in π.

5.4   PROJECTIVITIES IN DESARGUESIAN PLANES

In this section two classical theorems are proved concerning projectivities in Desarguesian planes. The first theorem is an equivalent formulation of the Desargues Theorem in terms of perspectivities; the second establishes an upper bound on the number of perspectivities necessary to generate a projectivity in a Desarguesian plane. Both theorems are interesting in themselves; the second has an important application that will be discussed in Chapter 6.

THEOREM 5.4.1   *The projective plane π satisfies the Desargues Theorem if and only if π has the following property: If f is the product of two perspectivities g and h where $g = L \overset{p}{\barwedge} L'$, $h = L' \overset{q}{\barwedge} L''$, $L \neq L'$, and L, L', and L'' are concurrent, then f is a perspectivity.*

Proof.   Suppose that π satisfies the Desargues Theorem and suppose that $f = L \overset{p}{\barwedge} L' \overset{q}{\barwedge} L''$ where L, L', and L'' are concurrent and where $L \neq L''$. Clearly, if $p = q$, then f is a perspectivity with center p. Suppose that $p \neq q$ and adopt the following definitions: $g = L \overset{p}{\barwedge} L'$, $h = L' \overset{q}{\barwedge} L''$, $s = L \cap L' \cap L''$, $a \in L$ such that $a \neq s$ and $a \notin pq$, $a' = g(a)$, $a'' = h \circ g(a)$, $r = pq \cap aa''$, and $f' = L \overset{r}{\barwedge} L''$ (see Figure 43). The remainder of the proof shows that $f' = f$, that is, that $f'(t) = f(t)$ for all $t \in L$. Clearly, $f'(a) = f(a) = a''$; $f'(s) = f(s) = s$; and $f'(pq \cap L) = f(pq \cap L) = pq \cap L''$. Suppose that $b \neq a, s$ or $pq \cap L$. Let $b' = g(b)$ and $b'' =$

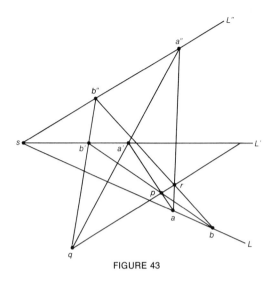

FIGURE 43

$h \circ g(b)$; therefore, $b'' = f(b)$. Now $aa'a''$ and $bb'b''$ are triangles centrally perspective from s. Furthermore, $aa' \cap bb' = p$ and $a'a'' \cap b'b'' = q$. Since π is Desarguesian, $aa'' \cap bb''$ is collinear with pq. Since $aa'' \cap pq = r$, we have $r \in bb''$, and so $f'(b) = b''$. Therefore, $f(t) = f'(t)$ for all points $t \in L$. Proof of the converse of this theorem is left as an exercise.

In general an upper bound cannot be established on the number of perspectivities needed to generate a projectivity. In a Desarguesian plane, every projective collineation is the product of at most five central collineations (by Theorem 4.5.18 and Corollary 5.2.2(iv)). Since projectivities and perspectivities share a relationship analogous to that of projective collineation and central collineation, it is not unexpected that such an upper bound can be established for Desarguesian planes.

First let us consider the cases of $\pi(2)$ and $\pi(3)$.

THEOREM 5.4.2 In $\pi(3)$, if f is a projectivity mapping L to L' and $L \neq L'$, then f is the product of at most two perspectivities.

Proof. Let L and M be two distinct lines intersecting at point p: $L = \{p,q,r,s\}$, $M = \{p,a,b,c\}$. There exist six possible maps from L to M that fix p. There are also six points in $\pi(3)$ that are not on $L \cup M$.

Each perspectivity fixes p and each of these six points determines a unique perspectivity from L to M. Therefore, each of the six possible maps is associated with exactly one perspectivity.

Let L and L' be two distinct lines, let $L = \{p,q,r,s\}$ and $L' = \{p,t,u,v\}$, and let f be an arbitrary map from L to L'. If $f(p) = p$, then, by the preceding statements, f is a perspectivity. Suppose that $f(p) \neq p$. Let $M = \{f(p),q,a,b\}$. Note that $L \cap M = q$ and that $M \cap L' = f(p)$. Define g and h as follows:

$$g: p \to f(p)$$
$$q \to q$$
$$r \to a$$
$$s \to b$$

$$h: f(p) \to f(p)$$
$$q \to f(q)$$
$$a \to f(r)$$
$$b \to f(s)$$

Thus $f = h \circ g$. Since g and h fix one point, each is a perspectivity.

THEOREM 5.4.3   In $\pi(2)$, if f is a projectivity from L to L' where $L \neq L'$, then f can be represented by the product of at most two perspectivities.

Proof.   The proof is analogous to that of Theorem 5.4.2.

LEMMA 5.4.4   If π is a Desarguesian plane and $f = L \overset{p}{\barwedge} L' \overset{q}{\barwedge} L''$ where $L \neq L''$, then for any line M such that $M \neq L$, $q \notin M$ and $M \in L \cap L'$, there exists a point r such that $f = L \overset{r}{\barwedge} M \overset{q}{\barwedge} L''$.

Proof.   The proof is left as an exercise.

THEOREM 5.4.5   If π is a Desarguesian plane, then every projectivity f that maps L to L', where $L \neq L'$, can be represented by the product of at most two perspectivities.

Proof.   By Theorems 5.4.2 and 5.4.3, we need consider only planes with five or more points on each line. Let f be the following projectivity $L = L_0 \overset{p_1}{\barwedge} L_1 \overset{p_2}{\barwedge} L_2 \overset{p_3}{\barwedge} L_3 \ldots \overset{p_n}{\barwedge} L_n = L'$. Assume that $n \geq 3$ and reduce the length of this chain from n to $n - 1$; the desired result will follow by induction. Notice that it will be sufficient to reduce $L_0 \stackrel{p_1}{\barwedge} L_1 \stackrel{p_2}{\barwedge} L_2 \stackrel{p_3}{\barwedge} L_3$ by one perspectivity.

First it will be shown that if two of the four lines L_0, L_1, L_2, and L_3 are the same, the previously mentioned chain may be reduced by an equivalent chain having the property that all four lines are distinct. We may assume that no two succeeding lines are identical, for if they were, one of our perspectivities would be the identity and could automatically be dispensed with.

Case 1. Suppose that $L_0 = L_3$. (Here $n \geq 4$ since $L \neq L'$.) Let $g = L_2 \stackrel{p_3}{\barwedge} L_3 \stackrel{p_4}{\barwedge} L_4$ and let M be a line through $L_2 \cap L_3$ such that $M \neq L_2$, $M \neq L_3$, and $p_4 \notin M$. By Lemma 5.4.4 there exists a point q such that $g = L_2 \stackrel{q}{\barwedge} M \stackrel{p_4}{\barwedge} L_4$. So we replace L_3 by M. Since $L_3 = L_0$, we have $M \neq L_0$, and therefore our chain begins $L_0 \stackrel{p_1}{\barwedge} L_1 \stackrel{p_2}{\barwedge} L_2 \stackrel{p_3}{\barwedge} M$.

Case 2. Suppose that $L_0 = L_2$. Let $g = L_1 \stackrel{p_2}{\barwedge} L_2 \stackrel{p_3}{\barwedge} L_3$. Then L_2 may be replaced by another line M as was done in Case 1 where $M \neq L_1$, $M \neq L_2$, $L_1 \cap L_2 \in M$, and $g = L_1 \stackrel{q}{\barwedge} M \stackrel{p_3}{\barwedge} L_3$ for some point q. If $M = L_3$, then we may dispense with a perspectivity. In any event, $L_0 \stackrel{p_1}{\barwedge} L_1 \stackrel{q}{\barwedge} M \stackrel{p_3}{\barwedge} L_3$ may be used to begin our chain.

Case 3. Suppose that $L_1 = L_3$. As in Case 1, we may replace L_1 by a line M distinct from L_0 and L_1; therefore, $L_0 \stackrel{q}{\barwedge} M \stackrel{p_2}{\barwedge} L_2 \stackrel{p_3}{\barwedge} L_3$ may be used. If $M = L_2$, we may simply drop one perspectivity. The preceding remarks lead to the conclusion that lines L_0, L_1, L_2, and L_3 may be assumed to be distinct.

The second step is to show that if any three of these four lines are concurrent, the chain is replaced by an equivalent chain where no three of the four lines are concurrent. We may assume that no three successive lines are concurrent, for if they were, then, by Theorem 5.4.1, the chain of three perspectivities could immediately be reduced to two perspectivities.

Case 1. Suppose that L_0, L_2, and L_3 are concurrent. By Lemma 5.4.4, L_2 can be replaced by M where $M \neq L_1$, $M \neq L_2$, $L_1 \cap L_2 \in M$, $p_3 \notin M$, and $L_1 \stackrel{p_2}{\barwedge} L_2 \stackrel{p_3}{\barwedge} L_3 = L_1 \stackrel{q}{\barwedge} M \stackrel{p_3}{\barwedge} L_3$ for some point q.

Now $M \neq L_0$ and $M \neq L_3$ because M does not pass through $L_0 \cap L_3$ (since L_2 does pass through $L_0 \cap L_3$). Thus $L_0 \overset{p_1}{\barwedge} L_1 \overset{q}{\barwedge} M \overset{p_3}{\barwedge} L_3$ is an equivalent chain that does not have three concurrent lines (see Figure 44).

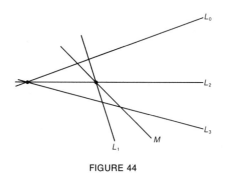

FIGURE 44

Case 2. Suppose that L_0, L_1, and L_3 are concurrent and proceed exactly as in the preceding case.

Thus we may assume not only that the four lines are distinct but also that no three of them are concurrent. On this assumption, we may now reduce the length of our chain from four perspectivities to three as follows.

Let $s = L_0 \cap L_1$, $t = L_2 \cap L_3$, and $M = st$ (see Figure 45); recall that $f = L_0 \overset{p_1}{\barwedge} L_1 \overset{p_2}{\barwedge} L_2 \overset{p_3}{\barwedge} L_3$.

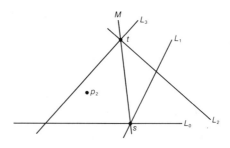

FIGURE 45

Case 1. Suppose that p_2 is not on M. Let $g = L_0 \overset{p_1}{\barwedge} L_1 \overset{p_2}{\barwedge} M \overset{p_2}{\barwedge} L_2 \overset{p_3}{\barwedge} L_3$. Clearly, $f = g$. By Theorem 5.4.1, $L_0 \overset{p_1}{\barwedge} L_1 \overset{p_2}{\barwedge} M$ is a perspectivity $L_0 \overset{q}{\barwedge} M$. Also by Theorem 5.4.1, $M \overset{p_2}{\barwedge} L_2 \overset{p_3}{\barwedge} L_3$ is a perspectivity $M \overset{r}{\barwedge} L_3$. Therefore, $f = L_0 \overset{q}{\barwedge} M \overset{r}{\barwedge} L_3$. Thus we have reduced the length of the chain of perspectivities.

Case 2. Suppose that p_2 is on M (see Figure 46). Let N be any line such that $L_1 \cap L_2 \in N$, $N \neq L_1$, $N \neq L_2$, $p_1 \notin N$, and $L_0 \cap L_3 \notin N$. (Assume the existence of five lines through $L_1 \cap L_2$.)

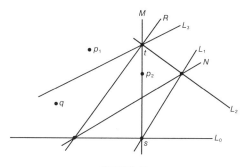

FIGURE 46

Clearly, $f = L_0 \overset{p_1}{\barwedge} N \overset{p_1}{\barwedge} L_1 \overset{p_2}{\barwedge} L_2 \overset{p_3}{\barwedge} L_3$. By Theorem 5.4.1, $N \overset{p_1}{\barwedge} L_1 \overset{p_2}{\barwedge} L_2$ is a perspectivity $N \overset{q}{\barwedge} L_2$, so $f = L_0 \overset{p_1}{\barwedge} N \overset{q}{\barwedge} L_2 \overset{p_3}{\barwedge} L_3$. We know that L_0, L_2, and L_3 are not concurrent, and since $L_1 \cap L_2 \in N$ and since L_0, L_1, and L_2 and L_1, L_2, and L_3 are not concurrent triples, it follows that L_0, N, and L_2 are not concurrent and that N, L_2, and L_3 are not concurrent. Thus L_0, N, L_2, and L_3 are four distinct lines, no three of which are concurrent. Define R as the line joining $L_0 \cap N$ and $L_2 \cap L_3$. Notice that $R \neq M$ because $N \cap L_0 \neq L_1 \cap L_0$. Furthermore, $q \notin R$ because $f: s \to t$; if $q \in R$, we would have $f: L_0 \cap N \to t$, and we know that $L_0 \cap N \neq s$. So we have reduced this case to the preceding one. This completes the proof.

COROLLARY 5.4.6 *In a Desarguesian plane* π, *if* f *is a projectivity mapping* $L \to L$, *then* f *can be represented by the product of at most three perspectivities.*

EXERCISES

1. Complete the proof of Theorem 5.4.1.

2. Prove Theorem 5.4.3.

3. Prove Lemma 5.4.4.

4. Complete the proof of Theorem 5.4.5 for planes π_F using the following outline:

   i. Assume that there are at least four points per line.
   ii. Let f be a projectivity mapping $L \to L'$ and choose a four-point p,q,r,s where $p,q \in L$, $r,s \in L'$, and $f\colon p,q \to r,s$.
   iii. Assume that p,q,r,s is the four-point u,v,o,e where $u = [1,0,0]$, $v = [0,1,0]$, $o = [0,0,1]$, and $e = [1,1,1]$.
   iv. Let g be a projective collineation that extends f; thus $g = f_R$ for some matrix R and

$$R\begin{pmatrix} 0 \\ 0 \\ 1 \end{pmatrix} = \begin{pmatrix} 0 \\ c \\ 0 \end{pmatrix} \quad R\begin{pmatrix} 1 \\ 0 \\ 0 \end{pmatrix} = \begin{pmatrix} d \\ d \\ d \end{pmatrix} \quad R\begin{pmatrix} 1 \\ 0 \\ 1 \end{pmatrix} = \begin{pmatrix} x \\ y \\ x \end{pmatrix}$$

   for some $c,d,x,y \in F$ such that $c,d,x \neq 0$ and $x \neq y$.
   v. Let

$$M = \begin{pmatrix} xd^{-1} & 0 & 0 \\ 0 & xd^{-1}-c & c \\ 0 & xd^{-1} & 0 \end{pmatrix} \quad N = \begin{pmatrix} d & 0 & 0 \\ d & \dfrac{y-x}{c} & 0 \\ d & 0 & \dfrac{y-x}{c} \end{pmatrix}$$

   vi. The mapping $f_N \circ f_M$ extends f; that is, $f_N \circ f_M \,|\, ou = f$.
   vii. The projectivity f is the product of two perspectivities.

5. In π_F find three perspectivities f_1, f_2, f_3 such that $f_3 \circ f_2 \circ f_1 = g$ where $g\colon \langle 0,1,0 \rangle \to \langle 0,1,0 \rangle$ and $g\colon [x,0,y] \to [-x,0,y]$. ($F$ does not have characteristic 2.)

5.5   QUADRANGULAR SETS

Planar configurations were studied in this text as early as Chapter 1 but no mention has yet been made of arrangements of points on a line. One arrangement in particular that is especially fruitful in Desarguesian planes is called a quadrangular set of points and is made up of the six intersections of opposite sides of a quadrangle with a seventh line (see Figure 47).

DEFINITION 5.5.1   *The unordered triple of ordered pairs of collinear points $\{(a,a'),(b,b'),(c,c')\}$ is said to be a* quadrangular set of points *if and only if $a = A \cap L$, $a' = A' \cap L$, $b = B \cap L$, $b' = B' \cap L$, $c = C \cap L$, and $c' = C' \cap L$ where A,A'; B,B'; and C,C' (respectively)*

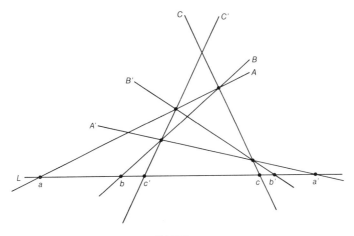

FIGURE 47

denote the opposite sides of a complete quadrangle, where A, B, and C are concurrent, and where L is a line that does not contain any vertex of the quadrangle.

Notation. If $\{(a,a'),(b,b'),(c,c')\}$ is a quadrangular set of points, the notation used is $\{(a,a'),(b,b'),(c,c')\} \in \gamma$.

The preceding definition indicates the following distinction of points: $a \neq b \neq c \neq a$, $a' \neq b' \neq c' \neq a'$, $a \neq b'$, $a \neq c'$, $b \neq a'$, $b \neq c'$, $c \neq a'$, $c \neq b'$. Also, any combination of the following three equalities may occur: $a = a'$, $b = b'$, $c = c'$. Note that if $a = a'$ and $b = b'$, then $c = c'$ if and only if π is a Fano plane. Notice also that there are other triples of concurrent lines, namely, (A,B',C'), (A',B',C), and (A',B,C'), and so the following sets are quadrangular: $\{(a,a'),\ (b',b),\ (c',c)\}$, $\{(a',a),(b',b),(c,c')\}$, $\{(a',a),(b,b'),(c',c)\}$. In other words, if $\{(a,a'),(b,b'),(c,c')\} \in \gamma$, then we may switch two of the three ordered pairs and produce another triple in γ.

Suppose that we were given five collinear points a, a', b, b', and c such that $a \neq b \neq c \neq a$, $a' \neq b'$, $a \neq b'$, $b \neq a'$, $c \neq a'$, and $c \neq b'$. We are interested in knowing whether there is a sixth point c' such that $\{(a,a'),(b,b'),(c,c')\} \in \gamma$, and if so, whether there is more than one such point. The following theorem establishes the existence of the point.

THEOREM 5.5.2 *Let a,b,c,a',b' be five collinear points on line L such that if $a \neq b \neq c \neq a$, $a' \neq b'$, $a \neq b'$, $a' \neq b$, $c \neq a'$, and $c \neq b'$, then there exists a point c' such that $\{(a,a'),(b,b'),(c,c')\} \in \gamma$.*

Proof. Let p be a point not on L and let $A = pa$, $B = pb$, and $C = pc$. Let A' be a line through a' and distinct from $a'p$ and L. Let $q = B \cap A'$, $r = C \cap A'$, and $s = b'r \cap A$, and let $c' = qs \cap L$. Now it must be shown that p,q,r,s is a four-point. This is easily proved, so it will be done here for only one case. Suppose that q, r, and s are collinear. We know that $b' \in rs$, so $b' \in qr$. But $qr = A'$, so $b' = a'$. This contradiction shows that q, r, and s are not collinear. The other cases are handled similarly.

The sixth point c' may or may not be unique, as Example 5.5.3 shows.

EXAMPLE 5.5.3

    i. Let $\pi = \pi(2)$. Let A,A',B,B',C,C' be six sides of a quadrangle. There exists only one more line L in $\pi(2)$ and it has three points. Thus $a = a'$, $b = b'$, and $c = c'$, and therefore the sixth point is unique.

    ii. Let $\pi = \pi(3)$. Let A,A',B,B',C,C' be the six sides of a quadrangle. Each vertex of the quadrangle has one more line passing through it since each point contains four lines. Thus there are $13 - 4 - 6$, or 3 eligible lines L. In one instance we may have $a = a'$, $b \neq b'$. Thus $c = c'$ is necessary since L has only four points. In another instance $a \neq a'$ and $b = b'$. Then, similarly, $c = c'$. In the third instance $a = a'$ and $b = b'$. Thus $c \neq c'$ because $\pi(3)$ is not a Fano plane and so c' is the unique fourth point on L. Therefore, given line L, there are only three permissible arrangements of points a, b, c, a', and b', and sixth point c' is unique in each.

    iii. Let $\pi = \pi_N(9)$ and let $a = a' = [0,0,1]$, $b = b' = [-1,0,1]$, and $c = [a,0,1]$. A check of Example 4.4.14 shows that by one construction $c' = [1,0,0]$ and that by another construction $c' = c$. Thus c' is not unique.

The nonuniqueness of c' in Example 5.5.3(iii) stems directly from the fact that $\pi_N(9)$ has different-order subplanes. Since Desarguesian planes have a characteristic, the following theorem is not unexpected.

THEOREM 5.5.4 *If* a,b,c,a',b' *are collinear points such that* $a \neq b$ $\neq c \neq a$, $a' \neq b'$, $a \neq b'$, $a' \neq b$, $c \neq a'$, *and* $c \neq b'$, *and if* π *is a Desarguesian plane, then there exists exactly one point* c' *such that* $\{(a,a'),(b,b'),(c,c')\} \in \gamma$.

Proof. By Example 5.5.3(i) and (ii), each line can be assumed to have at least five points. Let the two arbitrary quadrangles used in the construction have vertices p,q,r,s and p',q',r',s', respectively, and let $a = pq \cap L = p'q' \cap L$, $b = pr \cap L = p'r' \cap L$, $c = ps \cap L$, and $d = q'r' \cap L$. It must be shown that $c' = d$.

Case 1. Suppose that $p = p'$ (see Figure 48). Since qrs and $q'r's'$ are triangles centrally perspective from p, we know that $qs \cap q's'$, $rs \cap r's'$, and $qr \cap q'r'$ are collinear. Therefore, $qr \cap q'r'$ is on $a'b' = L$, and thus $c' = d$.

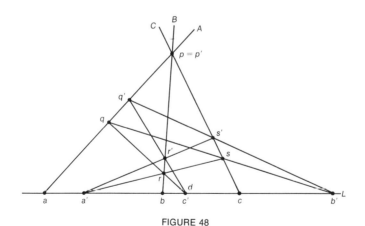

FIGURE 48

Case 2. Suppose that $p \neq p'$, $pa \neq p'a$, $pb \neq p'b$, $pc \neq p'c$, $rs \neq r's'$, and $qs \neq q's'$ (see Figure 49). Note that triangles pqs and $p'q's'$ are axially perspective from L, so by the dual of the Desargues Theorem, pp', qq', and ss' are concurrent. Similarly,

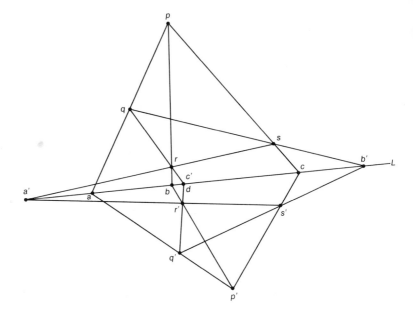

FIGURE 49

triangles *prs* and *p'r's'* are axially perspective, so *pp',rr',ss'* are concurrent. Thus *pp',qq',rr'* are concurrent, and so *pqr* and *p'q'r'* are centrally perspective triangles. By the Desargues Theorem, they are axially perspective, and so *qr* ∩ *q'r'* = *L*. Thus *c' = d*.

Case 3. Suppose that *p* ≠ *p'*, *pa* ≠ *p'a*, *pb* ≠ *p'b*, and *pc* ≠ *p'c* (see Figure 50). Thus *rs = r's'* or *qs = q's'*, possibilities that were excluded in Case 2. Choose a point *s"* on the line *ps* distinct from *p,s,c* and *q's'* ∩ *ps* (assume that there are at least five points on a line) and let *q" = pa* ∩ *b's"*, *r" = pb* ∩ *a's"*. By Case 1, the quadrangle with vertices *p,q",r",s"* yields the same arrangement on *L* as does *p,q,r,s*. Furthermore, the two four-points *p',q',r',s'* and *p,q",r",s"* satisfy the hypotheses of Case 2. Thus *c' = d*.

Case 4. Suppose that *p* ≠ *p'* and that either *pa = p'a* or *pb = p'b* or *pc = p'c*. Let *M* be a line passing through *p* that is distinct from *pa,pb,pc* and let *p"* be a point on *M* not incident with *pa, pb, pc*, and *L*; see Figure 51. (Here again assume the existence of at least five distinct points on a line.) Construct the quadrangle with vertices *p",q",r",s"* where *p"q"* ∩ *L = a*, *p"r"* ∩ *L = b*, *p"s"* ∩ *L*

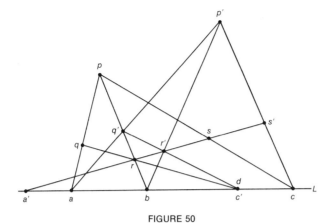

FIGURE 50

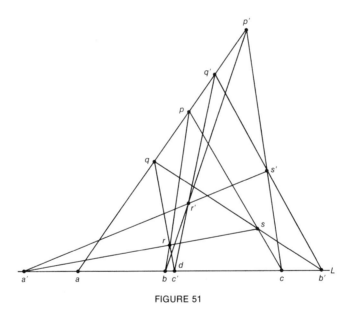

FIGURE 51

$= c$, $r''s'' \cap L = a'$, and $q''s'' \cap L = b'$. Now $p \neq p''$, $pa \neq p''a$, $pb \neq p''b$, and $pc \neq p''c$. Thus, by Case 3, $q''r'' \cap L = qr \cap L = c'$. Also $p' \neq p''$, $p'a \neq p''a$, $p'b \neq p''b$, and $p'c \neq p''c$, so $q'r' \cap L = q''r'' \cap L = d$. Thus $c' = d$.

All cases are now exhausted, so the proof is complete.

The following axiom formalizes this important property of uniqueness of the sixth quadrangular point.

AXIOM OF THE SIXTH QUADRANGULAR POINT   *If a,b,c,a',b' are collinear points such that $a \neq b \neq c \neq a$, $a' \neq b'$, $a \neq b'$, $a' \neq b$, $c \neq a'$, and $c \neq b'$, and if c' and d are such that $\{(a,a'),(b,b'),(c,c')\} \in \gamma$ and $\{(a,a'),(b,b'),(c,d)\} \in \gamma$, then $c' = d$.*

Theorem 5.5.4 shows that Desarguesian planes satisfy this axiom. Theorem 5.5.5 shows that the converse is also true; thus in a plane π, the Axiom of the Sixth Quadrangular Point is equivalent to the Desargues Theorem.

THEOREM 5.5.5   *The plane π satisfies the Axiom of the Sixth Quadrangular Point if and only if π is Desarguesian.*

Proof.   Because of Theorem 5.5.4 we need only show that the axiom implies the Desargues Theorem in π.

Suppose that π is not Desarguesian. Thus there exist triangles pqr and $p'q'r'$, which are axially perspective but not centrally perspective. Let $pq \cap p'q' = a$, $pr \cap p'r' = b$, $qr \cap q'r' = c$, and $a,b,c \in L$. Suppose that $pp' \cap qq' = s$. We may assume that $s \in L$ (by Theorem 5.2.5) and that p,q,r,s and p',q',r',s are four-points. Let $sp \cap L = d$, $sq \cap L = e$, $sr \cap L = f$, and $sr' \cap L = f'$. Since the triangles are not centrally perspective, $f \neq f'$. Therefore, $\{(d,c),(e,b),(f,a)\} \in \gamma$ and $\{(d,c),(e,b),(f',a)\} \in \gamma$. Thus π does not satisfy the axiom.

We also have the dual notion of a quadrangular set of lines. This is an arrangement of six concurrent lines, each passing through one of the six vertices of a quadrilateral (see Figure 52).

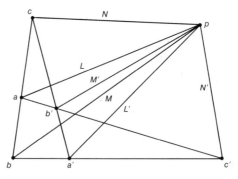

FIGURE 52

DEFINITION 5.5.6 *The (unordered) triple of ordered pairs of concurrent lines* $\{(L,L'),(M,M'),(N,N')\}$ *is said to be a* quadrangular set *of lines if and only if* $L = ap$, $L' = a'p$, $M = bp$, $M' = b'p$, $N = cp$, *and* $N' = c'p$ *where* a,a', b,b', *and* c,c', *respectively, denote the opposite vertices of a complete quadrilateral, where* a, b, *and* c *are collinear, and where* p *does not lie on any side of the quadrilateral.*

Notation. If $\{(L,L'),(M,M'),(N,N')\}$ is a quadrangular set of lines, the notation used is $\{(L,L'),(M,M'),(N,N')\} \in \beta$.

THEOREM 5.5.7 *If* L,M,N,L',M' *are concurrent lines such that* $L \neq M \neq N \neq L$, $L' \neq M'$, $L \neq M'$, $L' \neq M$, $N \neq L'$, *and* $N \neq M'$, *and if* π *is a Desarguesian plane, then there exists exactly one line* N' *such that* $\{(L,L'),(M,M'),(N,N')\} \in \beta$.

Proof. The proof follows from Theorem 5.5.4 and the principle of duality.

THEOREM 5.5.8 *Suppose that* π *is a Desarguesian plane. If six concurrent lines* A,A',B,B',C,C' *pass through a point* p *and meet a line* L *in six points* a,a',b,b',c,c', *respectively, and if* $\{(a,a'),(b,b'),(c,c')\}$ $\in \gamma$, *then* $\{(A',A),(B',B),(C',C)\} \in \beta$.

Proof. Let M be a line through a' and let $s = C \cap M$, $r = B \cap M$, and $q = b's \cap A$ (see Figure 53). It follows that p,q,r,s is a four-point and,

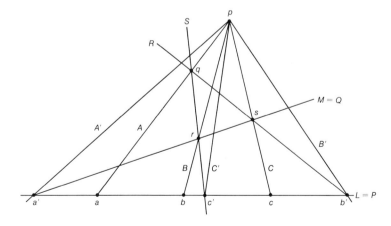

FIGURE 53

by construction, that $pq \cap L = a$, $pr \cap L = b$, $ps \cap L = c$, $rs \cap L = a'$, and $qs \cap L = b'$. Since $\{(a,a'),(b,b'),(c,c')\} \in \gamma$ and since c' is unique, $qr \cap L = c'$. Let $P = L$, $Q = M = rs$, $R = qs$, and $S = qr$. Clearly, P,Q,R,S are the four sides of a complete quadrilateral. Again by construction, $(P \cap Q)p = a'p = A'$, $(P \cap R)p = b'p = B'$, $(P \cap S)p = c'p = C'$, $(R \cap S)p = qp = A$, $(Q \cap S)p = rp = B$, $(Q \cap R)p = sp = C$. Thus $\{(A',A),(B',B),(C',C)\} \in \beta$.

THEOREM 5.5.9 *Suppose that π is a Desarguesian plane. If six collinear points a,a',b,b',c,c' lie on six concurrent lines A,A',B,B',C,C', respectively, and if $\{(A,A'),(B,B'),(C,C')\} \in \beta$, then $\{(a',a),(b',b),(c',c)\} \in \gamma$.*

Proof. The proof follows from Theorem 5.5.8 and the principle of duality.

The final theorem in this section shows that projectivities preserve the quadrangular arrangement of points on a line (and, dually, lines through a point).

THEOREM 5.5.10 *In a Desarguesian plane, if f is a projectivity from L to M such that f: $a,a',b,b',c,c' \rightarrow p,p',q,q',r,r'$ and if $\{(a,a'), (b,b'),(c,c')\} \in \gamma$, then $\{(p,p'),(q,q'),(r,r')\} \in \gamma$.*

Proof. It is sufficient to show that this theorem is true for perspectivities f. Suppose that $f = L \overset{O}{\barwedge} M$. Let $A = oa$, $B = ob$, $C = oc$, $A' = oa'$, $B' = ob'$, and $C' = oc'$. By Theorem 5.5.7, $\{(A',A),(B',B),(C',C)\} \in \beta$, and by Theorem 5.5.8, $\{(p,p'),(q,q'),(r,r')\} \in \gamma$ because $oa = op$, $ob = oq$, $oc = or$, $oa' = op'$, $ob' = oq'$, and $oc' = or'$.

EXERCISES

1. Let p,q,r,s be a four-point and let L be a line not containing p, q, r, or s. Generate a quadrangular set of points in π_F if $p = [0,1,0]$, $q = [0,1,1]$, $r = [1,1,1]$, $s = [1,1,0]$, and $L = \langle 0,1,0 \rangle$.

2. Let $x \in F$ such that $x \neq 0,1$ and let $a = [0,0,1] = a'$, $b = [1,0,1]$, $b' = [-1,0,1]$, and $c = [x,0,1]$ in π_F. Find c' such that $\{(a,a'),(b,b'),(c,c')\} \in \gamma$.

3. Suppose that $\{(p,p'),(q,q'),(r,r')\} \in \gamma$. Show that π is then a Fano plane if and only if $p = p'$ and $q = q'$ imply that $r = r'$.

4. In $\pi_H(9)$ show that if p,q,r,p',q' are collinear points such that $p \neq q \neq r \neq p$, $p' \neq q'$, $p \neq q'$, $p' \neq q$, $r \neq p'$, and $r \neq q'$, then there may exist more than one point r' such that $\{(p,p'),(q,q'),(r,r')\} \in \gamma$.

5. Complete a proof of Theorem 5.5.4 using Corollary 5.2.2(iv) and the following outline.
    i. Let p,q,r,s and p',q',r',s' be the two four-points and let a,a',b,b', c,c', and d be the same as in the proof of Theorem 5.5.4.
    ii. Let $f \in CC(a',pq)$ such that $f: b' \rightarrow c'$ and let $g \in CC(a,r's')$ such that $g: b \rightarrow d$. (Note that $f: c \rightarrow b$ and $g: c \rightarrow b'$).
    iii. Referring back to exercise 3.2.5 and Theorem 5.2.3, note also that $g \circ f = f \circ g$.
    iv. Conclude that $c' = d$.

6. Consider the following property: If f and g are projectivities mapping L onto L such that: (1) each is a product of two perspectivities, (2) the axial invariant point of f is the central invariant point of g, and (3) the axial invariant point of g is the central invariant point f (see exercise 3.4.2), then $f \circ g = g \circ f$. Show that π satisfies this property if and only if π satisfies the Axiom of the Sixth Quadrangular Point.

5.6   ARITHMETIC ON A LINE

The final two sections of this chapter consider useful special cases of quadrangular sets of points. This section develops a way of adding and multiplying points on a line with the help of quadrangular sets.

Let us first consider a purely geometric method for adding and multiplying points in the Cartesian plane. Let p and q be arbitrary points on the x-axis. Thus $p = (a,0)$ and $q = (b,0)$ for some real numbers a and b. In Figure 54, X and Y denote the two axes and o denotes the

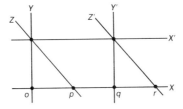

FIGURE 54

origin. Draw an arbitrary line $X' \parallel X$ and let Z denote the line joining $X' \cap Y$ to p. Then lines $Y' \parallel Y$ and $Z' \parallel Z$ can be constructed where Y' passes through q and Z' passes through $Y' \cap X'$. Let $r = Z' \cap X$; then $r = (c,0)$, and, by congruent triangles, $c = a + b$. If we define $p \boxplus q$ to be the point r thus generated, we see that this addition on X is isomorphic to addition on the real number line.

Multiplication is defined as follows. Let $p,q \neq o$, let $i = (1,0)$, and let L be a line through o distinct from X and Y (see Figure 55). Construct

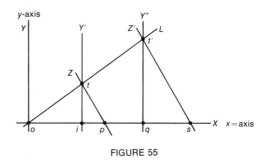

FIGURE 55

lines Y' and Y'' parallel to Y and passing through i and q, respectively, and construct parallel lines Z and Z' where Z connects $Y' \cap L$ to p and Z' passes through $Y'' \cap L$. Let $s = Z' \cap X$; then $s = (d,0)$ and, by the following observations, $d = ab$. Triangles opt and ost', where $t = L \cap Z$ and $t' = L \cap Z'$, are similar; therefore, $\overline{op}/\overline{ot} = \overline{os}/\overline{ot'}$. ($\overline{op}$ denotes the length of the line segment op.) Triangles oit and oqt' are also similar, so $\overline{oi}/\overline{ot} = \overline{oq}/\overline{ot'}$. Therefore, $\overline{oi}/\overline{op} = \overline{oq}/\overline{os}$; that is, $1/a = b/d$, and so $d = ab$. Thus if we define $p \boxdot q$ to be the point s and $o \boxdot p = p \boxdot o = o$, we see that multiplication on X is isomorphic to multiplication on the real number line.

Addition and multiplication on a line in a projective plane may now be defined using the appropriate extensions of these constructions. Let L be the line containing p and q, let L_∞ be the line at infinity, and let $u = L \cap L_\infty$. Addition is defined by $p \boxplus q = r$ where r is generated by the construction of Figure 56. Multiplication is defined by $p \boxdot q = s$ and $p \boxdot o = o \boxdot p = o$ where p, q, and s are related as in Figure 57. Figure 56 shows that $\{(o,r),(q,p),(u,u)\} \in \gamma$; Figure 57 shows that $\{(i,r),(q,p),(u,o)\} \in \gamma$.

The following definition can now be stated.

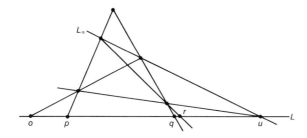

FIGURE 56

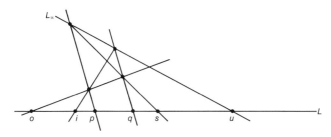

FIGURE 57

DEFINITION 5.6.1   *Let L be an arbitrary line in π and let o, i, u be three distinct arbitrary points on L.*

1. *If p and q are points distinct from o and u, then $p \boxplus q = r$ (with respect to o and u) if and only if $\{(o,r),(q,p),(u,u)\} \in \gamma$. Also, $p \boxplus o = o \boxplus p = p$ for $p \neq u$.*

2. *If p and q are points distinct from o, i, and u, then $p \boxdot q = s$ (with respect to o,i,u) if and only if $\{(i,s),(q,p),(u,o)\} \in \gamma$. Also $p \boxdot o = o \boxdot p = o$ and $i \boxdot p = p \boxdot i = p$ for all $p \neq u$.*

If π is a Desarguesian plane, the operations of addition and multiplication are well defined because r and s are unique by Theorem 5.5.4. Thus we have an algebraic system $(L - \{u\}, \boxplus, \boxdot)$. Since Desarguesian planes are (p, L) transitive for all p and L, they are eligible for algebraic representations as proposed in Section 4.5. Let $p \oplus q = f_q \circ f_p(o)$ and $p \odot q = g_p \circ g_p(i)$ as in Definition 4.5.6. It is then appropriate to compare $(L - \{u\}, \boxplus, \boxdot)$ with $(S, \oplus, \odot)$. We shall assume that o, u, and i refer to the same points in both discussions. Thus $S = L - \{u\}$. We shall also assume that $L_\infty = uv$.

THEOREM 5.6.2 $(S,\oplus) = (L - \{u\},\boxplus)$.

Proof.   We know that $o \boxplus p = p \boxplus o = p = o \oplus p = p \oplus o$. Suppose that $p \boxplus q = r$ where $p,q \neq o$. Then $\{(o,r),(q,p),(u,u)\} \in \gamma$, and we may construct a four-point a,b,c,d such that $ab = L_\infty$, $ac \cap ou = p$, $bd \cap ou = q$, $bc \cap ou = o$, $ad \cap ou = r$, and $cd \cap ou = u$ (see Figure 58). Since $f_q: o \to q$, it follows that $f_q: bo \to bq$, $c \to d$, and $p \to r$. Thus $(f_q \circ f_p)(o) = f_q(p) = r$, and so $p \oplus q = r$.

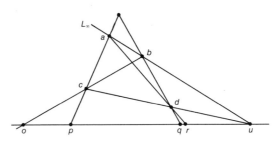

FIGURE 58

THEOREM 5.6.3 $(S - \{o\},\odot) \sim (L - \{u,o\},\boxdot)$.

Proof.   We know that $i \odot p = p \odot i = p = i \boxdot p = p \boxdot i$. Suppose that $p \boxdot q = s$ where $p,q \neq i$. Then $\{(i,s),(q,p),(u,o)\} \in \gamma$, and we may construct a four-point a,b,c,d such that $ab = L_\infty$, $ac \cap ou = p$, $bd \cap ou = q$, $bc \cap ou = i$, $ad \cap ou = s$, and $cd \cap ou = o$ (see Figure 59). Let v be an arbitrary point on L_∞ distinct from u. Since $\mathrm{Hom}(o,uv) \sim \mathrm{Hom}(u,ov)$ (because for Desarguesian planes π, $C(\pi)$ is transitive on four-points) and since $\mathrm{Hom}(u,ov) \sim (S - \{o\},\odot)$ by Theorem 4.5.7, we need show only that $(L - \{u,o\},\boxdot) \sim \mathrm{Hom}(o,uv)$. Denote by h_p the unique homology in $\mathrm{Hom}(o,uv)$ mapping $i \to p$ where $p \neq o$. The final step is to show that the mapping $\beta: h_p \to p$ is the sought-after isomorphism. Observe that $h_q: i \to q$ and thus that $h_q: bi \to bq$, $c \to d$ (since cd is held fixed), $ac \to ad$, and $p \to s$. Thus $h_q \circ h_p(i) = h_q(p) = s$ and so $\beta(h_q \circ h_p) = \beta(h_s) = s = p \boxdot q = \beta(h_p) \boxdot \beta(h_q)$.

For planes π_F, it is easily shown that $(S,\oplus,\odot) = (L - \{u\},\boxplus,\boxdot)$. For general Desarguesian planes, the algebra $(S,\oplus,\odot)$ defined on π is identical to the algebra $(L - \{u\},\boxplus,\boxdot)$ defined on π^d. More will be said about this in Chapters 7 and 8.

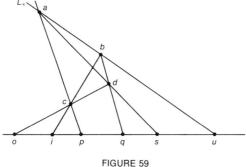

FIGURE 59

EXERCISES

1. Show that if π is a Desarguesian Fano plane, then $p \boxplus p = o$.

2. Show that for planes π_F, $(S,\oplus,\odot) = (L - \{u\},\boxplus,\boxdot)$, and thus (referring back to exercise 4.2.5) $(L - \{u\},\boxplus,\boxdot) \sim F$.

3. A way of adding points on a line (in a Desarguesian plane) can be formulated using projectivities. Consider the line L and the points $p \in L - \{t\}$ where t is a designated point on L. Also designate a point o on L different from t. Define addition as follows: Let M,N be two lines distinct from L and concurrent at t. Let c be a fixed point on N distinct from t and let $b = oc \cap M$. For any $q \in L - \{t\}$, let $d = qb \cap N$ and define $f_q = L \overset{c}{\underset{\wedge}{}} M \overset{d}{\underset{\wedge}{}} L$. Then define $p \oplus q$ as $f_q(p)$.
   a. Complete an analogous formulation for multiplication.
   b. Denoting your system by $(L - \{t\},\oplus,\odot)$, verify that it is identical to $(L - \{u\},\boxplus,\boxdot)$ of Definition 5.6.1 if the line L in both contexts is the same line and if $t = u$.
   c. Explain why your method of addition and multiplication would not be feasible for such non-Desarguesian planes as $\pi_N(9)$ and $\pi_H(9)$.

4. Let π be a Desarguesian plane and consider $\boxplus$ and $\boxdot$ as in $(L - \{u\},\boxplus,\boxdot)$. Let f_p be a mapping on L where $p \in L - \{u\}$ such that

$$f_p: u \to u$$
$$f_p: q \to p \boxplus q \text{ where } p \neq u$$

Let g_p be a mapping on L where $p \in L - \{u,o\}$ such that

$$g_p: o,u \to o,u$$
$$g_p: q \to p \boxdot q \text{ if } p \neq o,u$$

Let h_p be a mapping on L where $p \in L - \{u,o\}$ such that

$$h_p: o,u \to o,u$$
$$h_p: q \to q \boxdot p \text{ if } q \neq o,u$$

Show that f_p, g_p, and h_p are all projectivities on L.

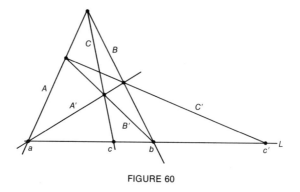

FIGURE 60

5.7 HARMONIC CONJUGATES

This section introduces another useful special case of a quadrangular set of points. If $\{(a,a'),(b,b'),(c,c')\} \in \gamma$ and $a = a'$, $b = b'$, then c' is the harmonic conjugate of c with respect to a and b. Since $\{(a,a'), (b,b'),(c,c')\} \in \gamma$ if and only if $\{(a' a),(b,b'),(c',c)\} \in \gamma$, a more symmetric definition of harmonic conjugate may be stated as follows.

DEFINITION 5.7.1 *The points c and c' are* harmonic conjugates *with respect to a and b on line L if and only if there exists a complete quadrangle with opposite sides A,A'; B,B'; and C,C' such that $A \cap L = A' \cap L = a$, $B \cap L = B' \cap L = b$, and $C \cap L = c$, $C' \cap L = c'$* (see Figure 60).

Notation. If c and c' are harmonic conjugates with respect to a and b, the notation used is $H(a,b; c,c')$. Since we are dealing with an ordered pair of (unordered) doubletons, we may write $H(a,b; c',c)$, $H(b,a; c,c')$, or $H(b,a; c',c)$.

For Fano planes, if a, b, and c are distinct points, then $H(a,b; c,c)$. For the plane $\pi_N(9)$, Example 5.5.3(iii) has shown that it is possible to have $H(a,b; c,c)$ and $H(a,b; c,d)$ where $d \neq c$. Evidently this is always possible in planes that contain a four-point with characteristic 2 and another with characteristic $k \neq 2$. The uniqueness of the fourth harmonic point is an important property of a plane. As with the sixth quadrangular point, this property will be given the status of an axiom and its validity will be tested in various planes.

AXIOM OF THE FOURTH HARMONIC POINT   *If $H(a,b; c,d)$, then $c \neq d$. Furthermore, if $H(a,b; c,d)$ and $H(a,b; c,d')$, then $d = d'$.*

Clearly, a plane must satisfy Fano's Axiom before it is eligible to satisfy the Axiom of the Fourth Harmonic Point. Thus $\pi_N(9)$, $(\pi_N(9))^d$, and $\pi_H(9)$ do not qualify. By Theorem 5.5.4, the Axiom of the Fourth Harmonic Point is satisfied in all Desarguesian planes with characteristics other than 2. This axiom also holds in planes satisfying both the Little Desargues Theorem and Fano's Axiom; it is proved with the help of two lemmas.

LEMMA 5.7.2   *In a Little Desarguesian plane π, if $f,g \in El(p)$ and $f(q) = g(q) \neq q$ for some point q, then $f|pq = g|pq$.*

Proof.   Suppose that $f \in El(p,L)$ and that $f: q,r \to a,b$ where $p \in qr$. Suppose also that $g \in El(p,M)$ and that $g: q,r \to a,c$. We must show that $b = c$. Let $N = pq$; then $p,q,r,a,b,c \in N$. Since π satisfies the Little Desargues Theorem, there exists $h \in El(p,N)$ such that $h: L \to M$. It is easily verified that $h \circ f \circ h^{-1} \in El(p,M)$ and that it maps $q \to a, r \to b$. Also, $g \in El(p,M)$ and g maps $q \to a, r \to c$. Since an elation is completely determined by its center, its axis, and the image of one appropriate point, we have proved that $b = c$.

LEMMA 5.7.3   *In a Little Desarguesian plane π, if $H(p,q; r,s)$, then there exists a line M and an elation $f \in El(p,M)$ such that $f: r,q \to q,s$.*

Proof.   Let $L = pq$. Since $H(p,q; r,s)$, there is a four-point a,b,c,d such that $ab \cap L = p$, $ac \cap L = q$, $bc \cap L = r$, and $ad \cap L = s$ (see Figure 61). Let $M = ab$ and $f \in El(p,M)$ such that $f: r \to q$. Then $f: rb \to qb$, $c \to d$ (because cd is held fixed), $ca \to da$, $q \to s$ (because L is held fixed).

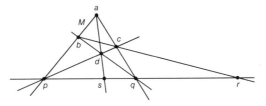

FIGURE 61

THEOREM 5.7.4 *If π is a Little Desarguesian plane that satisfies Fano's Axiom, then the Axiom of the Fourth Harmonic Point holds.*

Proof. Suppose that $H(p,q; r,s)$ and that $H(p,q; r,t)$. Since π satisfies Fano's Axiom, $r \neq s$ and $r \neq t$. By Lemma 5.7.3, there exists a line M, and $f \in \text{El}(p,M)$ such that $f: r,q \rightarrow q,s$. There also exists a line N, and $g \in \text{El}(p,N)$ such that $g: r,q \rightarrow q,t$. By Lemma 5.7.2, $s = t$.

The converse of Theorem 5.7.4 is also true. This is an unexpected result that provides a convenient characterization of the class of Little Desarguesian planes satisfying Fano's Axiom. Some standard properties of harmonic conjugates will be mentioned before the converse of Theorem 5.7.4 is proved.

THEOREM 5.7.5 *If π is a Little Desarguesian plane and if f is a projectivity mapping $a,b,c,d \rightarrow p,q,r,s$ where $H(a,b; c,d)$, then $H(p,q; r,s)$.*

Proof. The proof follows from Theorem 5.5.9.

THEOREM 5.7.6 *If π is a Little Desarguesian plane and if $H(a,b; c,d)$ and $H(p,q; r,s)$, then there exists a projectivity f such that $f: a,b,c,d \rightarrow p,q,r,s$.*

Proof. The proof is left as an exercise.

THEOREM 5.7.7 *If π is a Little Desarguesian plane satisfying Fano's Axiom and if $H(a,b; c,d)$, then $H(c,d; a,b)$.*

Proof. Let L be the line containing a,b,c,d and let p,q,r,s be a four-point such that $pq \cap L = rs \cap L = a$; $pr \cap L = qs \cap L = b$; $ps \cap L = c$; and $qr \cap L = d$ (see Figure 62). Since π satisfies Fano's Axiom, $c \neq d$. Therefore, t and u may be defined as follows: $t = pc \cap qd$, $u = cr \cap pd$. Since p,t,r,u is a four-point, $H(c,d; ut \cap L,b)$. Because qts and pur are triangles that are axially perspective from L and because $a = rs \cap pq$, we see that $a \in ut$ since π satisfies the Little Desargues Theorem. Thus $a = ut \cap L$ and therefore $H(c,d; a,b)$.

The term "harmonic" was used in Chapter 3 to refer to a special type of homology. It is natural to ask whether harmonic conjugates are

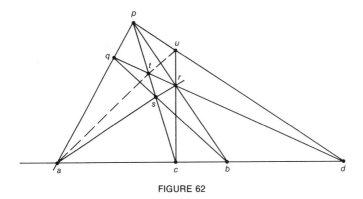

FIGURE 62

related to harmonic homologies. The following theorem explains this relationship.

THEOREM 5.7.8 *Suppose that π satisfies the Axiom of the Fourth Harmonic Point and suppose that $f \in \text{Hom}(p, L)$. Then f is a harmonic homology if and only if $H(p, q; r, f(r))$ where q is an arbitrary point on L and r is an arbitrary point on pq distinct from p and q.*

Proof. Suppose that f is a harmonic homology. Let $q \in L$ and $r \in pq$ such that $r \neq p, q$ and let M be an arbitrary line through p distinct from pq (see Figure 63). Also let N be an arbitrary line through r distinct from pq and let $M \cap L = a$, $M \cap N = c$, $N \cap L = b$, $f(r)b \cap M = e$, and $f(r)c \cap L = d$. Since $f: r \to f(r)$, we have $f: rc \to f(r)c$, $b \to d$ (since L is fixed). Also $f: f(r) \to f^2(r) = r$, so $f: f(r) e \to re$, and thus $b \to re$ $\cap L$. It follows that $d = re \cap L$. The four-point $r, b, f(r), d$ yields the relationship $H(c, e; p, a)$, so if $g = M \overset{b}{\barwedge} pq$, then $H(g(c), g(e); g(p), g(a))$, that is, $H(r, f(r); p, q)$. By Theorem 5.7.7, $H(p, q; r, f(r))$. Proof of the converse is left as an exercise.

For planes satisfying the Axiom of the Fourth Harmonic Point, Theorem 5.7.8 establishes the fact that there is at most one harmonic homology for a given p and L where $p \notin L$. Theorem 5.7.10 shows that there exists at least one harmonic homology for a given p and L where $p \notin L$. First it will be convenient to state a lemma.

LEMMA 5.7.9 *If $H(p, q; r, s)$ and $H(p, q'; r', s')$ where $pq \neq pq'$, then qq', rr', and ss' are concurrent.*

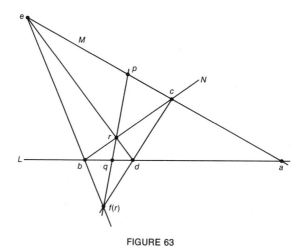

FIGURE 63

Proof. The proof is left as an exercise.

THEOREM 5.7.10 *Suppose that π satisfies the Axiom of the Fourth Harmonic Point. If p and L are arbitrary except that $p \notin L$, there exists a harmonic homology $f \in$ Hom(p,L).*

Proof. Define $f: \mathcal{P} \dashrightarrow \mathcal{P}$ as follows: $f(x) = x$ if $x = p$ or $x \in L$; if $x \neq p$ and $x \notin L$, then $f(x) = y$ where $H(p,t; x,y)$ and $t = px \cap L$. Clearly, f is well defined and one-to-one because fourth harmonics are unique. The substantial part of the proof involves showing that f preserves incidence. It is obvious that f preserves incidence on line L and through point p, so consider a line M distinct from L and not incident with p. Let $s = L \cap M$, let q be any point on M distinct from s, and let $r = pq \cap L$ (see Figure 64). Then $H(p,r; q,f(q))$. If $x \in M$ is distinct from s and q, we must prove that $f(x) \in f(q)f(s)$. Let $t = px \cap L$ as defined earlier in the proof; then $H(p,t; x,f(x))$. By Lemma 5.7.9, rt, qx, and $f(q)f(x)$ are concurrent. The point of concurrence is s, and since $f(s) = s$, we have $f(s) \in f(q)f(x)$.

Since f preserves incidence, we can see that f is an involution simply by observing that $H(p,t; x,f(x))$ implies both $H(p,t; f(x),x)$ and $H(f(p),f(t); f(x),f^2(x))$, that is, $H(p,t; f(x),f^2(x))$. Since fourth harmonics are unique, $x = f^2(x)$.

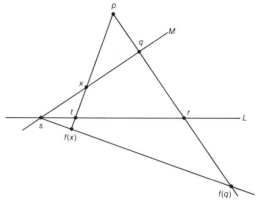

FIGURE 64

Because f is an involution, it must be an onto map, and thus f is a collineation. We know from the definition of f that it is a homology with center p and axis L.

THEOREM 5.7.11 *If π satisfies the Axiom of the Fourth Harmonic Point and if p and L are arbitrary except that $p \notin L$, there exists one and only one harmonic homology with center p and axis L.*

Proof. The proof follows from Theorems 5.7.8 and 5.7.10.

This chapter is concluded with a proof of the converse of Theorem 5.7.4.

THEOREM 5.7.12 *If π satisfies the Axiom of the Fourth Harmonic point, then π is a Little Desarguesian plane that satisfies Fano's Axiom.*

Proof. It is easily seen that π satisfies Fano's Axiom. It must be shown that π contains all possible elations; that is, that π is (p,L) transitive for all p and L such that $p \in L$. Suppose that p, L, a, and a' are given such that $p \in L$, $p \neq a$, $p \neq a'$, $pa \neq L$, and $a' \in pa$. Let q be such that $H(p,q;a,a')$, let g be the harmonic homology in $\text{Hom}(q,L)$, and let h be the harmonic homology in $\text{Hom}(a',L)$. Letting $f = h \circ g$, we see that f holds L pointwise fixed, so $f \in CC(L)$. Also $f: a \to a'$, so the center of f is on aa'. Let b be a point such that $b \notin aa'$ and $b \notin L$ and

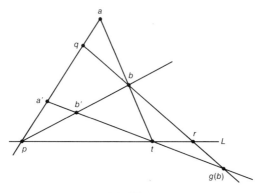

FIGURE 65

let $ab \cap L = t$ and $qb \cap L = r$ (see Figure 65). By Theorem 5.7.8, $H(q,r; b,g(b))$, and so, by Theorem 5.7.7, $H(g(b),b; r,q)$. The following argument shows that a', t, and $g(b)$ are collinear. We know that g: $a \rightarrow a'$ because $H(q,p; a,a')$, so g: $ab \rightarrow a'g(b)$. Since g holds L pointwise fixed, $t = g(t) = g(ab \cap L) = a'g(b) \cap L$. Therefore, $H(g(b),b'; t,a')$ for some b' on ta'. It follows from Lemma 5.7.9 that bb', rt, and qa' are concurrent. This point of intersection must be p since $rt \cap qa' = p$; thus p, b, and b' are collinear. We easily obtain $H(a',t; b',g(b))$ and $H(a',t; b',h(b'))$, and thus $h(b') = g(b)$. But $b' = h^2(b') = h(g(b)) = h \circ g(b) = f(b)$, so the center of f is on bb' as well as on aa'. Since $aa' \cap bb' = p$, we may conclude that f is the sought-after elation in $\mathrm{El}(p,L)$ mapping a onto a'.

We should observe that there are Little Desarguesian planes π that do not satisfy Fano's Axiom—for example, $\pi = \pi_F$ where $F = \mathrm{GF}(2'')$. Chapter 14 will show that all Little Desarguesian planes failing to satisfy Fano's Axiom are Fano planes. Thus, in all Little Desarguesian planes, the fourth (not necessarily distinct) harmonic point is unique. Theorem 5.7.4 has shown this to be true for planes satisfying Fano's Axiom; it is clearly also true for Fano planes.

Harmonic conjugates have not yet been discussed in the affine context. As will be pointed out in the exercises at the end of this section, such a concept generalizes the idea of midpoint.

EXERCISES

1. Prove Theorem 5.7.6.

2. Complete the proof of Theorem 5.7.8.

3. Prove Lemma 5.7.9.

4. Show that the harmonic conjugate relationship between four collinear points in π_R is related to the midpoint relationship between three collinear points in α_R as follows: The point r is the midpoint of the line segment joining p to q in α_R if and only if $H(p,q; r,s)$ in $\pi(\alpha_R)^+$ where $s = pq \cap L_\infty$. Hint: First show that this is true for $p = (a,0)$, $q = (b,0)$, and $r = (c,0)$; then complete the proof using perspective mappings.

5. a. Define the concept of harmonic conjugacy for four concurrent lines.
   b. Show that if N and N' are harmonic conjugates with respect to L and M, then $H(p,q; r,r')$ where $p = L' \cap L$, $q = L' \cap M$, $r = L' \cap N$, $r' = L' \cap N'$, and L' is any line not concurrent with L and M.
   c. Formulate an Axiom of the Fourth Harmonic Line.
   d. Verify that if π satisfies Fano's Axiom, then π is a Little Desarguesian plane if and only if π satisfies the Axiom of the Fourth Harmonic Line.
   e. Attempt exercise 4 in the context of harmonic conjugacy for lines.

6. Let π be a Little Desarguesian plane that satisfies Fano's Axiom.
   a. Let h be a harmonic homology in $\text{Hom}(p,L)$ and let $f \in CC(L)$. Show that if $h \circ f$ is a harmonic homology, then f is an elation.
   b. Using exercise 6(a), show that the composite of two harmonic homologies having a common axis is an elation.

7. Let π be a Little Desarguesian plane and let h be the harmonic homology with center p and axis L. Show that $h \circ f = f \circ h$ for all $f \in \text{Hom}(p,L)$.

8. Show that π is a Little Desarguesian plane if and only if π satisfies the following theorems:
   a. The Desargues Theorem is true for any pair of triangles pqr and $p'q'r'$ where $pr \cap p'r' \in qq'$.
   b. The Desargues Theorem is true for any pair of triangles pqr and $p'q'r'$ such that $r' \in pq$.

9. Formulate and prove a statement equivalent to the Axiom of the Fourth Harmonic Point. Hint: See the statement that is equivalent to the Axiom of the Sixth Quadrangular Point given in exercise 5.5.6.

CHAPTER 6

PAPPIAN PLANES

This chapter examines the second of the classical projective planes, the Pappian plane. Like the Desarguesian plane, the Pappian plane is defined in terms of the embeddability of a configuration—the Pappus configuration, $\Sigma(9_3)$. This chapter will show, as did the previous one, that a simple configuration theorem can lead to unexpected and far-reaching results.

6.1   THE PAPPUS THEOREM

The embeddability requirement for the Pappus configuration may be stated as follows.

PAPPUS THEOREM   *In a projective plane π, if p,q,r and p',q',r' are triples of distinct collinear points on distinct lines L and L', respectively, and if $L \cap L'$ is different from all six points, then the points $pq' \cap p'q, pr' \cap p'r, qr' \cap q'r$ are collinear (see Figure 66).*

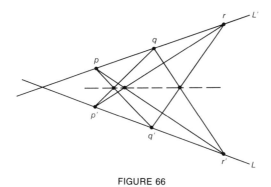

FIGURE 66

DEFINITION 6.1.1   *If a plane satisfies the Pappus Theorem, then it is a* Pappian plane.

Note that the class of Pappian planes satisfies the principle of duality.

THEOREM 6.1.2   *If π is a Pappian plane, if L, M, N and L', M', N' are triples of concurrent lines passing through distinct points p and p', and if pp' is different from all six lines, then the lines formed by the following point pairs are concurrent: $L \cap M', L' \cap M; L \cap N', L' \cap N; M \cap N', M' \cap N$.*

Proof.   The proof is left as an exercise.

COROLLARY 6.1.3   *The class of Pappian planes satisfies the principle of duality.*

Unlike the Desargues Theorem, which can apply to configurations other than the Desargues configuration, the hypotheses of the Pappus Theorem require the existence of at least nine distinct points and nine distinct lines; the theorem is true if and only if the Pappus configuration, $\Sigma(9_3)$, occurs everywhere (see Section 2.8).

EXAMPLE 6.1.4

i. $\pi(2)$ satisfies the Pappus Theorem because the hypotheses of the theorem cannot be met.

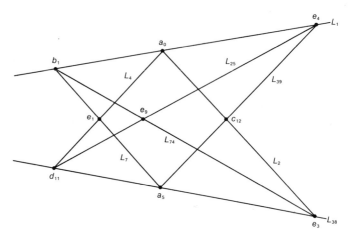

FIGURE 67

ii. π_F satisfies the Pappus Theorem. Let $p = [0,1,0]$, $q = [1,1,0]$, $r = [x,1,0]$ where $x \neq 0,1$; $p' = [0,0,1]$, $q' = [1,0,1]$, and $r' = [y,0,1]$ where $y \neq 0,1$. Notice that p,q,r lie on $L = \langle 0,0,1 \rangle$ and p',q',r' lie on $L' = \langle 0,1,0 \rangle$. Since $L \cap L' = [1,0,0] \neq p,q,r,p',q',r'$, the hypotheses of the Pappus Theorem have been satisfied. It was verified in the partial proof of Theorem 2.8.7 that $pq' \cap p'q = [1,1,1]$, that $pr' \cap p'r = [xy,y,x]$, that $qr' \cap q'r = [xy - 1, y - 1, x - 1]$, and that these three points lie on the line $\langle x - y, xy - x, y - xy \rangle$.

Suppose that a,b,c and a',b',c' are an arbitrary pair of three collinear points satisfying the hypotheses of the Pappus Theorem. Since p,q,p',q' and a,b,a',b' are four-points and since $C(\pi_F)$ is transitive on four-points, there exists a collineation $f: a,b,a',b' \rightarrow p,q,p',q'$. Since r and r', as defined previously, represent arbitrary third points on L and L', respectively, we may let $r = f(c)$ and $r' = f(c')$. Thus f^{-1} maps the Pappus configuration generated from p,q,r,p',q',r' into another Pappus configuration. It follows that $ab' \cap a'b$, $ac' \cap a'c$, $bc' \cap b'c$ are collinear. Thus π_F satisfies the Pappus Theorem.

iii. $\pi_H(9)$ does not satisfy the Pappus Theorem. Let b_1,a_0,e_4 and d_{11},a_5,e_3 be a pair of collinear triples. They satisfy the hypotheses of the Pappus Theorem because $b_1,a_0,e_4 \in L_1$; $d_{11},a_5,e_3 \in L_{38}$; and $L_1 \cap L_{38} = g_8$. It is a simple matter to show that

$b_1 a_5 \cap d_{11} a_0 = e_1$, that $d_{11} e_4 \cap b_1 e_3 = e_9$, and that $a_0 e_3 \cap a_5 e_4 = c_{12}$ (see Figure 67). But $e_1 e_9 = L_{60}$ and $c_{12} \notin L_{60}$, so the conclusion of the Pappus Theorem does not hold.

iv. $\pi_N(9)$ does not satisfy the Pappus Theorem. Proof of this fact is left as an exercise.

v. $\pi = \Sigma_0^+$ where Σ_0 is not a projective plane. Here π does not satisfy the Pappus Theorem because π does not contain a confined configuration outside of Σ_0, and the Pappus configuration is a confined configuration that must occur everywhere in a Pappian plane.

EXERCISES

1. Prove Theorem 6.1.2.

2. Establish Example 6.1.4(iv).

3. Two triangles abc and $a'b'c'$ are *k-ply perspective* if there exist k points $p_1, \ldots, p_k$ such that the triangles, after a suitable rearrangement of letters, are centrally perspective from each p_i, $i = 1, \ldots, k$.
   a. Construct a figure showing two triangles that are doubly perspective.
   b. Construct a figure showing two triangles that are triply perspective.
   c. Show that in a Pappian plane, if two triangles are doubly perspective, then they are triply perspective.

6.2   SPECIALIZED PAPPUS THEOREMS
AND RELATED DESARGUES THEOREMS

The Pappus Theorem does not lend itself naturally to the many special cases that the Desargues Theorem does. Three cases are mentioned here; one will prove to be especially useful to us.

PAPPUS (M, N, L) THEOREM   *If a, b, c and a', b', c' are triples of collinear points on M and N, respectively, such that $a, b, c, a', b', c' \notin L$, $ab' \cap a'b \in L$, and $ac' \cap a'c \in L$, then $bc' \cap b'c \in L$.*

PAPPUS (M, N) THEOREM   *π satisfies the Pappus (M, N) Theorem if and only if π satisfies the Pappus (M, N, L) Theorem for all lines $L \neq M, N$.*

PAPPUS (L) THEOREM   *π satisfies the Pappus (L) Theorem if and only if π satisfies the Pappus (M,N,L) Theorem for all distinct lines M and N different from L.*

Notation.   As with the specialized Desargues theorems, the specialized Pappus theorems are often invoked for specific lines. If the lines are X, Y, and Z, then "the Pappus (X,Y,Z) Theorem" is the shorthand notation for "the Pappus (M,N,L) Theorem where $M = X$, $N = Y$, and $L = Z$." Similarly, "the Pappus (X,Y) Theorem" and "the Pappus (Z) Theorem" denote "the Pappus (M,N) Theorem" and "the Pappus (L) Theorem where $M = X$, $N = Y$, and $L = Z$."

A few inferences can easily be drawn from these theorems. If $π$ satisfies the Pappus (M,N,L) Theorem, then it satisfies the Pappus (N,M,L) Theorem; if $π$ satisfies the Pappus (M,N) Theorem, then it satisfies the Pappus (N,M) Theorem; if $π$ satisfies the Pappus (M,N) Theorem for all distinct lines M and N, then it satisfies the Pappus Theorem; and if $π$ satisfies the Pappus (L) Theorem for all lines L, then it satisfies the Pappus Theorem. Actually, much stronger statements can be made in this regard. If $π$ satisfies the Pappus (M,N) Theorem for two fixed lines M and N, then $π$ satisfies the Pappus Theorem (Pickert, 1959; Burn, 1968; see also Theorem 11.5.13). Also, if $π$ satisfies the Pappus (L) Theorem for one fixed line L, then $π$ satisfies the Pappus Theorem (see Theorem 8.2.8). In this text the useful specialization is the Pappus (L) Theorem. Theorem 6.2.1 shows that this specialized Pappus Theorem implies both of the specialized Desargues theorems.

THEOREM 6.2.1   *If π satisfies the Pappus (L) Theorem, then π satisfies the Second Desargues (L) Theorem.*

Proof.   Since $π(2)$ satisfies the Pappus Theorem, we may assume that every line has at least four points. Suppose that abc and $a'b'c'$ are centrally perspective from point p, that $p \notin L$, that $ab \cap a'b' = d \in L$, and that $ac \cap a'c' = e \in L$ (see Figure 68). We must show that $bc \cap b'c' = f \in L$.

Let $q = ab \cap pe$, $r = b'q \cap a'c'$, and $s = ar \cap pc$. It can be shown that none of these points is on L. If $q \in L$, then $q = d$, and so p,d,e are collinear. But this implies that $p \in L$, which contradicts the hypothesis

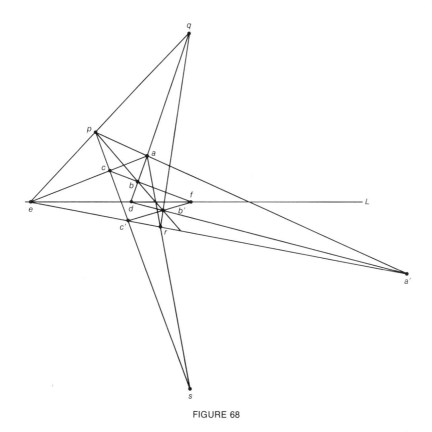

FIGURE 68

that $p \in L$. If $r \in L$, then $r = e$, and so b', e, q are collinear. Let M be the line containing these points. Now $p \in M$ because $p \in eq$, and so $b \in M$ because $b \in b'p$, and $a \in M$ because $a \in bq$. This is a contradiction because p, b, a are not collinear. Consider the triples q, r, b' and a', p, a. These are collinear points such that $qp \cap ra' = e \in L$ and $qa \cap b'a' = d \in L$. Therefore, by the Pappus (L) Theorem, $ra \cap pb' \in L$. If $s \in L$, it follows that $pc \cap pb' \in L$; thus $s = p$ and $p \in L$, a contradiction.

Consider the collinear triples p, c, s and a, q, b. Notice that $pq \cap ca = e \in L$. Also $sa \cap pb \in L$ because $ra = sa$, $pb = pb'$, and $ra \cap pb' \in L$ (by the preceding argument). Thus, by the Pappus (L) Theorem, $sq \cap cb \in L$. Similarly, consider the collinear triples p, c', s and r, q, b'. We know that $pq \cap c'r = e \in L$ and $pb' \cap sr = pb' \cap ra \in L$; therefore, $sq \cap b'c' \in L$, and we may conclude that $bc \cap b'c' \in L$.

COROLLARY 6.2.2   *If π satisfies the Pappus (L) Theorem, then π satisfies the First Desargues (L) Theorem.*

Proof.   The proof follows from Theorems 6.2.1 and 5.2.4.

THEOREM 6.2.3   *If π satisfies the Pappus Theorem, then π satisfies the Desargues Theorem.*

Proof.   If π satisfies the Pappus Theorem, then it satisfies the Pappus (L) Theorem for all L. By Theorem 6.2.1, it satisfies the Second Desargues (L) Theorem for all L, and thus by Theorem 5.2.4, it satisfies both the First and Second Desargues (L) theorems for all L. Therefore, π satisfies the Desargues Theorem.

The converse is not true. Chapter 7 will give examples of Desarguesian planes that are not Pappian.

Since Pappian planes are Desarguesian, a number of properties can be listed for Pappian planes that were proved for Desarguesian planes in Chapter 5.

THEOREM 6.2.4   *If π satisfies the Pappus Theorem, then*

    i. *π is (p, L) transitive for all p and L.*

    ii. *$PC(\pi) = CC(\pi)$.*

    iii. *$CC(\pi)$ is transitive on four-points.*

    iv. *π has a characteristic.*

    v. *$El(L)$ is Abelian.*

Proof.   The first property follows from Theorem 6.2.3 and Corollary 5.2.2(i); the second and third properties follow from Theorem 5.2.6; the fourth follows from Corollary 5.2.7; and the fifth follows from Theorem 5.2.3.

The following theorem states another property, which, however, is peculiarly Pappian.

THEOREM 6.2.5   *If π satisfies the Pappus (L) Theorem, then $Hom(p, L)$ is Abelian for all p such that $p \notin L$.*

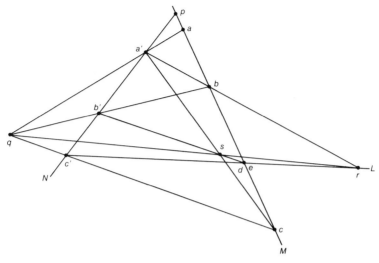

FIGURE 69

Proof. Let p,L,a be given where $a \neq p$ and $a \notin L$ and suppose that $f,g \in \text{Hom}(p,L)$ such that $f: a \to b$ and $g: a \to c$ (see Figure 69). Note that the homologies f and g are now completely defined. (We shall assume that a, b, and c are distinct; if this assumption is incorrect the proof is easily established.) Let M be the line containing p,a,b,c and let N be any other line through p. Let a' be an arbitrary point on N distinct from p and $N \cap L$ and let $b' = f(a')$ and $c' = g(a')$. It is easily seen that aa', bb', cc' all meet at a point $q \in L$.

Construct $g \circ f(a)$ and $f \circ g(a)$ as follows. Let $a'b \cap L = r$ and let $c'r \cap M = d$. Since $g: a' \to c'$, $g: a'r \to c'r$ and thus $g: b \to d$. But $b = f(a)$, so $d = g \circ f(a)$.

Let $a'c \cap L = s$ and let $b's \cap M = e$. Since $f: a' \to b'$, $f: a's \to b's$ and thus $f: c \to e$. But $c = g(a)$, so $e = f \circ g(a)$.

Consider the triples of collinear points b,c,d and c',b',a'. Notice that $bb' \cap cc' = q \in L$ and $ba' \cap dc' = r \in L$. Let $ca' \cap db' = t$; then $t \in L$ because π satisfies the Pappus (L) Theorem. But $ca' \cap L = s$, so $s = t$; thus b',d,s are collinear and therefore $b's \cap M = d$. It follows that $d = e$. Thus $g \circ f(a) = f \circ g(a)$. Since a was an arbitrary point, it follows that $\text{Hom}(p,L)$ is Abelian.

This may be seen to be an important theorem when placed in the context of Section 4.5. Referring to Theorems 4.5.7, 4.5.9, and 4.5.10, we see that for Pappian planes π, the structure $(S,\oplus,\odot)$ is a field. This assertion is stated formally in Theorem 6.2.6. If we can prove that this field may be used to represent π, we shall have shown the exciting result that the class of Pappian planes is exactly the class of planes π_F. The details of this proof will not be given until Chapter 8, but the student is encouraged to attempt a proof now.

THEOREM 6.2.6   *If π is a Pappian plane, then the algebraic system $(S,\oplus,\odot)$ of Definition 4.5.6 is a field.*

Proof.   The proof is based on Theorems 4.5.7, 4.5.9, 4.5.10, 6.2.4(v), and 6.2.5.

EXERCISES

1. Let π be a Pappian plane. Suppose that $f \in \text{Hom}(p,L)$ and that f: $q,r \rightarrow q',r'$ where p,q,r are collinear and $q,r \notin L$. Suppose also that $g \in \text{Hom}(p,L)$ and $g: q \rightarrow r$. Show that $g: q' \rightarrow r'$.

2. Show that if π satisfies the Pappus (M,N,L) Theorem, then π satisfies the Pappus (X,Y,Z) Theorem where (X,Y,Z) is any permutation of the lines M, N, and L.

3. Show that if π is a Desarguesian plane, then π satisfies Specialized Pappus Theorem One, which states that if a,b,c and a',b',c' are triples of collinear points on L and L', respectively, such that aa', bb', cc' are concurrent, then $ab' \cap a'b$, $ac' \cap a'c$, and $bc' \cap b'c$ are collinear.

4. Specialized Pappus Theorem Two states that if a,b,c and a',b',c' are triples of collinear points on L and L', respectively, such that $ab' \cap a'b$, $ac' \cap a'c$, and $L \cap L'$ are collinear (and lie on line M), then $bc' \cap b'c$ also lie on M.
   a. Show that if π satisfies Specialized Pappus Theorem Two, then π satisfies the Pappus (M,N,L) Theorem for any three concurrent lines M, N, and L.
   b. Show that if π is a Desarguesian plane, then π satisfies Specialized Pappus Theorem Two.

5. a. Prove the following partial converse of Theorem 6.2.5: If π satisfies the Second Desargues (L) Theorem and $\text{Hom}(p,L)$ is Abelian for all $p \notin L$, then π satisfies the Pappus (L) Theorem.
   b. Show that this theorem cannot be strengthened by addition of the following statement: If $\text{Hom}(p,L)$ is a nontrivial Abelian group for all $p \notin L$, then π satisfies the Pappus (L) Theorem.

6. a. Let π be a Pappian plane. Show that if $\{(a,a'),(b,b'),(c,c')\} \in \gamma$, then $\{(a',a),(b',b),(c',c)\} \in \gamma$.
   b. Recall the system $(L - \{u\}, \boxplus, \boxdot)$ of Definition 5.6.1. Using exercise 6(a), show that if π is a Pappian plane, then multiplication in this system is commutative.

6.3   AFFINE PAPPUS THEOREM

Like the Desargues Theorem, the Pappus Theorem has no obvious analogue in the affine plane; but, like the specialized Desargues theorems, the specialized Pappus Theorem does have an affine analogue.

AFFINE PAPPUS THEOREM   *If a,b,c and a',b',c' are triples of collinear points in α such that $ab' \parallel a'b$ and $ac' \parallel a'c$, then $bc' \parallel b'c$ (see Figure 70).*

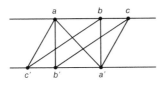

FIGURE 70

THEOREM 6.3.1   *The affine plane α satisfies the Affine Pappus Theorem if and only if $\pi(\alpha^+)$ satisfies the Pappus (L_∞) Theorem.*

Proof.   The proof follows directly from the statements of the theorems.

The following two theorems are direct consequences of previously stated theorems.

THEOREM 6.3.2   *If α satisfies the Affine Pappus Theorem, then α satisfies the First and Second Affine Desargues theorems.*

Proof.   The proof follows from Theorems 5.3.1 and 6.2.1.

THEOREM 6.3.3 *If α satisfies the Affine Pappus Theorem, then:*

  i. Trans(L) *is transitive for all L.*

  ii. Dil(p) *is transitive for all p.*

  iii. Trans(α) *is Abelian.*

  iv. Dil(p) *is Abelian for all p.*

Proof. The first three statements follow from Theorems 5.3.2, 5.3.3, and 6.3.2. The fourth statement follows from Theorem 6.2.5.

6.4 PROJECTIVITIES IN A PAPPIAN PLANE

As indicated in Section 5.4, the Desargues Theorem implies several facts concerning projectivities. Since all Pappian planes are Desarguesian, these facts are true for Pappian planes also. In this section the equivalence of the Pappus Theorem is proved with a proposition concerning projectivities. For the sake of convenience the following theorem will be referred to as "Theorem A."

THEOREM A *For any two distinct lines L and L', if f is a projectivity mapping $L \to L'$ such that f leaves $L \cap L'$ fixed, then f is a perspectivity.*

THEOREM 6.4.1 *If the projective plane π satisfies the Pappus Theorem, then π satisfies Theorem A.*

Proof. Suppose that π satisfies the Pappus Theorem and let f be a projectivity mapping $L \to L'$ such that f leaves $L \cap L'$ fixed. By Theorem 6.2.3, π satisfies the Desargues Theorem, and so by Theorem 5.4.5, $f = L \overset{p}{\barwedge} M \overset{q}{\barwedge} L'$. The rest of the proof consists of showing that this fact, along with the fact that $f: L \cap L' \to L \cap L'$, implies that f is a perspectivity.

Let $a = L \cap L'$, $m = L \cap M$, $n = M \cap L'$, $t = pa \cap M$, $g = L \overset{p}{\barwedge} M$, and $h = M \overset{q}{\barwedge} L'$ (see Figure 71). Thus $f = h \circ g$. Since $g: a \to t$, points p, a, t are collinear. $f: a \to a$ because $h: t \to a$, so p, t, q are collinear, and therefore p, a, q are collinear. Let $r = pn \cap qm$ and let $f' = L \overset{r}{\barwedge} L'$. The proof is completed by showing that $f = f'$. Let b be an arbitrary point on L. If $b = a$, then clearly $f(a) = f'(a) = a$. If $b = m$, then $f(m) = h(m)$

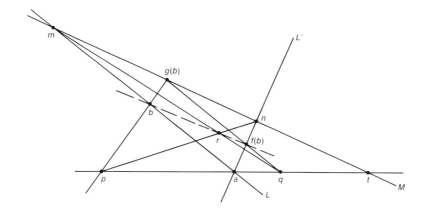

FIGURE 71

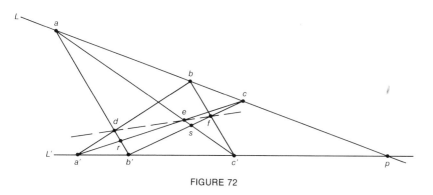

FIGURE 72

$= mq \cap L'$ and $f'(m) = rm \cap L'$. This is the same point, since r, m, q are collinear. Suppose that $b \neq a, m$ and consider points p, a, q and $m, g(b), n$ (on M). Applying the Pappus Theorem, we see that points $pg(b) \cap am = b$, $pn \cap qm = r$, and $an \cap qg(b) = h(g(b)) = f(b)$ are collinear. But since $b, f(b), r$ are collinear, $f(b) = f'(b)$.

THEOREM 6.4.2 *If π satisfies Theorem A, then π satisfies the Pappus Theorem.*

Proof. Let $a, b, c \in L$ and $a', b', c' \in L'$ satisfy the hypothesis of the Pappus Theorem. Let $ab' \cap a'b = d$, $ac' \cap a'c = e$, and $bc' \cap b'c = f$. We must show that d, e, f are collinear. Let $p = L \cap L'$, $r = ab' \cap a'c$, $s = b'c \cap ac'$, $g = ab' \overset{a'}{\underset{\wedge}{}} L$, and $h = L \overset{c'}{\underset{\wedge}{}} b'c$ (see Figure 72).

Thus g: $a,d,b',r \rightarrow a,b,p,c$; h: $a,b,p,c \rightarrow s,f,b',c$; and $h \circ g$: a,d,b',r $\rightarrow s,f,b',c$. By Theorem A, $h \circ g$ is a perspectivity. Since $h \circ g$: a,r $\rightarrow s,c$, the center of the perspectivity is $as \cap rc = ac' \cap a'c = e$. Since $h \circ g$: $d \rightarrow f$, we know that d,e,f are collinear.

6.5   THE PAPPUS THEOREM
AND THE FUNDAMENTAL THEOREMS

Section 4.4 introduced the two fundamental theorems of projective planes. This section establishes that the Fundamental Theorem of Projective Geometry (FT-I) is equivalent to the Pappus Theorem and states a partial result concerning the relationship of the Fundamental Theorem of Projective Collineations (FT-II) and the Pappus Theorem. The equivalence of these two will be established in Chapter 14. This section begins with two specialized theorems.

SPECIALIZED FUNDAMENTAL THEOREM   *There exist two ordered triples of three distinct collinear points a,b,c and a',b',c' such that exactly one projectivity f maps $a,b,c \rightarrow a',b',c'$.*

SPECIALIZED THEOREM A   *There exist two distinct lines L and L' such that if f is a projectivity mapping $L \rightarrow L'$ and f leaves $L \cap L'$ invariant, then f is a perspectivity.*

THEOREM 6.5.1   *If π satisfies Specialized Theorem A, then π satisfies the Specialized Fundamental Theorem.*

Proof.   Let L and L' be distinct lines such that if f: $L \cap L' \rightarrow L \cap L'$, then f is a perspectivity. Let $L \cap L' = a$, let b,c be distinct points on L other than a, and let b',c' be distinct points on L' other than a. Now there exists a projectivity f: $a,b,c \rightarrow a,b',c'$, which must be the perspectivity $L \overset{d}{\barwedge} L'$ where $d = bb' \cap cc'$. Thus f is unique and so it satisfies the Specialized Fundamental Theorem on L and L' for a,b,c.

THEOREM 6.5.2   *A projective plane π satisfies FT-I if and only if π satisfies the Specialized Fundamental Theorem.*

Proof.   Clearly, if π satisfies FT-I, it satisfies the Specialized Fundamental Theorem. Suppose that π satisfies the Specialized Fundamental

Theorem for a,b,c and a',b',c'. Let p,q,r and p',q',r' be an arbitrary pair of ordered triples of collinear points and let both f and g be projectivities mapping $p,q,r \to p',q',r'$. At least one such projectivity exists by Theorem 4.4.7. Also let h: $a,b,c \to p,q,r$ and let j: $p',q',r' \to a',b',c'$. Thus $j \circ f \circ h$: $a,b,c \to a',b',c'$ and $j \circ g \circ h$: $a,b,c \to a',b',c'$. Since π satisfies the Specialized Fundamental Theorem, $j \circ f \circ h = j \circ g \circ h$; therefore, $f = g$.

THEOREM 6.5.3 *If a projective plane π satisfies* FT-I, *then π satisfies Theorem* A.

Proof. Let L and L' be distinct lines and suppose that f is a projectivity mapping $L \to L'$ such that f leaves $L \cap L' = a$ fixed. Let $b,c \in L$, $b' = f(b)$, $c' = f(c)$, $d = bb' \cap cc'$, and $g = L \overset{d}{\barwedge} L'$. Since both f and g map $a,b,c \to a,b',c'$ and since π satisfies FT-I, it follows that $f = g$. Since g is a perspectivity, so is f.

THEOREM 6.5.4 *The following five theorems are equivalent:*

> *Pappus Theorem*
> *Theorem* A
> *Specialized Theorem* A
> *Fundamental Theorem of Projective Geometry*
> *Specialized Fundamental Theorem*

Proof. This proof follows from Theorems 6.4.1, 6.4.2, 6.5.1, 6.5.2, and 6.5.3, and from the obvious fact that Theorem A implies Specialized Theorem A.

The last two theorems in this section deal with the relationship between the Pappus Theorem and FT-II.

THEOREM 6.5.5 *If π satisfies the Pappus Theorem and if f is a projective collineation that leaves a four-point fixed, then f is the identity.*

Proof. Let p,q,r,s be a four-point fixed by f and let $pq \cap rs = t$. Since f is a collineation, we know that f leaves lines pq and rs invariant and point t invariant. Since f induces projectivities on lines, we know that $f|pq$ and $f|rs$ are projectivities that leave three points fixed. Since π is

Pappian, $f|pq$ and $f|rs$ are the identity projectivities. Thus f is a collineation that leaves two lines pointwise invariant and so, by Theorem 3.2.2(iii), f must be the identity.

THEOREM 6.5.6 *If π satisfies the Pappus Theorem, then π satisfies* FT-II.

Proof. Suppose that π satisfies the Pappus Theorem. Let p,q,r,s and p',q',r',s' be arbitrary four-points. Since $PC(\pi)$ is transitive on four-points, there exists $f \in PC(\pi)$ such that $f: p,q,r,s \rightarrow p',q',r',s'$. Also suppose that there exists $g \in PC(\pi)$ such that $g: p,q,r,s \rightarrow p',q',r',s'$. Then $g^{-1} \circ f$ leaves p,q,r,s fixed and, by Theorem 6.5.5, $g^{-1} \circ f$ is the identity. It follows that $g = f$.

The converse of Theorem 6.5.6 is proved in Chapter 14.

EXERCISES

1. Show that if f is a projective collineation fixing any four distinct points in a Pappian plane, then f is a central collineation.

2. The projectivity $f: L \rightarrow L$ is an *involutory projectivity* if and only if $f \circ f$ is the identity and f is not the identity.
   a. Refer back to Theorem 4.2.13. If $\pi = \pi_F$ and $f = f_M$ where

   $$M = \begin{pmatrix} a & b \\ c & d \end{pmatrix}$$

   then f is an involutory projectivity if and only if $a = -d$.

   For exercises 2(b) through (e), suppose that π is a Pappian plane and that f is an involutory projectivity on L.
   b. Show, by example, that f may hold fixed 2, 1, or 0 points.
   c. (1) Show that if f holds 2 points fixed, then there exists a harmonic homology h such that $h|L = f$.
      (2) Show that if f holds 1 point fixed, then there exists an involutory elation g such that $g|L = f$.
   d. Show that if f holds points p and q fixed, then $H(p,q; r,f(r))$.
   e. (1) Show that if π satisfies Fano's Axiom, then f may hold fixed 2 points or 0 points but not 1 point.
      (2) Show that if π is a Fano plane, then f may hold fixed 1 point or 0 points but not 2 points.

3. Refer back to the definitions of addition and multiplication on the line *ou* given in Definition 5.6.1. Let *f* be an involutory projectivity on *ou* and show that

   a. If f: $p \to q$ and f leaves u fixed, then f: $o \to p \boxplus q$.
   b. If f: $p,u \to q,o$, then f: $i \to p \boxdot q$.

4. In this exercise, consider only those projectivities that are products of two perspectivities and map L onto L (refer back to exercise 3.4.2). Show that if π satisfies the Second Desargues (L) Theorem, then a projectivity with a given central invariant point p and axial invariant point q is uniquely determined by the image of one point different from p and q. (p may equal q.)

CHAPTER 7

PLANES OVER DIVISION RINGS AND FIELDS

Chapter 2 introduced several different types of projective planes. The type that has been most receptive to examination is the plane π_F because of its algebraic definition. This chapter analyzes a general type of plane represented by equivalence classes of ordered triples of numbers. These numbers may come from a field or from a more general algebraic system called a division ring.

7.1 DIVISION RINGS AND FIELDS

DEFINITION 7.1.1 *A set S with two binary operations, $+$ and $\cdot$, is called a* division ring *or a* skew field *if and only if the following axioms hold:*

1. *If $a, b \in S$, then $a + b \in S$ and $a \cdot b \in S$.*

2. *If $a, b, c \in S$, then $a + (b + c) = (a + b) + c$ and $a \cdot (b \cdot c) = (a \cdot b) \cdot c$.*

3. *There exists* $o,e \in S$, $o \neq e$, *such that* $a + o = o + a = a$ *and* $e \cdot a = a \cdot e = a$ *for all* $a \in S$.

4. *For all* $a \in S$ *there exists* $b \in S$ *such that* $a + b = b + a = o$. *For all* $a \in S$, $a \neq o$, *there exists* $c \in S$ *such that* $a \cdot c = c \cdot a = e$.

5. *For all* $a,b \in S$, $a + b = b + a$.

6. *For all* $a,b,c \in S$, $a \cdot (b + c) = a \cdot b + a \cdot c$ *and* $(a + b) \cdot c = a \cdot c + b \cdot c$.

Notation. A general division ring $(S,+,\cdot)$ is denoted by D.

EXAMPLE 7.1.2 Let $S = \{a + bi + cj + dk: a,b,c,d$ are real numbers$\}$. Equality is defined as identity; that is, $a + bi + cj + dk = a' + b'i + c'j + d'k$ if and only if $a = a'$, $b = b'$, $c = c'$, and $d = d'$. Addition is defined in a coordinatewise fashion; that is, if $q = a + bi + cj + dk$ and $q' = a' + b'i + c'j + d'k$, then $q \oplus q' = a + a' + (b + b')i + (c + c')j + (d + d')k$. Multiplication is associative and distributive and is based on Table 17 for the indeterminates i, j, and k. Thus $q \circ q' = aa' - bb' - cc' - dd' + (ab' + ba' + cd' - dc')i + (ac' + ca' + db' - bd')j + (ad' + da' + bc' - cb')k$. This system $(S,\oplus,\circ)$ is a division ring, but proof of this fact is tedious, so it will be omitted.

TABLE 17

$\circ$	i	j	k
i	-1	k	$-j$
j	$-k$	-1	i
k	j	$-i$	-1

DEFINITION 7.1.3 *The division ring described in Example 7.1.2 is called the* ring of quaternions.

A representation of the ring of quaternions in which the axioms of a division ring (given in Definition 7.1.1) are easily verified will be included in the exercises at the end of this section.

Some new examples of fields are introduced here to generate more examples of division rings. The fields GF(4) and GF(9) displayed in Tables 1 and 2 (p. 39) may be exhibited in another way. Let $S = \{rx + s: r,s \in J/2\}$, that is, the set of all polynomials of degree 0 or 1 with coefficients in $J/2$. Define addition as ordinary polynomial addition with coefficients modulo 2 and define multiplication as ordinary polynomial multiplication modulo $x^2 + x + 1$ with coefficients modulo 2. Thus $(rx + s) \oplus (tx + u) = vx + w$ where $v = r + t$ mod 2 and where $w = s + u$ mod 2; $(rx + s) \circ (tx + u) = mx + n$ where $m = m'$ mod 2, where $n = n'$ mod 2, and where $m'x + n'$ is the remainder yielded by the division of $rtx^2 + (ru + st)x + su$ by $x^2 + x + 1$. It is easily checked that $(S, \oplus, \circ)$ as defined here is the field GF(4). Similarly, GF(9) may be represented by the set of all polynomials of degree 0 and 1 with coefficients from $J/3$, where addition and multiplication of polynomials are modulo $x^2 + x + 2$ and coefficients are modulo 3. In general, finite fields may be generated as described in Construction 7.1.4.

CONSTRUCTION 7.1.4 Let F be an arbitrary field and suppose that $P(x)$ is an irreducible polynomial of degree $n > 1$, that is, $P(x) \neq Q(x)R(x)$ for polynomials $Q(x)$ and $R(x)$ of degree less than n. (It should be noted that some fields — for example, the field of complex numbers — do not admit irreducible polynomials.) Let $S = \{Q(x): Q(x)$ is a polynomial of degree less than $n\}$. Define $\oplus$ and $\circ$ on S as follows: $Q(x) \oplus R(x) = S(x)$ where $S(x)$ is the result of ordinary polynomial addition with the understanding that addition of coefficients is carried out in the field F. Also, $Q(x) \circ R(x) = T(x)$ where $T(x)$ is the result of polynomial multiplication modulo $P(x)$, that is, $T(x)$ is the remainder resulting from the division of $Q(x)R(x)$ by $P(x)$. The coefficients of the product $Q(x)R(x)$ and of the quotient are calculated in F. The system $(S, \oplus, \circ)$ constructed here is a field; proof of this fact can be found in any advanced algebra textbook.

EXAMPLE 7.1.5

i. Let $F = R$, the field of real numbers, and let $P(x) = x^2 + 1$. It is easily checked that $(S, \oplus, \circ)$ is isomorphic to the field of complex numbers, the isomorphism being $\gamma: a + bx \to a + bi$.

ii. Let $F = J/2$ and $P(x) = x^3 + x + 1$. Then $(S, \oplus, \circ)$ is a Galois Field with eight elements.

iii. Let $F = $ GF(4) and $P(x) = x^2 + x + a$. Then $(S, \oplus, \circ)$ is a Galois Field with sixteen elements.

iv. Let $F = J/p$ and let $P(x)$ be any irreducible polynomial of degree n. Such polynomials always exist and, except for $p = n = 2$, they are not unique. Nevertheless, for a given p and n, all the fields $(S,\oplus,\circ)$ generated by the irreducible polynomials of degree n over J/p are isomorphic. Hence the following definition.

DEFINITION 7.1.6   *The field $(S,\oplus,\circ)$ of Example 7.1.5(iv) is called the* Galois Field *of order p^n.*

Notation.   The Galois Field of order p^n is denoted by $GF(p^n)$.

For any prime number p and any positive integer n, there exists one and only one field of order p^n, namely, $GF(p^n)$. Observe that the fields of Example 7.1.5(iii) and $GF(16)$ are isomorphic although each was generated in a slightly different manner.

Using the idea of polynomials, another type of field may be generated: the *quotient field* of a polynomial ring. The details of this construction are outlined in Construction 7.1.7; a more complete presentation can be found in any standard advanced algebra text.

CONSTRUCTION 7.1.7   Let F be an arbitrary field and consider the set T of all polynomials of n indeterminates $x_1, \ldots, x_n$. Thus a typical element of T is the sum of terms of the form $cx_1^{k_1}x_2^{k_2} \ldots x_n^{k_n}$ where $c \in F$ and $k_i \geq 0$; if $k_i = 0$, we assume that $x_i^{k_i} = 1$. Let $R = (T,+,\cdot)$ be the system of polynomials under usual addition and multiplication. This system, called a *polynomial ring*, lacks multiplicative inverses. To make amends for this deficiency, R is embedded in a field as follows. Let $T' = \{(a,b): a,b \in T\}$ and let Q be the set of all equivalence classes of T' under the relation $\sim$ where $(a,b) \sim (c,d)$ if and only if $ad = bc$. Letting $[a,b]$ denote the equivalence class containing (a,b), we may define $\oplus$ and $\odot$ on Q by $[a,b] \oplus [c,d] = [ad + cb, bd]$ and $[a,b] \odot [c,d] = [ac,bd]$. The student will find it a valuable exercise to show that the resulting system $(Q,\oplus,\odot)$ is a field and that $(Q',\oplus,\odot)$, where $Q' = \{[a,1]: a \in T\}$, is a subsystem isomorphic to R.

Although the preceding construction may seem formidable to the uninitiated, what has been done is to supplement the system R with fractions a/b. In the notation a/b was denoted by $[a,b]$. The equivalence relation was introduced because of the redundancy of notation for fractions, that is, $a/b = c/d$ if and only if $ad = bc$. This fact could have simply been noted, as is done in the ordinary rational number system, without mentioning equivalence relations; however, it is customary in mathematics to identify formally all equivalent objects and thus avoid possible ambiguity.

Notation. The field of Construction 7.1.7 is denoted by $F(x_1, \ldots, x_n)$.

EXAMPLE 7.1.8

    i. Let $F = R$, the field of real numbers. The field $F(x)$ is called the field of rational functions and its elements are of the form $P(x)/Q(x)$ where $P(x)$ and $Q(x)$ are polynomials of arbitrary degree with coefficients in R. Equality in $F(x)$ is defined as follows: $P(x)/Q(x) = R(x)/S(x)$ if and only if $P(x)S(x) = Q(x)R(x)$.

    ii. Let $F = J/p$. Then $F(x)$ is the infinite field whose elements are of the form $P(x)/Q(x)$ where $P(x)$ and $Q(x)$ are polynomials of arbitrary degree with coefficients in F. Equality is defined as follows: $P(x)/Q(x) = R(x)/S(x)$ if and only if $P(x)S(x) = Q(x)R(x)$.

A class of division rings, all of which are called *quaternion algebras*, is defined in Construction 7.1.9.

CONSTRUCTION 7.1.9 Let F be an arbitrary field and consider the field $F(x,y)$. Let $S = \{a + bi + cj + dk: a,b,c,d \in F(x,y)\}$. Define equality as identity; define addition in a coordinatewise fashion: I: $q = a + bi + cj + dk$ and $q' = a' + b'i + c'j + d'k$, then $q \oplus q' = a + a' + (b + b')i + (c + c')j + (d + d')k$. Multiplication is associative and distributive and is based on Table 18 for i,j,k. Thus $q \circ q' = (a + b + cj + dk)(a' + b'i + c'j + d'k) = aa' + ab'i + ac'j + ad'k + ba'i + bb'i + bc'ij + bd'ik + ca'j + cb'ji + cc'j^2 + cd'jk + da'k + db'ki + dc'kj + dd'k^2 = (aa' + bb'x + cc'y + cd'y - dd'xy) + (ab' + ba' + bb' - cd'y - db'x + dc'y)i + (ac' + bd'x + ca' + cb')j + (ad' + bc' + bd' - cb' + da')k$.

TABLE 18

$\circ$	i	j	k
i	$i + x$	k	$k + xj$
j	$j - k$	y	$y - yi$
k	$-xi$	yi	$-xy$

As with the quaternions (see Example 7.1.2), the proof that $(S,\oplus,\circ)$ of Construction 7.1.9 is a division ring is tedious and is therefore omitted here.

Notation. The division ring $(S,\oplus,\circ)$ is denoted by Q_F.

The definition of characteristic of a division ring is the same as that of the characteristic of a field.

DEFINITION 7.1.10 *A division ring D has* characteristic n *if and only if n is the least number of multiplicative identities that must be added together to obtain 0 (the additive identity). If there is no such (finite) number, then the characteristic is 0.*

Note that the characteristic of a division ring is either a prime number p or 0. It is easily seen that both $F(x_1, \ldots , x_n)$ and Q_F have the same characteristic as F. Thus there are division rings of any permissible characteristic.

In projective geometry, the most important theorem concerning the relationship of division rings and fields is the celebrated theorem of Wedderburn (1905):

THEOREM 7.1.11 *A finite division ring is a field.*

Proof. A proof of this theorem can be found in Hall (1959), p. 375.

EXERCISES

1. Find an irreducible polynomial of degree n in the field F where:
   a. $n = 3, F = J/3$.
   b. $n = 2, F = J/7$.
   c. $n = 4, F = J/2$.
   d. $n = 4, F = J/3$.
   e. $n = 2, F = \text{GF}(9)$.

2. a. Find two irreducible polynomials of degree 2 in $J/3$ other than $x^2 + x + 2$.
   b. Show that the Galois Fields obtained from these polynomials are isomorphic to the GF(9) displayed in Table 2 (p. 39).

3. Write out the addition and multiplication tables for GF(8) using the irreducible polynomial $x^3 + x + 1$ over $J/2$.

4. Show that the configuration $\Sigma(8_3)$ occurs in all planes π_F where $F = GF(p^{2n})$.

5. Consider the set S of 2×2 matrices over the complex numbers that have the form
$$\begin{pmatrix} a + bi & c + di \\ -c + di & a + bi \end{pmatrix}$$
where a, b, c, d are arbitrary real numbers. Define addition and multiplication on S by ordinary matrix addition and multiplication.

   a. Show that $(S, +, \cdot)$ is a division ring but not a field. Hint: Show that:
      i. Multiplication is closed.
      ii. Multiplicative inverses exist for all nonzero elements.
      iii. Multiplication is not commutative.
   All other division ring axioms are easily verified.
   b. Exhibit an isomorphism γ from $(S, +, \cdot)$ to the ring of quaternions.

6. Show that for any finite field, there exists an irreducible polynomial of degree n where:
   a. $n = 2$ (refer back to exercise 3.4.6(d)).
   b. $n = 3$.
   c. n is an arbitrary positive integer.

7.2   PLANES DEFINED BY DIVISION RINGS

A plane over a division ring is defined in much the same way that a plane over a field was defined in Section 2.1.

CONSTRUCTION 7.2.1   Let D be an arbitrary division ring. The plane $\pi_D = (\mathcal{P}, \mathcal{L}, \mathcal{I})$ is formed as follows. Let $\mathcal{P} = \{[x, y, z]: x, y, z \in D$ and $x = y = z = 0$ is not true$\}$, where $[x, y, z] = [x', y', z']$ if and only if there exists $r \in D$, $r \neq 0$, such that $x = x'r$, $y = y'r$, $z = z'r$. Let $\mathcal{L} = \{\langle a, b, c \rangle:$ $a, b, c \in D$ and $a = b = c = 0$ is not true$\}$, where $\langle a, b, c \rangle = \langle a', b', c' \rangle$ if and only if there exists $s \in D$, $s \neq 0$, such that $a = sa'$, $b = sb'$, $c = sc'$. Let $\mathcal{I}$ be defined as follows: $([x, y, z], \langle a, b, c \rangle) \in \mathcal{I}$ if and only if $ax + by + cz = 0$.

If D is a field, this construction is exactly the same as Construction 2.1.6 since multiplication is commutative in a field. The notation for points and lines in π_D is given in Construction 7.2.1; $\in$ denotes incidence.

THEOREM 7.2.2   *The plane π_D is a projective plane.*

Proof.   Let $p = [x,y,z]$ and $p' = [x',y',z']$ be distinct points. Not all of $x,y,z = 0$, so we may assume without loss of generality that $x \neq 0$. Furthermore, we may assume that $x = 1$. We know that $yx' - y' = 0$ and that $zx' - z' = 0$ is impossible because if these statements were true, then $1 \cdot x' = x'$, $yx' = y'$, $zx' = z'$, and thus $p = p'$. So we may assume without loss of generality that $yx' - y' \neq 0$.

First, to show that there exists a line containing p and p', let $L = \langle a,b,c \rangle$ where $a = -(by + z)$, $b = (zx' - z')(y' - yx')^{-1}$, $c = 1$. Now $ax + by + cz = -(by + z)\cdot 1 + by + 1\cdot z = 0$.

Also $ax' + by' + cz' = -(by + z)x' + by' + z' = b(y' - yx') - zx' + z' = (zx' - z')(y' - yx')^{-1}(y' - yx') - zx' + z' = 0$. Thus L contains p and p'.

Next, to show uniqueness, suppose that $L' = \langle a',b',c' \rangle$ also contains p and p'. If $c' = 0$, then $a' + b'y = 0$ and $a'x' + b'y' = 0$. Thus $-b'yx' + b'y' = 0$, and therefore $b'(y' - yx') = 0$. Since $y' - yx' \neq 0$, we have $b' = 0$. Now $a' + b'y = 0$, so $a' = 0$, and thus $a' = b' = c' = 0$. This contradiction establishes that $c' \neq 0$; we may assume that $c' = 1$. We now have the following equalities:

1. $a + by + z = 0$

   because $p,p' \in L$

2. $ax' + by' + z' = 0$

3. $a' + b'y + z = 0$

   because $p,p' \in L'$

4. $a'x' + b'y' + z' = 0$

Equalities (1) and (3) imply that $a + by = a' + b'y$; (2) and (4) imply that $ax' + by' = a'x' + b'y'$. So we have:

5. $a - a' = (b' - b)y$

6. $(a - a')x = (b' - b)y'$

Now suppose that $b \neq b'$. Then (5) and (6) imply that $(b' - b)yx = (b' - b)y'$. However, this implication contradicts the fact that $yx' - y' \neq 0$. Thus $b = b'$. Equality (5) implies that $a = a'$. Thus we conclude that $L = L'$. This establishes Pj1.

Pj2 may be shown in a similar manner. Finally, the points $[0,0,1]$, $[0,1,0]$, $[1,0,0]$, and $[1,1,1]$ form a four-point, thus establishing Pj3.

An affine plane can be formed in the usual fashion by deleting one line from π_D. If the line $\langle 0,0,1 \rangle$ is deleted, the resulting affine plane, denoted by α_D, may be described as follows. Points are pairs (x,y): $x,y \in D$, and lines are sets of points satisfying linear equations $ax + by + c = 0$ for some $a,b,c \in D$ such that $a = b = c = 0$ is not true.

EXERCISES

1. Let D be a division ring and define $\pi'_D = (\mathscr{P}', \mathscr{L}', \mathscr{I}')$ as follows:

   $\mathscr{P}' = \{[[x,y,z]]: x,y,z \in D, x = y = z = 0$ is not true, and $[[x,y,z]] = [[x',y',z']]$ if and only if $x' = rx$, $y' = ry$, $z' = rz$ for some $r \in D$, $r \neq 0\}$

   $\mathscr{L}' = \{\langle\langle a,b,c \rangle\rangle: a,b,c \in D, a = b = c = 0$ is not true, and $\langle\langle a,b,c \rangle\rangle = \langle\langle a',b',c' \rangle\rangle$ if and only if $a' = as$, $b' = bs$, $c' = cs$ for some $s \in D$, $s \neq 0\}$

   $\mathscr{I} = \{([[x,y,z]], \langle\langle a,b,c \rangle\rangle): xa + yb + zc = 0\}$

   Show that π'_D is isomorphic to $(\pi_D)^d$.

2. Let D be a division ring.

   a. Define $\pi_D{}^{(1)} = (\mathscr{P}^{(1)}, \mathscr{L}^{(1)}, \mathscr{I}^{(1)})$ as follows:

   $\mathscr{P}^{(1)} = \{[x,y,1]: x,y \in D\} \cup \{[1,x,0]: x \in D\} \cup \{[0,1,0]\}$

   $\mathscr{L}^{(1)} = \{\langle m,1,k \rangle: m,k \in D\} \cup \{\langle 1,0,k \rangle: k \in D\} \cup \{\langle 0,0,1 \rangle\}$

   $\mathscr{I}^{(1)} = \{([x,y,z], \langle m,n,k \rangle): mx + ny + kz = 0\}$

   Show that $\pi_D{}^{(1)}$ is isomorphic to π_D.

   b. Define $\pi_D{}^{(2)} = (\mathscr{P}^{(2)}, \mathscr{L}^{(2)}, \mathscr{I}^{(2)})$ as follows:

   $\mathscr{P}^{(2)} = \{[[x,y,1]]: x,y \in D\} \cup \{[[1,x,0]]: x \in D\} \cup \{[[0,1,0]]\}$

   $\mathscr{L}^{(2)} = \{\langle\langle m,1,k \rangle\rangle: m,k \in D\} \cup \{\langle\langle 1,0,k \rangle\rangle: k \in D\} \cup \{\langle\langle 0,0,1 \rangle\rangle\}$

   $\mathscr{I}^{(2)} = \{([[x,y,z]], \langle\langle m,n,k \rangle\rangle): xm + yn + zk = 0\}$

   Show that $\pi_D{}^{(2)}$ is isomorphic to $(\pi_D)^d$.

   c. Let $f: [x,y,z] \to [[x,y,z]]$ and $F: \langle m,n,k \rangle \to \langle\langle m,n,k \rangle\rangle$. Show that (f,F) is an isomorphism from $\pi_D{}^{(1)}$ to $\pi_D{}^{(2)}$ if and only if D is a field.

3. Let $D = (R,+,\cdot)$ be a division ring. The system $(R,+,\circ)$ where $a \circ b = b \cdot a$ is called the *dual division ring* of D. This system is denoted by D^d.

   a. Verify that D^d is, itself, a division ring.

   b. Show that π_{D^d} is isomorphic to $(\pi_D)^d$.

7.3   PROPERTIES OF π_D

It has already been established in Example 6.1.4(ii) that planes of type π_F satisfy the Pappus Theorem and therefore the Desargues Theorem.

This study is now continued with an examination of the relationship between π_D and the Pappus and Desargues theorems.

THEOREM 7.3.1   Let $p = [x,y,z]$, $p' = [x',y',z']$, and $p'' = [x'',y'',z'']$ be distinct points in π_D. Then $p'' \in pp'$ if and only if $x'' = xr + x's$, $y'' = yr + y's$, and $z'' = zr + z's$ for some $r,s \in D$ where $r,s \neq 0$.

Proof.   Let $pp' = \langle a,b,c \rangle$. We know that $ax + by + cz = 0$ and that $ax' + by' + cz' = 0$. Thus, for $r,s \neq 0$, $a(xr + x's) + b(yr + y's) + c(zr + z's) = axr + byr + czr + ax's + by's + cz's = (ax + by + cz)r + (ax' + by' + cz')s = 0$. Therefore, if $x'' = xr + x's$, $y'' = yr + y's$, and $z'' = zr + z's$, then $p'' \in pp'$.

Suppose that $p'' \in pp'$. Since $x = y = z = 0$ is impossible, we may, without loss of generality, assume that $x \neq 0$. Furthermore, we may assume that $x = 1$. Using the same argument as in Theorem 7.2.2, we may assume that $yx' - y' \neq 0$, and thus we may solve the equations $xr + x's = x''$ and $yr + y's = y''$ for r and s. It is easily checked that $s = (y' - yx')(y'' - yx'')$ and that $r = x'' - x's$. If $c = 0$, then $a + by = 0$ and $ax' + by' = 0$. Thus $-byx' + by' = b(y' - yx') = 0$. Now $y' - yx' \neq 0$, so $b = 0$. Therefore, $a = 0$, and we reach a contradiction. Thus $c \neq 0$, and we shall assume that $c = 1$. Since p,p',p'' are collinear, it follows that $a + by + z = 0$, $ax' + by' + z' = 0$, and $ax'' + by'' + z'' = 0$. Thus $z'' - zr - z's = -ax'' - by'' - (-a - by)(x'' - x's) - (-ax' - by')s = -ax'' - by'' + ax'' - ax's + byx'' - byx's + ax's + by's = b(y(x'' - x's) + (-y'' + y's)) = b(-y'' + y's + yr) = b \cdot 0 = 0$. Thus $z'' = zr + z's$. Clearly, if $r = 0$ or $s = 0$, then p'' would not be distinct from p and p'. Thus $x'' = xr + x's$, $y'' = yr + y's$, $z'' = zr + z's$, and $r,s \neq 0$.

COROLLARY 7.3.2   If p,q,r are collinear, there exist coordinates p_i, q_i, $i = 1,2,3$, of p,q, respectively, such that $r_i = p_i + q_i$ for given coordinates r_i of r.

THEOREM 7.3.3   If D is a division ring, then π_D is a Desarguesian plane.

Proof.   Let p,q,r and p',q',r' be vertices of two triangles centrally perspective from a seventh point s. Let $p = [p_1,p_2,p_3]$, $q = [q_1,q_2,q_3]$, $r = [r_1,r_2,r_3]$, $s = [s_1,s_2,s_3]$, $p' = [p'_1,p'_2,p'_3]$, $q' = [q'_1,q'_2,q'_3]$, $r' =$

$[r'_1, r'_2, r'_3]$. Since s, p, p' are collinear by Corollary 7.3.2, we may assume that the coordinates of p and p' are such that $p'_i = p_i + s_i$, $i = 1,2,3$. Similarly, we may assume that the coordinates of q and q' are such that $q'_i = q_i + s_i$ and that the coordinates of r and r' are such that $r'_i = r_i + s_i$, $i = 1,2,3$.

Let $a = pq \cap p'q'$, $b = pr \cap p'r'$, $c = qr \cap q'r'$. It is then easily checked that $a = [p_1 - q_1, p_2 - q_2, p_3 - q_3]$, $b = [p_1 - r_1, p_2 - r_2, p_3 - r_3]$, and $c = [q_i - r_1, q_2 - r_2, q_3 - r_3]$. Therefore, a, b, c are collinear by Theorem 7.3.1.

It is not true that all planes of the form π_D satisfy the Pappus Theorem. In fact, π_D satisfies that theorem if and only if D is a field. As mentioned previously, if D is a field, π_D satisfies the Pappus Theorem by Example 6.1.4(ii). The converse is stated in Theorem 7.3.4.

THEOREM 7.3.4   *If π_D is a Pappian plane, then D is a field.*

Proof.   Let D be a division ring and let $x, y \in D$ where $x, y \neq 0, 1$. Let $p = [1, -1, 0]$, $q = [q_1, q_2, q_3] = [x, -1, 0]$, $r = [r_1, r_2, r_3] = [y, -1, 0]$; also, let $p' = [0, 1, 1]$, $q' = [q'_1, q'_2, q'_3] = [0, x, 1]$, $r' = [r'_1, r'_2, r'_3] = [0, y, 1]$. Clearly, p, q, r lie on $\langle 0, 0, 1 \rangle$ and p', q', r' lie on $\langle 1, 0, 0 \rangle$. Let $a = pq' \cap p'q$, $b = pr' \cap p'r$, $c = qr' \cap q'r$. It is then easily checked that $a = [x, 0, 1]$ and $b = [y, 0, 1]$. Now let $c = [c_1, c_2, c_3]$. Since $c \notin \langle 1, 0, 0 \rangle$, we have $c_1 \neq 0$, so we may choose $c_1 = x$. Since $c \in q'r$, it follows from Theorem 7.3.1 that there exists $s, t \in D$ where $s, t \neq 0$ such that $c_i = r_i s + q'_i t$. Thus $s = y^{-1}x$ and $c = [x, -y^{-1}x + xt, t]$. Similarly, since $c \in qr'$, there exists $u, v \in D$ where $u, v \neq 0$ such that $c_i = q_i u + r'_i v$. Thus $u = 1$ and $c = [x, yv - 1, v]$. It follows that $t = v$, and so $xt - y^{-1}x = yt - 1$.

Since the Pappus Theorem holds, a, b, c are collinear. Since $ab = \langle 0, 1, 0 \rangle$, we have $c_2 = 0$. Thus $yt = 1$ and so $t = y^{-1}$. Also $xt = y^{-1}x$, so $y^{-1}x = xy^{-1}$. Thus $y^{-1}xy = x$ or $xy = yx$. This proves that multiplication is commutative and that D is a field.

THEOREM 7.3.5   *The plane π_D is Pappian if and only if D is a field.*

Proof.   The proof follows immediately from Theorem 7.3.4 and Example 6.1.4(ii).

Theorem 7.3.5 may be used to cite Desarguesian planes that are not Pappian—for example, π_D where D is the ring of quaternions. However, Theorems 7.1.1 and 7.3.5 imply that a finite plane π_D must necessarily be Pappian.

Since π_D is Desarguesian, every affine restriction of π_D is isomorphic. Thus, without ambiguity, any restriction of π_D may be denoted by α_D. Also α_D satisfies both Affine Desargues theorems and α_D satisfies the Affine Pappus Theorem if and only if D is a field.

There is also a convenient relationship between the division ring D of the plane π_D and Fano's Axiom and Fano planes, as stated in Theorems 7.3.6 and 7.3.7.

THEOREM 7.3.6 π_D is a Fano plane if and only if D has characteristic 2.

Proof. Let π_D be a Fano plane and consider the quadrangle with vertices $p = [1,0,0]$, $q = [0,1,0]$, $r = [0,0,1]$, $s = [1,1,1]$. Clearly, $pq \cap rs = t = [1,1,0]$, $pr \cap qs = u = [1,0,1]$, $ps \cap qr = v = [0,1,1]$. Let t,u,v be on the line $\langle a,b,c \rangle$; thus $a + b = 0$, $a + c = 0$, $b + c = 0$. Since at least one of $a,b,c \neq 0$, we may assume without loss of generality that $a \neq 0$. Furthermore, we may assume that $a = 1$. Thus $b = -1$, $c = 1$, and $1 + 1 = 0$. Therefore, D is of characteristic 2. Proof of the converse is left as an exercise.

THEOREM 7.3.7 *The plane π_D satisfies Fano's Axiom if and only if D does not have characteristic 2.*

Proof. The proof is left as an exercise.

The last theorem in this section, which was established by Singer (1938), states that every finite plane of type π_D is a cyclic plane. Although the proof of this theorem (see Hall, 1959, p. 401) makes use of a few elementary facts about Galois Fields that will not be proved here, the ingenuity of the argument can still be appreciated. Notice first that for a finite plane π_D, D must be a finite division ring and thus a field. Furthermore, D must be a Galois Field $GF(p^n)$ for some prime p and positive integer n. The essential fact concerning Galois Fields that will be used here is that the multiplicative group of the nonzero elements of $GF(p^n)$ is cyclic.

THEOREM 7.3.8   *A finite plane of type π_F is a cyclic plane.*

Proof.   Let $F = \mathrm{GF}(p^n)$ and let $F' = \mathrm{GF}(p^{3n})$. Elements of F' may be considered to be of the form $a + bx + cx^2$ where $a,b,c \in F$, where addition is ordinary polynomial addition with coefficient addition in F, and where polynomial multiplication is modulo an irreducible cubic in F as explained in Construction 7.1.4. Also we may assume that the element x is a generator of G', the multiplicative group of nonzero elements of $\mathrm{GF}(p^{3n})$. Thus we may represent uniquely every element of F' by the triple (a,b,c) where $(a,b,c) = a + bx + cx^2$, $a,b,c \in F$. The multiplicative group G of nonzero elements of $\mathrm{GF}(p^n)$ may be generated by x^N where $N = k^2 + k + 1$ and $k = p^n$. This may be seen by noting that G is a cyclic subgroup of G', that G is of order $k - 1$, and that G' is of order $k^3 - 1$. An equivalence relation on triples (a,b,c) may be established as follows: $(a,b,c) \sim (a',b',c')$ iff $r = s \bmod N$ where $(a,b,c) = x^r$ and $(a',b',c') = x^s$. This equivalence relation is the same relation that was used to define points in π_F. (Proof of this fact is left as an exercise.) Therefore, the notation $[a,b,c]$ or $[x^r]$ will be used to denote the class containing (a,b,c) where $(a,b,c) = x^r$.

Let f be a mapping on the points of π_F defined as follows: $f: [a,b,c] \to [a',b',c']$ where $(a',b',c') = (a,b,c)x$. It is easily checked that f is well defined. To show that f is a collineation, recall that lines in π_F may be considered to be sets of points $L_{p,p'}$ where p and p' are the points $[a,b,c]$ and $[a',b',c']$, respectively, and where $L_{p,p'} = \{[ra + sa', rb + sb', rc + sc']: r,s \in F, r,s \neq 0\} = \{[rx^i + sx^j]: x^i = (a,b,c), x^j = (a',b',c')\}$. Clearly, $f(L_{p,p'}) = \{[rx^{i+1} + sx^{j+1}]\} = L_{f(p),f(p')}$ since $f(p) = [x^{i+1}]$ and $f(p') = [x^{j+1}]$. Therefore, f preserves collinearity and is thus a collineation. Furthermore, f is a generator for a cyclic transitive subgroup of $C(\pi_p)$. This is easily seen by noting that $f^i: [x^0] \to [x^i]$ and that every point of π_F is of the form $[x^i]$, $i = 0, \ldots, N - 1$. By Theorem 4.6.2, it follows that π_F is a cyclic plane.

EXERCISES

1. Complete the proof of Theorem 7.3.6.

2. Prove Theorem 7.3.7.

3. Construct a Fano plane that is not Pappian.

4. Prove that the equivalence relation defined on triples of elements of F in the proof of Theorem 7.3.8 is the same equivalence relation as S where $((x,y,z), (x',y',z')) \in S$ if and only if there exists $s \in F, s \neq 0$, such that $x' = sx, y' = sy$, and $z' = sz$.

5. Suppose that $F = \mathrm{GF}(p^n)$ and that $p^{2n} + p^n + 1$ is a prime number.

   a. Show that if f is a collineation that leaves no points fixed in π_F, then f generates a group of collineations that is transitive on the points of π_F.

   b. Find a collineation f in π that generates a transitive collineation group on π where $\pi = \pi(2)$; where $\pi = \pi(3)$; where $\pi = \pi(5)$.

*6. Use the ideas of the proofs of Theorems 4.6.2 and 7.3.8 to generate your own perfect difference sets. You will want to use a computer for orders n, such that $n > 9$.

7.4   CROSS RATIO

Corollary 7.3.2 tells us that if a,b,c are distinct collinear points, there exist coordinates a_i, b_i, c_i, $i = 1,2,3$, for a,b,c, respectively, such that $a_i + b_i = c_i$. A fourth collinear point, $d = [d_1, d_2, d_3]$, distinct from a and b, may be located on the line ab with the help of an element $r \in D, r \neq 0$, as follows: $d_i = a_i r + b_i$. The point d is said to be in the ratio r to c with respect to a and b. This ratio is a useful concept in projective planes π_D, and an affine analogue can be used to represent a generalization of the ratio of two line-segment lengths.

DEFINITION 7.4.1   *Let a,b,c be three distinct collinear points such that the respective coordinates a_i, b_i, c_i, $i = 1,2,3$, are in the relation $c_i = a_i + b_i$. Then a fourth collinear point $d = [d_1, d_2, d_3]$ is said to be in ratio $[r]$ to c with respect to a and b if and only if $r = st^{-1}$ where $d_i = a_i s + b_i t$, $i = 1,2,3$, for some $s,t \neq 0$. We know that such an s,t exist. Here $[r]$ denotes $\{k^{-1}rk : k \in D\}$. Also $[r]$ is called the* cross ratio *of d to c with respect to a and b.*

Notation.   This cross ratio $[r]$ is denoted by $R(a,b; c,d)$.

First note that $[r]$ is an equivalence class that is generated by the equivalence relation T on D where $(r,s) \in T$ if and only if $rp = ps$ for some $p \in D$. If D is a field, the cross ratio $[r]$ is the singleton set $\{r\}$. Then the set notation is suppressed and the cross ratio is said to be r.

Second, note that the cross ratio is well defined. Suppose that a'_i, b'_i, c'_i, d'_i, $i = 1,2,3$, are also coordinates of a, b, c, d, respectively, which satisfy the condition that $c'_i = a'_i + b'_i$. Since $c'_i = a'_i + b'_i$, there exists $k \in D$ such that $a_i k = a'_i$, $b_i k = b'_i$, $c_i k = c'_i$. Letting $d'_i = d_i n = a_i sn + b_i tn = a'_i k^{-1} sn + b'_i k^{-1} tn$, we conclude that the ratio is $[k^{-1} sn \ n^{-1} t^{-1} k] = [k^{-1} rk] = [r]$.

Third, note that in a field, if $R(a,b; c,d) = R(a,b; c,e)$, then $d = e$; that is, given three points and a cross ratio, the fourth point is unique. This is not true for division rings. If $R(a,b; c,d) = R(a,b; c,e)$, then we can only say that if $a_i + b_i = c_i$, $d_i = a_i r + b_i$, and $e_i = a_i s + b_i$, then $r = k^{-1} sk$ for some k.

The cross ratio of four concurrent lines may be defined in an analogous fashion.

DEFINITION 7.4.2 *Let L, M, N be three concurrent lines such that the respective coordinates l_i, m_i, n_i, $i = 1,2,3$, are in relation $n_i = l_i + m_i$. Then a fourth concurrent line $O = \langle o_1, o_2, o_3 \rangle$ is said to be in ratio $[s]$ to N with respect to L and M if and only if $s = u^{-1} t$ where $o_i = t l_i + u m_i$, $i = 1,2,3$, for some $t, u \neq 0$. Also $[s]$ is called the cross ratio of O to N with respect to L and M.*

THEOREM 7.4.3 *If a, b, c, d are distinct collinear points and L, M, N, O are distinct concurrent lines such that $a \in L$, $b \in M$, $c \in N$, $d \in O$, then $R(a,b; c,d) = R(L,M; N,O)$.*

Proof. Let $R(a,b; c,d) = [r]$ and $R(L,M; N,O) = [s]$. Then

$$0 = \sum_{i=1}^{3} n_i c_i = \sum_{i=1}^{3} (l_i + m_i)(a_i + b_i)$$

$$= \sum_{i=1}^{3} l_i a_i + \sum_{i=1}^{3} m_i a_i + \sum_{i=1}^{3} l_i b_i + \sum_{i=1}^{3} m_i b_i$$

$$= 0 + \sum_{i=1}^{3} m_i a_i + \sum_{i=1}^{3} l_i b_i + 0$$

Let

$$t = \sum_{i=1}^{3} m_i a_i$$

Then

$$\sum_{i=1}^{3} l_i b_i = -t$$

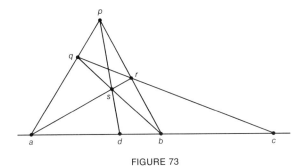

FIGURE 73

Notice that $t \neq 0$ since $a \notin M$. Also,

$$0 = \sum_{i=1}^{3} o_i d_i = \sum_{i=1}^{3} (sl_i + m_i)(a_i r + b_i)$$

$$= \sum_{i=1}^{3} sl_i a_i r + \sum_{i=1}^{3} m_i a_i r + \sum_{i=1}^{3} sl_i b_i + \sum_{i=1}^{3} m_i b_i = 0 + tr - st + 0$$

Thus $tr = st$, and therefore $[r] = [s]$.

THEOREM 7.4.4 *If $R(a,b; c,d) = [r]$ and if f is a projectivity such that $f: a,b,c,d \to a',b',c',d'$, then $R(a',b'; c',d') = [r]$.*

Proof. It is sufficient to show that cross ratio is preserved under perspectivity, so we may assume that f is a perspectivity. Let $L = aa'$, $M = bb'$, $N = cc'$, $O = dd'$. Since f is a perspectivity, L,M,N,O are concurrent. By Theorem 7.4.3, $R(a,b; c,d) = R(L,M; N,O) = R(a',b'; c',d') = [r]$.

THEOREM 7.4.5 *Let a,b,c,d be four collinear points with coordinates a_i, b_i, c_i, d_i, $i = 1,2,3$, such that $c_i = a_i + b_i$ and $H(a,b; c,d)$. Then $R(a,b; c,d) = [-1]$.*

Proof. Since $[1,0,0]$, $[0,1,0]$, $[0,0,1]$ are noncollinear, at least one of these points is not collinear with a,b,c,d. We may assume without loss of generality that $p = [1,0,0]$ is one such point. Let $q = [a_1 + 1, a_2, a_3]$, $r = [b_1 - 1, b_2, b_3]$. Clearly, $q \in ap$, $r \in bp$, $q \neq a,p$, and $r \neq b,p$. Let $s = qb \cap ra$ (see Figure 73); then $s = [a_1 + 1 - b_1, a_2 - b_2, a_3 - b_3]$.

Let $d = ps \cap ab$; then $d = [a_1 - b_1, a_2 - b_2, a_3 - b_3]$. Now it is clear that $H(a,b; c,d)$ and $R(a,b; c,d) = [-1]$. Notice that if the division ring has characteristic 2, then $-1 = 1$ and $c = d$.

THEOREM 7.4.6   *If $R(a,b; c,d) = [r]$, then:*

$$R(b,a; d,c) = R(c,d; a,b) = R(d,c; b,a) = [r]$$
$$R(a,b; d,c) = R(b,a; c,d) = R(c,d; b,a) = R(d,c; a,b) = [r^{-1}]$$
$$R(a,c; b,d) = R(c,a; d,b) = R(b,d; a,c) = R(d,b; c,a) = [1 - r]$$
$$R(a,c; d,b) = R(c,a; b,d) = R(b,d; c,a) = R(d,b; a,c) = [(1 - r)^{-1}]$$
$$R(a,d; c,b) = R(d,a; b,c) = R(b,c; d,a) = R(c,b; a,d) = [-r(1 - r)^{-1}]$$
$$R(a,d; b,c) = R(d,a; c,b) = R(b,c; a,d) = R(c,b; d,a) = [1 - r^{-1}]$$

Proof.   The proof is left as an exercise.

Cross ratio can also be considered from a one-dimensional standpoint. As explained in Section 3.4, in one dimension points may be considered as equivalence classes of ordered pairs of elements. Formally, $(x,y) \sim (x',y')$ if and only if there exists $r \in D$, $r \neq 0$, such that $x' = xr$, $y' = yr$. Although any line may be coordinatized in this way, the line $\langle 0,0,1 \rangle$ in π_D already contains points of the form $[x,y,0]$, and these points may conveniently be denoted by $[x,y]$. So for the sake of convenience, let us restrict our attention to the line $\langle 0,0,1 \rangle$ in π_D.

Cross ratio on pairs may be defined in the same way that cross ratio was defined on triples. Notice first that since any three points a,b,c on $\langle 0,0,1 \rangle$ have coordinates $a_i, b_i, c_i, i = 1,2,3$, so that $a_i + b_i = c_i$, the same may be said for any three points a,b,c whose coordinates are pairs $a_i, b_i, c_i, i = 1,2$.

DEFINITION 7.4.7   *Let a,b,c,d be four distinct points on $\langle 0,0,1 \rangle$ and let $a_i, b_i, c_i, i = 1,2$, be coordinates of a,b,c, respectively, such that $a_i + b_i = c_i$. Let s and t be such that $d_i = a_i s + b_i t$ and let $r = st^{-1}$ and $[r] = \{k^{-1}rk\}$. Then the cross ratio of d to c with respect to a,b is $[r]$.*

Notation.   This cross ratio is denoted by $R(a,b; c,d)$.

If D is a field, this cross ratio may be written in another way:

THEOREM 7.4.8 *If D is a field and a = $[a_1, a_2]$, b = $[b_1, b_2]$, c = $[c_1, c_2]$, and d = $[d_1, d_2]$, then*

$$R(a,b;\,c,d) = \frac{\begin{vmatrix} a_1 & a_2 \\ c_1 & c_2 \end{vmatrix} \begin{vmatrix} b_1 & b_2 \\ d_1 & d_2 \end{vmatrix}}{\begin{vmatrix} a_1 & a_2 \\ d_1 & d_2 \end{vmatrix} \begin{vmatrix} b_1 & b_2 \\ c_1 & c_2 \end{vmatrix}}$$

Proof. The proof is left as an exercise.

If we delete the point [1,0] from our line, there remain only those points of the form [s,1] where $s \in D$. These are points of an affine line. If we take D to be the real numbers, the cross ratio $R(a,b;\,c,d)$ is simply the ratio of distances. For example, let $a = (a',1)$, $b = (b',1)$, $c = (c',1)$, $d = (d',1)$. Then

$$R(a,b;\,c,d) = \frac{(a' - c')(b' - d')}{(a' - d')(b' - c')}$$

These four differences represent distances between points on the real line. If we consider the deleted point in our cross ratio (for example, if we let $a = [1,0]$, then we see that

$$R(a,b;\,c,d) = \frac{(b' - d')}{(b' - c')}$$

—that is, the cross ratio is the directed distance from b to d divided by the directed distance from b to c. Here we may think of a as the point at infinity. For a further examination of cross ratio in the affine plane, see the exercises at the end of this section.

EXERCISES

1. Verify that if $R(a,b;\,c,d) = [r]$, $R(a,b;\,d,e) = [s]$, and $R(a,b;\,c,e) = [t]$, then $[rs] = [t]$.

2. Prove Theorem 7.4.6.

3. Prove Theorem 7.4.8.

4. Find the cross ratio of the following sets of points in π_F:
   a. [1,1,0],[−1,1,0]; [1,0,0],[0,1,0].
   b. [−2,1,1],[1,−3,2]; [−1,0,1],[−1,1,0] where $F = R$.

  c. $[1,a,0],[1,0,0]$; $[1,b,0],[0,1,0]$ where $F = \mathrm{GF}(4)$ (see Section 2.1).

  d. $[-1,0,1],[-1,1,0]$; $[c,1,d],[-1,d,b]$ where $F = \mathrm{GF}(9)$ (see Section 2.1).

  e. p_0,p_1; p_3,p_9 in π_{S_3} (see Section 2.3).

5. a. Prove that the point pairs p,p'; q,q'; r,r' correspond under an involutory projectivity on L in π_F if and only if $R(p,q;r,r') = R(p',q';r',r)$ (see exercise 6.5.2).

  b. Prove that if $\{(a,a'),(b,b'),(c,c')\} \in \gamma$, then there exists an involutory projectivity f on line ab such that $f\colon a,b,c \to a',b',c'$.

6. Let p,q,r be collinear points in α_F. Then the ratio of p,q to q,r (denoted by $p,q/q,r$) is the number $-R(p,r;\ q,s)$ where $R(p,r;\ q,s)$ is the cross ratio in the plane $\pi(\alpha_F{}^+)$ and $s = pq \cap L_\infty$. Verify that in α_F, $p,q/q,r$ is the ratio of the directed length of segment pq to the directed length of segment qr.

7. Prove the following theorem (Menelaus' Theorem): Let a_1,a_2,a_3 be vertices of a triangle in α_F and let b_1,b_2,b_3 be three points different from these vertices such that $b_1 \in a_2a_3$, $b_2 \in a_1a_3$, and $b_3 \in a_1a_2$. Then b_1,b_2,b_3 are collinear if and only if $r_1 \cdot r_2 \cdot r_3 = -1$ where $r_1 = a_2,b_1/b_1,a_3$; $r_2 = a_3,b_2/b_2,a_1$; and $r_3 = a_1,b_3/b_3,a_2$. Hint: First prove that in π_F, if $a_1 = [0,0,1]$, $a_2 = [1,0,1]$, $a_3 = [0,1,1]$, $b_1 = [z,1-z,1]$, $b_2 = [0,y,1]$, $b_3 = [x,0,1]$, $c_1 = [-1,1,0]$, $c_2 = [0,1,0]$, and $c_3 = [1,0,0]$, then $t_1 \cdot t_2 \cdot t_3 = 1$ if and only if b_1,b_2,b_3 are collinear where $t_1 = R(a_1,a_2;\ b_3,c_3)$, $t_2 = R(a_2,a_3;\ b_1,c_1)$, and $t_3 = R(a_3,a_1;\ b_2,c_2)$.

8. Prove the following theorem (Ceva's Theorem): Let a_1,a_2,a_3 be vertices of a triangle in α_F and let b_1,b_2,b_3 be three points different from these vertices such that $b_1 \in a_2a_3$, $b_2 \in a_1a_3$, and $b_3 \in a_1a_2$. Then a_1b_1, a_2b_2, a_3b_3 are concurrent or parallel if and only if $r_1 \cdot r_2 \cdot r_3 = 1$ where $r_1 = a_2,b_1/b_1,a_3$; $r_2 = a_3,b_2/b_2,a_1$; and $r_3 = a_1,b_3/b_3,a_2$.

7.5 COLLINEATIONS ON π_D

The last two sections of this chapter deal with collineations on π_D. A fairly thorough treatment of this subject has already been given for planes π_F, chiefly in Example 3.1.2, Theorem 3.1.4, Example 3.2.4(ii), Theorems 3.2.5, 3.2.6, and 3.2.7, Example 3.2.12(ii) (all of which deal with matrix-representable collineations), and in Section 4.2. The two types of collineations examined in this section are matrix-representable collineations and automorphic collineations.

  All the results for matrix-representable collineations mentioned in the preceding paragraph are true in π_D except Theorem 3.1.4. The proofs are exactly the same for both planes if we assume that a 3×3 matrix over D is nonsingular by definition if and only if it represents a one-to-one mapping from D_3 onto D_3. It should be noted that for fields the concept of an $n \times n$ nonsingular matrix M has several equivalent formulations. For example:

There exists a matrix M^{-1} such that $M^{-1}M = MM^{-1} = I$, the identity matrix.
The n row vectors are linearly independent.
The determinant of $M \neq 0$.

If the entries of M are from a division ring, only the first statement is equivalent. The concepts of "linear dependence" and "determinant" are not easily formulated for a division ring.

DEFINITION 7.5.1 *Let M be a 3 × 3 nonsingular matrix over D; then*

$$M = \begin{pmatrix} a & b & c \\ d & e & f \\ g & h & i \end{pmatrix}$$

The mapping f: $[x, y, z] \rightarrow [ax + by + cz, dx + ey + fz, gx + hy + iz]$ *is called a* matrix-representable transformation.

Notation. f is denoted by f_M.

THEOREM 7.5.2 *The mapping f_M is well defined on the points of π_D.*

Proof. The proof is analogous to the proof in Example 3.1.2.

THEOREM 7.5.3 *The mapping f_M is a collineation on π_D.*

Proof. The proof is analogous to the proof in Example 3.1.2.

THEOREM 7.5.4 *Let M and N be nonsingular matrices over D. Then $f_M = f_N$ if and only if there exists $r \in D$ such that r commutes with all members of D and $M = rN$.*

Proof. The proof is left as an exercise.

Notation. The group of matrix-representable collineations on D is denoted by $\mathbf{PGL}(2,D)$.

THEOREM 7.5.5 *The collineation f_M leaves $\langle 0,0,1 \rangle$ pointwise fixed if and only if*

$$M = \begin{pmatrix} 1 & 0 & r \\ 0 & 1 & s \\ 0 & 0 & t \end{pmatrix}$$

The collineation f_M leaves $\langle 0,1,0 \rangle$ pointwise fixed if and only if

$$M = \begin{pmatrix} 1 & r & 0 \\ 0 & s & 0 \\ 0 & t & 1 \end{pmatrix}$$

The collineation f_M leaves $\langle 1,0,0 \rangle$ pointwise fixed if and only if

$$M = \begin{pmatrix} r & 0 & 0 \\ s & 1 & 0 \\ t & 0 & 1 \end{pmatrix}$$

Proof. The proof is the same as the proofs of Theorems 3.2.5 and 3.2.6.

THEOREM 7.5.6 *An elation with axis $\langle 0,0,1 \rangle$ may be represented by f_M where*

$$M = \begin{pmatrix} 1 & 0 & r \\ 0 & 1 & s \\ 0 & 0 & 1 \end{pmatrix}$$

for some $r,s \in D$. A homology with axis $\langle 0,0,1 \rangle$ and center $[0,0,1]$ is the map f_M where

$$M = \begin{pmatrix} 1 & 0 & 0 \\ 0 & 1 & 0 \\ 0 & 0 & t \end{pmatrix}$$

for some $t \in D$, $t \neq 0$. If D does not have characteristic 2, then f_M is a harmonic homology where

$$M = \begin{pmatrix} 1 & 0 & 0 \\ 0 & 1 & 0 \\ 0 & 0 & -1 \end{pmatrix}$$

Proof. The proof is the same as the proof in Example 3.2.12.

The automorphic collineation in π_D, which is introduced in the following paragraphs, should not be totally unfamiliar to the student, since it was included in exercises 3.1.6, 4.1.5, and 4.2.7 (see also Example 4.3.4(iii)).

DEFINITION 7.5.7 *Let D be a division ring. Then* $\gamma: D \rightarrow D$ *is an* automorphism *if and only if* γ *is a one-to-one map from D onto itself such that* $\gamma(a + b) = \gamma(a) + \gamma(b)$ *and* $\gamma(a \cdot b) = \gamma(a) \cdot \gamma(b)$.

EXAMPLE 7.5.8

i. Let C be the field of complex numbers and let $\gamma: a + bi \rightarrow a - bi$. It is easily checked that γ is an automorphism.

ii. Let $D = \mathrm{GF}(p^n)$ and let $\gamma: a \rightarrow a^p$. Clearly, $(a \cdot b)^p = a^p \cdot b^p$, so $\gamma(a \cdot b) = \gamma(a) \cdot \gamma(b)$. Also $(a + b)^p = a^p + pa^{p-1}b + \ldots + pab^{p-1} + b^p$. The binomial theorem states that all the coefficients of $a^{p-k}b^k$, $1 \leq k \leq p - 1$, are multiples of p. Since $\mathrm{GF}(p^n)$ has characteristic p, all the terms except a^p and b^p equal 0. Therefore, $(a + b)^p = a^p + b^p$ and so $\gamma(a + b) = \gamma(a) + \gamma(b)$.

If $\gamma(a) = \gamma(b)$, then $a^p = b^p$, and so $a^p - b^p = 0$. Thus $(a - b)^p = 0$ and $a = b$. It follows that γ is one-to-one.

Since γ is a one-to-one map of a finite set into itself, it must be an onto map. Thus we have shown that $a \rightarrow a^p$ is an automorphism.

It should be noted that since the composition of automorphisms is an automorphism, there are n automorphisms $I, \gamma, \gamma^2,$ $\ldots, \gamma^{n-1}$ on $\mathrm{GF}(p^n)$ where $\gamma^k: a \rightarrow a^{p^k}$. These automorphisms are in fact the only automorphisms on $\mathrm{GF}(p^n)$.

iii. Let D be a division ring and let a be a fixed nonzero element of D. Then the mapping $\gamma: x \rightarrow axa^{-1}$ for all $x \in D$ is easily shown to be an automorphism. An automorphism of this form is called an *inner automorphism*. If D is not a field and a is such that $ab \neq ba$ for all $b \in D$, then γ is not the identity.

THEOREM 7.5.9 *The identity is the only automorphism of the real field R onto itself.*

Proof. Let γ be an automorphism of R onto itself. The following equalities are true:

$\gamma(0) = \gamma(0) + \gamma(0)$, thus $\gamma(0) = 0$

$\gamma(1) = \gamma(1) \cdot \gamma(1)$, and since $\gamma(1) \neq 0$ (because γ is one-to-one), $\gamma(1) = 1$

$0 = \gamma(0) = \gamma(a - a) = \gamma(a) + \gamma(-a)$, so $\gamma(-a) = -\gamma(a)$

$1 = \gamma(1) = \gamma(a \cdot a^{-1}) = \gamma(a) \cdot \gamma(a^{-1})$, so $\gamma(a^{-1}) = (\gamma(a))^{-1}$ for $a \neq 0$

$\gamma(na) = \gamma(a + a + \ldots + a) = n\gamma(a)$

The field R satisfies the property that if $a > 0$, then $\sqrt{a} \in R$. So if $a > 0$, then $\gamma(a) = \gamma(\sqrt{a} \cdot \sqrt{a}) = \gamma(\sqrt{a}) \cdot \gamma(\sqrt{a}) = (\gamma(\sqrt{a}))^2 > 0$. Therefore, if $a > b$, then $a - b > 0$, so $\gamma(a - b) > 0$ and $\gamma(a) - \gamma(b) > 0$, and thus $\gamma(a) > \gamma(b)$.

In addition, R satisfies the Archimedian property of ordered fields, namely, for any $a, b > 0$ there exists an integer m such that $mb > a$.

Let $a > 0$ be an arbitrary real number and let n be an arbitrary integer. Clearly, there exists an integer m such that $m - 1 \leq na < m$. By the results previously stated in this proof, we also have $m - 1 \leq n(\gamma(a)) < m$. Thus $-1 < n(a - \gamma(a)) < 1$ for all n. It follows easily that $\gamma(a) = a$.

Since $\gamma(-a) = -\gamma(a)$, it follows that $\gamma(a) = a$ for $a < 0$ also. Therefore, $\gamma(a) = a$ for all $a \in R$ and so γ is the identity.

Let $\gamma: D \to D$ be an automorphism. Then γ induces a mapping f from $D_3 \to D_3$ as follows: $f: (x,y,z) \to (\gamma(x), \gamma(y), \gamma(z))$. Since $f: (xr, yr, zr) \to (\gamma(xr), \gamma(yr), \gamma(zr)) = (\gamma(x) \cdot \gamma(r), \gamma(y) \cdot \gamma(r), \gamma(z) \cdot \gamma(r))$, we see that f induces a well-defined mapping on the points of π_D. Similarly, f induces a well-defined mapping on the lines of π_D.

THEOREM 7.5.10 *Let γ be an automorphism on D and let f_γ and F_γ be mappings defined on the points and lines of π_D as follows:*

$$f_\gamma: [x, y, z] \to [\gamma(x), \gamma(y), \gamma(z)]$$
$$F_\gamma: \langle a, b, c \rangle \to \langle \gamma(a), \gamma(b), \gamma(c) \rangle$$

Then (f_γ, F_γ) is a collineation on π_D.

Proof. Clearly, f_γ and F_γ are well-defined, one-to-one, onto mappings.

Suppose that $p \in L$ where $p = [x,y,z]$ and $L = \langle a,b,c \rangle$. Then $ax + by + cz = 0$. Therefore, $0 = \gamma(0) = \gamma(ax + by + cz) = \gamma(a)\gamma(x) + \gamma(b)\gamma(y) + \gamma(c)\gamma(z)$, and it follows that $f_\gamma(p) \in F_\gamma(L)$. Thus (f_γ, F_γ) is a collineation.

DEFINITION 7.5.11   *If γ is an automorphism on D, the collineation (f_γ, F_γ) is called an* automorphic collineation.

THEOREM 7.5.12   *The set of automorphic collineations on π_D forms a group.*

Proof.   It need only be shown that the product of two automorphic collineations is an automorphic collineation. Let β and γ be automorphisms on D and let $\delta = \gamma \circ \beta$. It is easily seen that δ is an automorphism and that $f_\gamma \circ f_\beta = f_\delta$.

Notation.   The group of automorphic collineations on π_D is denoted by $\mathrm{Aut}(\pi_D)$.

The automorphic collineation is fundamentally different from the collineations that have previously been discussed.

In fact, no automorphic collineation other than the identity can be a central collineation, and only automorphic collineations defined by inner automorphisms can be projective collineations or matrix-representable collineations. A proof of this statement will be begun in this section and completed in the next.

THEOREM 7.5.13   *Let $f_\gamma \in \mathrm{Aut}(\pi_D)$. If f_γ is a central collineation, then f_γ is the identity.*

Proof.   Clearly, f_γ holds fixed the four-point $[1,0,0]$, $[0,1,0]$, $[0,0,1]$, $[1,1,1]$. Since f_γ holds a line pointwise fixed, it must also hold fixed two points off this line. Therefore, f_γ is the identity.

THEOREM 7.5.14   *If f is a projective collineation, then f preserves the cross ratio of collinear points.*

Proof.   The proof follows from Theorem 7.4.4.

THEOREM 7.5.15   *If f_γ is an automorphic collineation and $R(a,b;c,d)$*
$= [r]$, then $R(f_\gamma(a),f_\gamma(b); f_\gamma(c),f_\gamma(d)) = [\gamma(r)]$.

Proof.   Since $R(a,b;c,d) = [r]$, there exist coordinates a_i,b_i,c_i,d_i of a,b,c,d, respectively, $i = 1,2,3$, where $a_i + b_i = c_i$ and $a_i r + b_i = d_i$. It follows that $\gamma(a_i) + \gamma(b_i) = \gamma(c_i)$ and $\gamma(a_i)\gamma(r) + \gamma(b_i) = \gamma(d_i)$; therefore, $R(f_\gamma(a),f_\gamma(b); f_\gamma(c),f_\gamma(d)) = [\gamma(r)]$.

COROLLARY 7.5.16   *If f_γ is an automorphic collineation and γ is an inner automorphism, then f_γ preserves the cross ratio of collinear points.*

COROLLARY 7.5.17   *If f_γ is an automorphic collineation and $H(a,b;c,d)$, then $H(f_\gamma(a),f_\gamma(b); f_\gamma(c),f_\gamma(d))$.*

COROLLARY 7.5.18   *If $f_\gamma \in \text{Aut}(\pi_F)$ and γ is not the identity, then f_γ does not preserve the cross ratio of collinear points.*

EXERCISES

1. Prove Theorem 7.5.4.

2. Show that the identity is the only automorphism on the following fields:
   a. J/p.
   b. Q, the field of rational numbers.

3. Let $CR(L)$ denote the set of all one-to-one mappings from L onto L that preserve cross ratio in the plane π_F. Show that:
   a. $\text{Proj}(L) \subseteq CR(L)$.
   b. $CR(\langle 0,0,1 \rangle) \subseteq \text{Proj}(\langle 0,0,1 \rangle)$. Hint: Denoting points on $\langle 0,0,1 \rangle$ by ordered pairs as was done in Section 3.4, let $f \in CR(\langle 0,0,1 \rangle)$ such that f fixes $[0,1]$, $[1,0]$, and $[1,1]$. Define g from F onto F by $g(x) = x'$ where $f([x,1]) = [x',1]$ and show that g is the identity.
   c. $CR(L) \subseteq \text{Proj}(L)$.
   d. $CR(L) = \text{Proj}(L)$.

4. Let $\text{Hrm}(L)$ denote the set of all one-to-one mappings from L onto L that preserve harmonic conjugacy in π_D. Show that:
   a. $\text{Proj}(L) \subseteq \text{Hrm}(L)$.
   b. (1) $H([1,0],[0,1]; [x,1],[-x,1])$.
      (2) $H([x,1],[y,1]; [(x+y)/2,1],[1,0])$.
      (3) $H([x,1],[-x,1]; [1,1],[x^2,1])$.

c. If $D = J/p$, then $\mathrm{Hrm}(\langle 0,0,1 \rangle) \subseteq \mathrm{Proj}(\langle 0,0,1 \rangle)$. Hint: Let $f \in$ $\mathrm{Hrm}(\langle 0,0,1 \rangle)$ such that f fixes $[0,1]$, $[1,0]$, and $[1,1]$. Define g from D onto D by $g(x) = x'$ where $f([x,1]) = [x',1]$. Establish that $(g/2)(x) = g(x/2)$, that $g(x + y) = g(x) + g(y)$, and that $g(nx) = ng(x)$, then deduce that f is the identity.

d. If $D = Q$, the field of rational numbers, then $\mathrm{Hrm}(\langle 0,0,1 \rangle) \subseteq \mathrm{Proj}$ $(\langle 0,0,1 \rangle)$.

e. If $D = R$, the field of real numbers, then $\mathrm{Hrm}(\langle 0,0,1 \rangle) \subseteq \mathrm{Proj}(\langle 0,0,1 \rangle)$. Hint: Following the hint given in 4(c), note that $g(x^2) = (g(x))^2$. Thus g preserves order on R and fixes Q, and therefore, f is the identity.

f. If $D = J/p$, Q, or R, then $\mathrm{Hrm}(L) = \mathrm{Proj}(L)$.

g. If D is a field that admits an automorphism other than the identity, then $\mathrm{Proj}(L) \subset \mathrm{Hrm}(L)$.

7.6 TRANSFORMATION GROUPS IN π_D

This section examines, in the context of π_D, the results of Chapter 4 that concern π_F. First recall Theorem 4.2.10, which establishes the isomorphisms of $\mathrm{El}(p,L)$ with $(F,+)$ and $\mathrm{Hom}(p,L)$ with $(F - \{0\},\cdot)$ in π_F.

THEOREM 7.6.1 *Let $(D,+,\cdot)$ be a division ring. Then the group $\mathrm{El}(p,L)$ in π_D is isomorphic to $(D,+)$ and the group $\mathrm{Hom}(p,L)$ in π_D is isomorphic to $(D - \{0\},\cdot)$.*

Proof. It follows from Theorems 4.1.5 and 5.2.6 that all groups of the forms $\mathrm{El}(p,L)$ and $\mathrm{Hom}(p,L)$ are isomorphic. The remainder of the proof is the same as the proof of Theorem 4.2.10.

Establishment of these isomorphisms led to the idea of defining an algebraic structure on a general projective plane with transitive central collineation groups, which was done here in Definitions 4.5.6 and 5.6.1. Both structures are constructed on the points of one line of the plane, and for the special case of the plane π_F, both structures are isomorphic to the field F (see exercises 4.2.5 and 5.6.2). As indicated by Theorems 7.6.3 and 7.6.4, only one of these isomorphisms carries over to the plane π_D.

The following three theorems show the relationships between the division ring and the two algebraic structures. In each of the theorems it will be assumed that $(D,+,\cdot)$ is a given division ring and that in π_D,

$L = \langle 0,1,0 \rangle$, $u = [1,0,0]$, $v = [0,1,0]$, $o = [0,0,1]$, $e = [1,1,1]$, and $i = [1,0,1]$. Also recall the structures $(S,\oplus,\odot)$ and $(L - \{u\},\boxplus,\boxdot)$ of Definitions 4.5.6 and 5.6.1, respectively.

THEOREM 7.6.2   *Let $\times$ and $\otimes$ be defined on S as follows: $p \times q = q \odot p$ and $p \otimes q = h_q \circ h_p(i)$ where $p,q \neq o$, $h_p \in \mathrm{Hom}(o,uv)$, and $h_p: i \to p$. Then $(S - \{o\},\times) = (S - \{o\},\otimes)$.*

Proof.   Let $p = [a,0,1]$ and $q = [b,0,1]$. Recall that $p \odot q = g_q \circ g_p(i)$ where $g_p \in \mathrm{Hom}(u,ov)$ and $g_p(i) = p$. Clearly, $g_p = f_M$ and $h_p = f_N$ where

$$M = \begin{pmatrix} a & 0 & 0 \\ 0 & 1 & 0 \\ 0 & 0 & 1 \end{pmatrix} \quad \text{and} \quad N = \begin{pmatrix} 1 & 0 & 0 \\ 0 & 1 & 0 \\ 0 & 0 & a^{-1} \end{pmatrix}$$

Thus $p \times q = q \odot p = g_p \circ g_q(i) = [ab,0,1] = h_q \circ h_p(i) = p \otimes q$.

THEOREM 7.6.3   *The algebras $(D,+,\cdot)$ and $(S,\oplus,\otimes)$ are isomorphic.*

Proof.   Let $f_x \in \mathrm{El}(u,uv)$ such that $f_x: o \to [x,0,1]$ and let $h_x \in \mathrm{Hom}(o,uv)$ such that $h_x: i \to [x,0,1]$ where $x \neq 0$. Theorems 4.2.10 and 7.6.2 imply that $f_y \circ f_x(o) = f_{x+y}(o)$ and that $h_y \circ h_x(i) = h_{xy}(i)$. Let β be a mapping from D onto S such that $\beta: x \to [x,0,1]$. It follows that $\beta(x + y) = [x + y,0,1] = f_{x+y}(o) = f_y \circ f_x(o) = [x,0,1] \oplus [y,0,1] = \beta(x) \oplus \beta(y)$, and that $\beta(x \cdot y) = [xy,0,1] = h_{xy}(i) = h_y \circ h_x(i) = [x,0,1] \otimes [y,0,1] = \beta(x) \otimes \beta(y)$ where $x,y \neq 0$. If x or $y = 0$, then $\beta(x \cdot y) = 0 = \beta(x) \otimes \beta(y)$ (assuming that $o \otimes p = p \otimes o = o$ by definition). Thus β is an isomorphism from $(D,+,\cdot)$ to $(S,\oplus,\otimes)$.

THEOREM 7.6.4   *The algebras $(S,\oplus,\otimes)$ and $(L - \{u\},\boxplus,\boxdot)$ are identical.*

Proof.   The proof is analogous to the proofs of Theorems 5.6.2 and 5.6.3.

For planes π_F, we know from Corollary 4.2.9 that $CC(\pi_F) = \mathrm{PGL}(2,F) = PC(\pi_F) \subseteq C(\pi_F)$. This, in fact, is true for π_D, as will be shown in the following paragraphs. The equality $CC(\pi_D) = PC(\pi_D)$ follows from Theorem 5.2.5 because π_D is Desarguesian. To show that $CC(\pi_D) = \mathrm{PGL}(2,D)$, it must first be established that $\mathrm{PGL}(2,D)$ is transitive on four-points. In Section 4.2 this was done for $PGL(2,F)$ with the help of two lemmas. These lemmas are proved again here for D. Recall that a matrix M is nonsingular if and only if M has both a left and a right inverse (clearly, the two inverses are equal), or, equivalently, if M represents a one-to-one mapping from D_3 onto D_3.

LEMMA 7.6.5   *The points* $p = [x_1,x_2,x_3]$, $q = [y_1,y_2,y_3]$, $r = [z_1,z_2,z_3]$ *are noncollinear if and only if the matrix*

$$M = \begin{pmatrix} x_1 \ y_1 \ z_1 \\ x_2 \ y_2 \ z_2 \\ x_3 \ y_3 \ z_3 \end{pmatrix}$$

is nonsingular.

Proof.   Suppose that M is nonsingular. If p,q,r are collinear, they all belong to a common line $\langle a,b,c \rangle$. Thus $ax_1 + bx_2 + cx_3 = ay_1 + by_2 + cy_3 = az_1 + bz_2 + cz_3 = 0$, or equivalently, $(a,b,c)M = (0,0,0)$. Since M is nonsingular, it has a right inverse M^{-1}, so $(a,b,c)MM^{-1} = (0,0,0)M^{-1}$, that is, $(a,b,c) = (0,0,0)$. This contradiction shows that p,q,r are noncollinear.

Suppose that p,q,r are noncollinear. Then there is no (a,b,c) except $(0,0,0)$ such that $(a,b,c)M = (0,0,0)$. Therefore, M represents a one-to-one transformation and thus has a right inverse. Letting $pq = \langle a_1,a_2,a_3 \rangle$, $pr = \langle b_1,b_2,b_3 \rangle$, and $qr = \langle c_1,c_2,c_3 \rangle$, and recalling that $r \notin pq$, $q \notin pr$, and $p \notin qr$, we may choose these coordinates so that

$$\sum_{i=1}^{3} a_i z_i = \sum_{i=1}^{3} b_i y_i = \sum_{i=1}^{3} c_i x_i = 1$$

Let

$$N = \begin{pmatrix} c_1 \ c_2 \ c_3 \\ b_1 \ b_2 \ b_3 \\ a_1 \ a_2 \ a_3 \end{pmatrix}$$

Then $NM = I$, so M has a left inverse.

LEMMA 7.6.6   *If* p,q,r,s *is a four-point where* p,q,r *have the co-ordinates described in the preceding proof and* $s = [w_1,w_2,w_3]$, *then there exists* $a,b,c \in F$ *such that* $w_i = ax_i + by_i + cz_i$, $i = 1,2,3$.

Proof.   The proof is analogous to that of Lemma 4.2.2.

THEOREM 7.6.7 *The group* $PGL(2,D)$ *is transitive on four-points.*

Proof. The proof is analogous to the existence proof of Theorem 4.2.3.

The group $PGL(2,F)$ is sharply transitive on four-points, as indicated by the proof of Theorem 4.2.3; however, this statement is not necessarily true for $PGL(2,D)$. Thus the uniqueness part of the proof of Theorem 4.2.3 fails. It is true that if f_M fixes u,v,o,e, then

$$M = \begin{pmatrix} a & 0 & 0 \\ 0 & a & 0 \\ 0 & 0 & a \end{pmatrix}$$

but it is not necessarily true that f_M must then be the identity. If a does not commute with every element of D, then f_M is not the identity because $f_M([x,y,z]) = [ax,ay,az]$, and this is not necessarily the point $[x,y,z]$. This remark will be returned to later in the section.

THEOREM 7.6.8 $CC(\pi_D) \subseteq PGL(2,D)$.

Proof. The proof is analogous to that of Theorem 4.2.4.

THEOREM 7.6.9 $PGL(2,D) \subseteq CC(\pi_D)$.

Proof. Let $f \in PGL(2,D)$ and let $f: u,v,o,e \rightarrow p,q,r,s$. By Theorem 5.2.5, $CC(\pi_D)$ is transitive on four-points, so there exists $g \in CC(\pi_D)$ such that $g: u,v,o,e \rightarrow p,q,r,s$. Since $CC(\pi_D) \subseteq PGL(2,D)$, we have $g \in PGL(2,D)$, so $g^{-1} \circ f = f_M$ where

$$M = \begin{pmatrix} a & 0 & 0 \\ 0 & a & 0 \\ 0 & 0 & a \end{pmatrix}$$

But $M = M_1 M_2 M_3$ where

$$M_1 = \begin{pmatrix} a & 0 & 0 \\ 0 & 1 & 0 \\ 0 & 0 & 1 \end{pmatrix} \qquad M_2 = \begin{pmatrix} 1 & 0 & 0 \\ 0 & a & 0 \\ 0 & 0 & 1 \end{pmatrix} \qquad M_3 = \begin{pmatrix} 1 & 0 & 0 \\ 0 & 1 & 0 \\ 0 & 0 & a \end{pmatrix}$$

and so f_{M_i}, $i = 1,2,3$, is a central collineation. Thus $f_M = f_{M_3} \circ f_{M_2} \circ f_{M_1} = h \in CC(\pi_D)$, and therefore, $g^{-1} \circ f = h$ or $f = g \circ h$ for some $g,h \in CC(\pi_D)$. Thus $f \in CC(\pi_D)$.

THEOREM 7.6.10   *In π_D we have* $CC(\pi_D) = PC(\pi_D) = PGL(2,D)$.

Proof.   The proof is based on Theorems 5.2.5, 7.6.8, and 7.6.9.

The remark preceding Theorem 7.5.14 suggests that $f_\gamma \in \text{Aut}(\pi_D)$ $\cap$ $PGL(2,D)$ if and only if γ is an inner automorphism. Note that since $CC(\pi_D) = PC(\pi_D) = PGL(2,F)$, we can replace $PGL(2,F)$ by $CC(\pi_D)$ or $PC(\pi_D)$ in the following theorem.

THEOREM 7.6.11   *Let $f_\gamma \in \text{Aut}(\pi_D)$. Then f_γ is a matrix-representable collineation if and only if γ is an inner automorphism.*

Proof.   Clearly, f_γ holds fixed the four-point u,v,o,e. Suppose that $f_\gamma = f_M$ for some M. It easily follows that

$$M = \begin{pmatrix} a & 0 & 0 \\ 0 & a & 0 \\ 0 & 0 & a \end{pmatrix}$$

for some $a \in D$, and so f_γ: $[x,y,z] \to [ax,ay,az] = [axa^{-1}, aya^{-1}, aza^{-1}]$. Thus γ: $x \to axa^{-1}$. Suppose that γ: $x \to axa^{-1}$. Then $f_\gamma = f_M$ where M is the preceding matrix.

Theorem 7.6.11 is based on the obvious fact that if f is an automorphic collineation, then f fixes the four-point u,v,o,e. The converse of this statement is also true but is not obvious.

THEOREM 7.6.12   *If g is a collineation that fixes $u = [1,0,0]$, $v = [0,1,0]$, $o = [0,0,1]$, and $e = [1,1,1]$, then there exists an automorphism β on D such that $f_\beta = g$.*

Proof.   Define β on D as follows: β: $x \to x'$ where $x' \in D$ such that g: $[x,0,1] \to [x',0,1]$. Notice that $g([x,0,1])$ lies on $\langle 0,1,0 \rangle$ because g fixes u and o and thus the line ou, which is $\langle 0,1,0 \rangle$. Clearly, $\beta(0) = 0$ because $g(o) = o$. Also $\beta(1) = 1$ because $g(ou \cap ve)$ is held fixed and $ou \cap ve = i = [1,0,1]$. Assume that $x,y \neq 0,1$ and let $a = [x,0,1]$, $b = [y,0,1]$, $c = [x + y,0,1]$, $d = [xy,0,1]$, $a' = [x',0,1]$, $b' = [y',0,1]$, $c' = [\beta(x + y),0,1]$, and $d' = [\beta(xy),0,1]$.

From Theorems 7.6.3 and 7.6.4, we know that $\{(o,c),(b,a),(u,u)\}$ $\in \gamma$ and $\{(i,d),(b,a),(u,o)\} \in \gamma$. Since g is a collineation, $\{(o,c'),(b',a'),$

$(u,u)\} \in \gamma$ and $\{(i,d'),(b',a'),(u,o)\} \in \gamma$, and therefore $a' \boxplus b' = c'$ and $a' \boxdot b' = d'$. It follows that $\beta(x) \boxplus \beta(y) = \beta(x + y)$ and that $\beta(x) \boxdot \beta(y) = \beta(xy)$; therefore, β is an isomorphism on D.

That $f_\beta = g$ remains to be shown. By the definition of β, we know that $f_\beta(a) = g(a)$ for all points $a = [x,0,1]$ on $\langle 0,1,0 \rangle$. Also $f_\beta(u) = g(u) = u$, $f_\beta(o) = g(o) = o$, and $f_\beta(e) = g(e) = e$. Thus $f_\beta - g$ leaves fixed all points on $\langle 0,1,0 \rangle$ and points o and e not on $\langle 0,1,0 \rangle$. It follows that $f_\beta - g$ is the identity, and so $f_\beta = g$.

Theorem 7.6.12 allows us to conclude that every collineation on π_D is either a matrix-representable collineation, an automorphic collineation, or a product of the two.

DEFINITION 7.6.13   *The mapping g is a* semilinear collineation *if and only if* $g = f_M \circ f_\gamma$ *for some* $f_M \in \text{PGL}(2,D)$ *and* $f_\gamma \in \text{Aut}(\pi_D)$.

THEOREM 7.6.14   *If f is a collineation on π_D, then f is a semilinear collineation.*

Proof.   Let $f: u,v,o,e \rightarrow a,b,c,d$. Since $\text{PGL}(2,D)$ is transitive on four-points, there exists $f_M \in \text{PGL}(2,D)$ such that $f_M: a,b,c,d \rightarrow u,v,o,e$. Thus $f_M^{-1} \circ f$ fixes u,v,o,e, and so, by Theorem 7.6.12, there exists an automorphism γ on D such that $f_M^{-1} \circ f = f_\gamma$. It follows that $f = f_M \circ f_\gamma$.

THEOREM 7.6.15   *If D admits only inner automorphisms, then* $\text{CC}(\pi_D) = \text{PGL}(2,D) = \text{PC}(\pi_D) = \text{C}(\pi_D)$.

Proof.   The proof follows from Theorems 7.6.10, 7.6.11, and 7.6.14.

COROLLARY 7.6.16   *In π_R the groups* $\text{CC}(\pi_R)$, $\text{PGL}(2,R)$, $\text{PC}(\pi_R)$, *and* $\text{C}(\pi_R)$ *are identical.*

THEOREM 7.6.17   *Every collineation of π_D preserves harmonic conjugacy on lines.*

Proof.   The proof follows from Corollary 7.5.18 and Theorem 7.6.14.

Section 3.4 introduced the idea of representing points on $\langle 0,0,1 \rangle$ in π_F by equivalence classes of pairs rather than by triples. That representation can also be utilized for division rings D. Furthermore, the mapping f_M where M is a nonsingular 2×2 matrix over D represents a projectivity on $\langle 0,0,1 \rangle$; and if we let PGL(1,D) denote this group, we have the analogue of Theorem 4.2.13, that is, PGL(1,D) = Proj($\langle 0,0,1 \rangle$). Proof of this fact is left as an exercise. What will be proved here is that Theorem 4.2.12 may not be generalized to division rings.

THEOREM 7.6.18 *If D is not a field, the group* Proj($\langle 0,0,1 \rangle$) *in π_D is not sharply 3-transitive.*

Proof. Let $a,b \in D$ such that $ab \neq ba$. Also let $p = [1,0]$, $q = [0,1]$, $r = [1,1]$, and

$$M = \begin{pmatrix} a & 0 \\ 0 & a \end{pmatrix}$$

It is easily checked that f_M holds p,q,r fixed; but f_M: $[b,1] \to [ab,a]$ $\neq [ba,a] = [b,1]$, so f_M is not the identity.

A similar theorem holds for PGL(2,D), as was mentioned in the discussion following the proof of Theorem 7.6.7.

THEOREM 7.6.19 *If D is not a field, then* PC(π_D) *is not sharply transitive on four-points.*

Proof. Let $a \in D$ such that $ab \neq ba$ for some $b \in D$ and let γ: $x \to axa^{-1}$. Then the collineation f_γ holds the four-point u,v,o,e fixed but is not the identity. Recall Theorem 7.6.11 and note that $f_\gamma = f_M$ where

$$M = \begin{pmatrix} a & 0 & 0 \\ 0 & a & 0 \\ 0 & 0 & a \end{pmatrix}$$

Therefore, $f_\gamma \in$ PC(π_D), and so PC(π_D) is not sharply transitive on four-points.

Theorems 7.6.18 and 7.6.19 should not surprise us. Theorem 6.5.4 indicated that the Pappus Theorem is equivalent to the Specialized

Fundamental Theorem. If D is a division ring that is not a field, then, by Theorems 6.5.4 and 7.3.5, Proj(L) in π_D is not sharply 3-transitive. Furthermore, since FT-II is equivalent to the Pappus Theorem (a fact stated in Chapter 6 but not fully proved there), PC(π_D) is not sharply transitive on four-points. Thus we were already aware that the planes π_D contain a surplus of projectivities and projective collineations not found in planes of the form π_F. The proofs of Theorems 7.6.18 and 7.6.19 provide an explicit documentation of this surplus.

EXERCISES

1. Show that if π_D satisfies FT-II, then π_D is a Pappian plane.

2. Show that the algebras $(L - \{u\}, \boxplus, \boxdot)$ on π_D and $(S, \oplus, \odot)$ on π_{D^d} are isomorphic (refer back to exercise 7.2.3).

3. Show that in π_D:
   a. $\text{PGL}(1,D) = \text{Proj}(\langle 0,0,1 \rangle)$.
   b. $\text{PGL}(1,D) \sim \text{Proj}(L)$ for all L.

4. a. Let D be a division ring and let $F(D) = \{x \in D : xy = yx \text{ for all } y \in D\}$. Show that the involutory projectivities in π_D may be represented by f_M where
   $$M = \begin{pmatrix} a & 0 \\ 0 & -a \end{pmatrix}$$
   and $a \in F(D)$, or
   $$M = \begin{pmatrix} a & 0 \\ 0 & a \end{pmatrix}$$
   and $a \notin F(D)$, $a^2 \in F(D)$ (refer back to exercise 6.5.2).
   b. Find an involutory projectivity on $\langle 0,0,1 \rangle$ in π_D (where D is the ring of quaternions) that leaves infinitely many points fixed.

5. a. Show that if $f \in \text{PGL}(2,F)$, then f is uniquely determined by the images of five suitably chosen points.
   b. Show that if $f: L \to M$ is a projectivity in π_D, then f is uniquely determined by the images of four suitably chosen points on L.

CHAPTER 8

COORDINATIZING PLANES

In summary, the first seven chapters have shown that the projective planes defined over an underlying algebra are the only planes amenable to detailed analysis. This chapter shows that the two geometrically defined planes, the Desarguesian plane and the Pappian plane, have an algebraic representation. Specifically, every Desarguesian plane π is isomorphic to a plane π_D and every Pappian plane is isomorphic to a plane π_F. These two theorems are the culmination of the analysis included in Parts 1 and 2. They also mark the beginning of the study of the algebraic representation of geometric planes, which is continued in Part 3.

8.1 THE DIVISION RING D_π

As indicated in Chapter 7, for every division ring D, there is an associated Desarguesian plane π_D. The following paragraphs show that, given a Desarguesian plane π (actually a plane satisfying the Second Desargues (L) Theorem for some L), there is an associated division ring D_π.

LEMMA 8.1.1 *Suppose that π satisfies the Second Desargues (L) Theorem. If $f, g \in El(p,L)$ and q is an arbitrary point not on L, then there exists a unique $h \in Hom(q,L)$ such that $g = h \circ f \circ h^{-1}$.*

Proof. The proof is left as an exercise.

LEMMA 8.1.2 *Let π satisfy the Second Desargues (L) Theorem; let $f \in El(p,L)$, $g \in El(q,L)$ (p may equal q) be given elations that are different from the identity; and let o be an arbitrary fixed point not on L. Then the mapping γ from $El(p,L)$ onto $El(q,L)$ is an isomorphism where γ is defined as follows: $\gamma: h \rightarrow j \circ g \circ j^{-1}$ where j is the unique homology in $Hom(o,L)$ such that $h = j \circ f \circ j^{-1}$.*

Proof. Proof that γ is one-to-one and onto is left as an exercise. The proof that γ preserves compositions is divided into two cases:

Case 1. Suppose that $p \neq q$. Let $M = op$, $M' = oq$, $f(o) = r$, $g(o) = s$, and $t = rs \cap L$ (see Figure 74). Also let j be the unique element of $El(t,L)$ such that $j: r \rightarrow s$. Thus $g = j \circ f$, since the two elations g and $j \circ f$ agree at point o. Let k, m be arbitrary elations in $El(p,L)$ and let $x,y,z \in Hom(o,L)$ such that $k = x \circ f \circ x^{-1}$, $m = y \circ f \circ y^{-1}$, $m \circ k = z \circ f \circ z^{-1}$. Thus $\gamma(k) = x \circ g \circ x^{-1} = x \circ j \circ f \circ x^{-1}$ $= x \circ j \circ x^{-1} \circ x \circ f \circ x^{-1} = (x \circ j \circ x^{-1}) \circ k$, $\gamma(m) = y \circ g \circ y^{-1} = (y \circ j \circ y^{-1})$ $\circ m$, and $\gamma(m \circ k) = z \circ g \circ z^{-1} = (z \circ j \circ z^{-1}) \circ (m \circ k)$. Since $El(L)$ is Abelian, $\gamma(m) \circ \gamma(k) = (y \circ j \circ y^{-1}) \circ m \circ (x \circ j \circ x^{-1}) \circ k = y \circ j \circ y^{-1} \circ x$ $\circ j \circ x^{-1} \circ (m \circ k)$. The following argument shows that $z \circ j \circ z^{-1}$ $= y \circ j \circ y^{-1} \circ x \circ j \circ x^{-1}$. Note first that both mappings are in $El(t,L)$. Suppose that $z \circ j \circ z^{-1}(o) = a$, $y \circ j \circ y^{-1} \circ x \circ j \circ x^{-1}(o) = b$. Since both maps are in $El(t,L)$, a,b,t are collinear. Let $o' = (m \circ k)^{-1}(o)$; then $z \circ j \circ z^{-1}(m \circ k(o')) = a$ and $y \circ j \circ y^{-1} \circ x \circ j \circ x^{-1}(m \circ k(o')) = b$. Since $\gamma(m \circ k), \gamma(m), \gamma(k) \in El(q,L)$, we conclude that $\gamma(m \circ k)(o')$, $(\gamma(m) \circ \gamma(k))(o')$, and q are collinear. But this means that a,b,q are collinear, and this fact, combined with the fact that a,b,t are collinear, implies that $a = b$. Therefore, the maps agree at point o. Since they are elations with the same axis, they are identical. We may conclude that in Case 1, $\gamma(m \circ k) = \gamma(m) \circ \gamma(k)$.

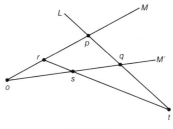

FIGURE 74

Case 2. Suppose that $p = q$. Let $r \in L$ such that $r \neq p$ and let j be a given elation in $\text{El}(r, L)$. By Case 1, the map γ is an isomorphism where γ maps $\text{El}(p, L)$ onto $\text{El}(r, L)$, and $\gamma\colon h \to k \circ j \circ k^{-1}$ where $k \in \text{Hom}(o, L)$ and $h = k \circ f \circ k^{-1}$. Also the map β is an isomorphism where $\beta\colon \text{El}(r, L) \to \text{El}(q, L)$, and $\beta\colon m \to n \circ g \circ n^{-1}$ where $n \in \text{Hom}(o, L)$ and $m = n \circ j \circ n^{-1}$. Thus the map $\beta \circ \gamma\colon \text{El}(p, L) \to \text{El}(q, L)$ is an isomorphism and $\beta \circ \gamma$ is the right type of map; that is, $\beta \circ \gamma\colon h \to k \circ g \circ k^{-1}$ where $k \in \text{Hom}(o, L)$ and $h = k \circ f \circ k^{-1}$.

CONSTRUCTION 8.1.3   Let π be a projective plane satisfying the Second Desargues (L) Theorem. Let M be an arbitrary line of π distinct from L. Let $u = L \cap M$ and let o and i be distinct arbitrary points on M different from u. Let f_a denote the unique $f \in \text{El}(u, L)$ such that $f\colon o \to a$. Let h_a, $a \neq o$, denote the unique $h \in \text{Hom}(o, L)$ such that $h\colon i \to a$. Define the algebraic structure $D_\pi = (D, \oplus, \odot)$ as follows:

$$D = M - \{u\}$$
$$a \oplus b = f_b \circ f_a(o)$$
$$a \odot b = h_b \circ h_a(i) \text{ for all } a, b \neq o$$
$$a \odot o = o \odot a = o \text{ for all } a \in D$$

THEOREM 8.1.4   *The algebra D_π is a division ring.*

Proof.

i. If $a, b \in D$, then $a \oplus b$ and $a \odot b \in D$ because both $\text{El}(u, L)$ and $\text{Hom}(o, L)$ map points of D onto points of D.

ii. $a \oplus (b \oplus c) = (a \oplus b) \oplus c$; $a \odot (b \odot c) = (a \odot b) \odot c$. This follows because the composition of functions is associative.

iii. $o \oplus a = f_a \circ f_o(o) = f_a(o) = a$.
$a \oplus o = f_o \circ f_a(o) = f_o(a) = a$ because f_o is the identity.

$i \odot a = h_a \circ h_i(i) = h_a(i) = a.$

$a \odot i = h_i \circ h_a(i) = h_i(a) = a$ because h_i is the identity.

Thus o is the additive identity; i is the multiplicative identity.

iv. Let f_{-a} denote $(f_a)^{-1}$; let $h_{a^{-1}}$ denote $(h_a)^{-1}$ for $a \neq o$. Therefore,
$-a \oplus a = f_a \circ f_{-a}(o) = f_a \circ (f_a)^{-1}(o) = o$ and $a \oplus (-a) = f_{-a} \circ f_a(o)$
$= (f_a)^{-1} \circ (f_a)(o) = o$. Thus $-a$ is the additive inverse of a. Similarly, a^{-1} represents the multiplicative inverse of a.

v. $a \oplus b = f_b \circ f_a(o) = f_a \circ f_b(o) = b \oplus a$ because $El(L)$ is Abelian by
Theorem 4.5.11.

vi. Note that $f_{a \odot b} = h_b \circ f_a \circ h_b^{-1}$ for $b \neq o$. Clearly, both maps are in
$El(u,L)$. Also $f_{a \odot b}(o) = a \odot b$; $h_b \circ f_a \circ h_b^{-1}(o) = h_b \circ f_a(o) = h_b(a)$
$= a \odot b$. Thus the mappings agree at the point o and therefore
they must be identical. The right distributive law may now be
shown as follows: $(b \oplus c) \odot a = f_{(b \oplus c) \odot a}(o) = h_a \circ f_{(b \oplus c)} \circ h_a^{-1}(o)$
$= h_a \circ f_c \circ f_b \circ h_a^{-1}(o) = h_a \circ f_c \circ h_a^{-1} \circ h_a \circ f_b \circ h_a^{-1}(o) = f_{c \odot a} \circ f_{b \odot a}(o)$
$= f_{c \odot a}(b \odot a) = (b \odot a) \oplus (c \odot a)$.

Lemma 8.1.2 is used for the left distributive law. Let $p = q = u$,
$f = f_i$, and $g = f_a$. Thus the mapping γ from $El(u,L)$ onto $El(u,L)$
is an isomorphism where $\gamma: f_b \rightarrow j \circ f_a \circ j^{-1}$ and $f_b = j \circ f_i \circ j^{-1}$.
Notice that $j = h_b$ because $f_b(o) = h_b \circ f_i \circ h_b^{-1}(o) = b$. Thus γ:
$f_b \rightarrow h_b \circ f_a \circ h_b^{-1} = f_{a \odot b}$ is an isomorphism. Therefore, $\gamma(f_{b \oplus c})$
$= \gamma(f_c \circ f_b) = \gamma(f_c) \circ \gamma(f_b)$, that is, $f_{a \odot (b \oplus c)} = f_{a \odot c} \circ f_{a \odot b}$. So
$a \odot (b \oplus c) = f_{a \odot (b \oplus c)}(o) = f_{a \odot c} \circ f_{a \odot b}(o) = f_{a \odot c}(a \odot b) = (a \odot b)$
$\oplus (a \odot c)$.

It follows from steps (i) through (vi) that D_π is a division ring.

THEOREM 8.1.5 *If π satisfies the Pappus (L) Theorem, then D_π is
a field.*

Proof. Since π satisfies the Pappus (L) Theorem, it satisfies the
Second Desargues (L) Theorem, so D_π is a division ring. Also, by
Theorem 6.2.5, $Hom(o,L)$ is Abelian; therefore, $a \odot b = h_b \circ h_a(i)$
$= h_a \circ h_b(i) = b \odot a$ for all $a,b \neq o$. Thus D_π is a field.

Notice that the algebra D_π has, in fact, been defined in terms of
fixed lines L and M and fixed points o, i, and u. These items have been
omitted from the notation because any other admissible choices would
have yielded an isomorphic division ring, as will be shown now.

THEOREM 8.1.6 *In a Desarguesian plane π, the division ring D_π is unique up to isomorphism.*

Proof. Let L,M,o,u,i and L',M',o',u',i' be the constituents of two division rings, $D_\pi = (D,\oplus,\odot)$ and $D'_\pi = (D',+,\cdot)$, respectively. Let u,v,o,e and u',v',o',e' be two four-points such that $uv = L$, $ou = M$, $ve \cap M = i$ and $e'v' = L'$, $o'u' = M'$, $v'e' \cap M' = i'$. Clearly, such four-points exist. Finally, since π is Desarguesian, there exists a collineation g on π such that $g: u,v,o,e \to u',v',o',e'$.

To show that g is an isomorphism from D_π to D'_π, first note that the arithmetic in D'_π is defined as follows: $g(a) + g(b) = f'_{g(b)}(g(a))$ where $f'_{g(b)} \in \text{El}(u',L')$ and $f'_{g(b)}: o' \to g(b)$; $g(a) \cdot g(b) = h'_{g(b)}(g(a))$ for $a,b \neq o$, where $h'_{g(b)} \in \text{Hom}(o',L')$ and $h'_{g(b)}: i' \to g(b)$. It is easily checked that $f'_{g(b)} = g \circ f_b \circ g^{-1}$ and that $h'_{g(b)} = g \circ h_b \circ g^{-1}$. Therefore, $g(a \oplus b) = g \circ f_{a \oplus b}(o) = g \circ f_b \circ f_a(o) = g \circ f_b \circ g^{-1}(o') = g \circ f_b \circ g^{-1} \circ g \circ f_a \circ g^{-1}(o') = f'_{g(b)} \circ f'_{g(a)}(o') = g(a) + g(b)$. Similarly, $g(a \odot b) = g(a) \cdot g(b)$ for $a,b \neq o$. Clearly, if a or $b = o$, we also have equality. Thus g preserves addition and multiplication and therefore is an isomorphism.

Three algebras have previously been defined on projective planes: $(S,\oplus,\odot)$ of Definition 4.5.6, $(L - \{u\},\boxplus,\boxdot)$ of Definition 5.6.1, and $(S,\oplus,\otimes)$ of Theorem 7.6.2. Theorems 7.6.3 and 7.6.4 show the relationships between these algebras in the plane π_D. For Desarguesian planes, these algebras may be related to the algebra $D_\pi = (D,\oplus,\odot)$ as follows. Let u, o, and i correspond to the same point for all algebras; thus $S = L - \{u\} = D$. Also let the lines uv (of S), L_∞ (of $L - \{u\}$), and L (of D) be identical. Hence the following theorem.

THEOREM 8.1.7 *If π is a Desarguesian plane, then the algebras $(D,\oplus,\odot)$, $(L - \{u\},\boxplus,\boxdot)$, and $(S,\oplus,\otimes)$ are identical.*

Proof. The proof is left as an exercise.

EXERCISES

1. Prove Lemma 8.1.1.

2. Prove Theorem 8.1.7.

8.2   THE RELATIONSHIPS BETWEEN D AND D_{π_D}
AND BETWEEN π AND π_{D_π}

Now we face two important questions. First, if we are given a division ring D, form the Desarguesian plane π_D, and then form the new division ring D_{π_D}, what can we say about the relationship between D and D_{π_D}? Second, if we begin with a Second Desarguesian (L) plane π, form the division ring D_π, and then build the projective plane π_{D_π}, what can we say about the relationship between π and π_{D_π}? Our hope is that D will be isomorphic to D_{π_D} and that π will be isomorphic to π_{D_π}. The latter isomorphism is especially important since it will establish the theorem that every Second Desarguesian (L) plane is equivalent to a plane π_D for some D—informally stated, every Second Desarguesian (L) plane has an algebraic representation. The former isomorphism follows from Theorems 7.6.4 and 8.1.7. It is proved again here to emphasize the nature of the isomorphism.

THEOREM 8.2.1   Let $(D,+,\cdot)$ be a division ring. Then $(D,+,\cdot)$ is isomorphic to $D_{\pi_D} = (M - \{u\}, \oplus, \odot)$.

Proof.   Recall that the points of π_D are of the form $[a,b,c]$, $a,b,c \in D$. Let $L = \langle 0,0,1 \rangle$, $M = \langle 0,1,0 \rangle$. Note that $u = [1,0,0]$ and that the points of $M - \{u\}$ can be uniquely represented in the form $[a,0,1]$ for some $a \in D$. Let $o = [0,0,1]$ and $i = [1,0,1]$ and define the mapping γ from D onto D_{π_D} such that $\gamma(a) = [a,0,1]$. Clearly, γ is one-to-one and onto. Recall that arithmetic in D_{π_D} is accomplished using elements of $\text{El}(u,L)$ and $\text{Hom}(o,L)$; $f_{[a,0,1]}$ can be represented by

$$\begin{pmatrix} 1 & 0 & a \\ 0 & 1 & 0 \\ 0 & 0 & 1 \end{pmatrix}$$

and $h_{[a,0,1]}$ can be represented by

$$\begin{pmatrix} 1 & 0 & 0 \\ 0 & 1 & 0 \\ 0 & 0 & a^{-1} \end{pmatrix}$$

This notation is shortened to f_a and h_a. Thus $\gamma(a + b) = [a + b,0,1] = f_b[a,0,1] = f_b \circ f_a(o) = [a,0,1] \oplus [b,0,1] = \gamma(a) \oplus \gamma(b)$. If $a,b \neq 0$, then

$\gamma(a \cdot b) = [a \cdot b, 0, 1] = h_b[a, 0, 1] = h_b \circ h_a(i) = [a, 0, 1] \odot [b, 0, 1] = \gamma(a) \odot \gamma(b)$. If a or $b = 0$, then $\gamma(a \cdot b) = 0 = \gamma(a) \odot \gamma(b)$. Thus γ is an isomorphism.

THEOREM 8.2.2 *If π satisfies the Second Desargues (L) Theorem, then π is isomorphic to π_{D_π}.*

Proof. Let M and M' be two lines distinct from L and let $M \cap L = u$, $M' \cap L = v$, and $M \cap M' = o$. Also let i and i' be points distinct from o, u, and v on M and M', respectively. Finally, for $a \in M - \{o, u\}$, let $h_a \in \mathrm{Hom}(o, L)$ be such that $h_a(i) = a$. Define a partial mapping γ from $\mathcal{P}$ to $\mathcal{P}_{D_\pi}$ as follows (see Figure 75):

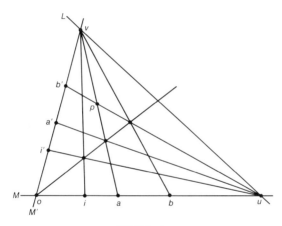

FIGURE 75

γ: $u \rightarrow [i, o, o]$
     $a \rightarrow [a, o, i]$ for $a \in M - \{u\}$
     $v \rightarrow [o, i, o]$
     $a' \rightarrow [o, a, i]$ where $a' = h_a(i')$
     $p \rightarrow [a, b, i]$ where $p \notin L, M, M'$ and $vp \cap M = a$, $up \cap M' = b'$

The mapping γ on L (aside from u and v) will be disregarded here. It is sufficient to establish an isomorphism between $\mathcal{P} - L$ and $\mathcal{P}_{D_\pi} - \langle o, o, i \rangle$ since the extensions to π and π_{D_π} are unique.
     Proof that γ is a one-to-one mapping from $\mathcal{P} - L$ onto $\mathcal{P}_{D_\pi} - \langle o, o, i \rangle$ is left to the student. The fact that γ preserves collinearity remains to

be proved. Notice that the point $[x,y,z]$ is on the line $\langle a,b,c \rangle$ in π_{D_π} iff $a \odot x \oplus b \odot y \oplus c \odot z = 0$.

Case 1.   Suppose that points of π lie on line N passing through u. Let $N \cap M' = a'$ and let p be any other point on N. Then $pu \cap M' = a'$ and $pv \cap M = b$ for some $b \in M$, $b \neq o$. Thus $\gamma(a') = [o,a,i]$, $\gamma(p) = [b,a,i]$, $\gamma(u) = [i,o,o]$. Since $[i,o,o][o,a,i] = \langle o,i,-a \rangle$ and $[b,a,i] \in \langle o,i,-a \rangle$, it follows that $\gamma(a')$, $\gamma(p)$, and $\gamma(u)$ are collinear in π_{D_π}.

Case 2.   Suppose that the points lie on a line passing through v. Then the argument used is analogous to that used in Case 1.

Case 3.   Suppose that the points of π lie on line N passing through o. Let p and q be on line N. We know that $\gamma(o) = [o,o,i]$. Let $\gamma(p) = [a,b,i]$ and let $\gamma(q) = [c,d,i]$. Now there exists $h \in \text{Hom}(o,L)$ such that $h: p \to q$. Thus $h: pv \to qv$ and therefore $h: a \to c$. Also $h: pu \to qu$, so $h: b' \to d'$, and thus $h: b \to d$. Since $h: a \to c$, $h = h_c \circ h_{a^{-1}}$, and since $h: b \to d$, it follows that $h_c \circ h_{a^{-1}}(b) = d$. Therefore, $b \odot a^{-1} \odot c = d$. Since $\gamma(o)\gamma(p) = [o,o,i][a,b,i] = \langle i,-a \odot b^{-1},o \rangle$ and since $c \oplus -a \odot b^{-1} \odot d = o$, it follows that $[c,d,i] \in \langle i,-a \odot b^{-1},o \rangle$. Thus $\gamma(o)$, $\gamma(p)$, and $\gamma(q)$ are collinear in π_{D_π}.

Case 4.   Suppose that the points lie on a line N and that N does not pass through u,v,o. Let $N \cap M = a$, $N \cap M' = b'$, and $N \cap L = w$, and let p be another point on N distinct from a, b', and w. Also let $N' = ow$ (see Figure 76) and let $\gamma(p) = [c,d,i]$. Recalling that f_{-a} denotes the member of $\text{El}(u,L)$ mapping $o \to -a$, note that $f_{-a}: a \to o$ and that therefore, $f_{-a}: N \to N'$. Thus $f_{-a}(a), f_{-a}(b')$, and $f_{-a}(p)$ are on N', and so, by Case 3, their images under γ are collinear. We must now show that $\gamma(f_{-a}(a)) = [o,o,i]$, $\gamma(f_{-a}(b')) = [-a,b,i]$, and $\gamma(f_{-a}(p)) = [c \oplus -a,d,i]$. The first equality follows because $f_{-a}(a) = o$. The latter two equalities follow because $f_{-a}: o = b'v \cap M \to (f_{-a}(b'))v \cap M = -a$ and $f_{-a}: c = pv \cap M \to (f_{-a}(p))v \cap M = c \oplus -a$ and because $(f_{-a}(b'))u \cap M' = b'$ and $(f_{-a}(p))u \cap M' = d'$.

Since $\gamma(f_{-a}(a)) \gamma(f_{-a}(b')) = [o,o,i][-a,b,i] = \langle i,a \odot b^{-1},o \rangle$ and since $\gamma(f_{-a}(p)) = [c \oplus -a,d,i] \in \langle i,a \odot b^{-1},o \rangle$ it follows that $c \oplus -a \oplus d \odot a \odot b^{-1} = o$. Also $\gamma(a),\gamma(b') = [a,o,i][o,b,i] =$

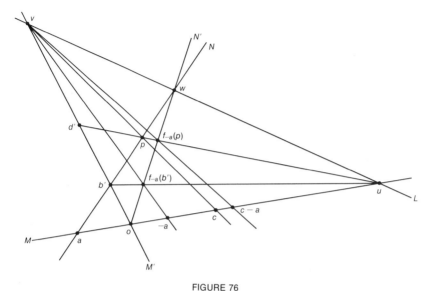

FIGURE 76

$\langle i, a \odot b^{-1}, -a \rangle$, and since $\gamma(p) = [c,d,i]$ and $c \oplus d \odot a \odot b^{-1} \oplus -a$
$= o$, it follows that $\gamma(p) \in \langle i, a \odot b^{-1}, -a \rangle$. Thus $\gamma(a)$, $\gamma(b')$, and
$\gamma(p)$ are collinear in π_{D_π}.

This completes the proof that collinearity is preserved on $\mathscr{P} - L$; the
preceding remarks indicate that γ is an isomorphic map from π to π_{D_π}.

Notice that we could have carried out the same coordinatization
process with affine planes α that satisfy the Second Affine Desargues
Theorem as we did with Second Desarguesian (L) planes π. The only
difference between D_α and D_π is that addition and multiplication in D_α
are defined in terms of translations and dilatations rather than elations
and homologies. These comments are formalized in the following
construction and theorem.

CONSTRUCTION 8.2.3   Let α be an affine plane satisfying the Second
Affine Desargues Theorem. Let M be an arbitrary line of α and let o
and i be distinct arbitrary points on M. Let $f_a \in \text{Trans}(M)$ such that

f_a: $o \rightarrow a$ and let $h_a \in \text{Dil}(o)$ such that h_a: $i \rightarrow a$ where $a \neq o$. Define the algebraic structure $D_\alpha = (M,+,\times)$ as follows:

$$a + b = f_b \circ f_a(o)$$
$$a \times b = h_b \circ h_a(i) \text{ for all } a,b \neq o$$
$$a \times o = o \times a = o \text{ for all } a \in M$$

THEOREM 8.2.4

    i. *The algebra D_α is a division ring.*

    ii. *If $(D,+,\cdot)$ is a division ring, then $(D,+,\cdot)$ is isomorphic to D_{α_D}.*

    iii. *If α satisfies the Second Affine Desargues Theorem, then α is isomorphic to α_{D_α}.*

    iv. *If α satisfies the Second Affine Desargues Theorem, then $\pi(\alpha^+)$ satisfies the Second Desargues (L_∞) Theorem and $D_{\pi(\alpha^+)}$ is isomorphic to D_α.*

Proof. The proof is left as an exercise.

The consequences of Theorem 8.2.2 are numerous and significant. One of the most celebrated results is the theorem stating that every finite Desarguesian plane is Pappian. This follows because all Desarguesian planes are of the form π_D and, by Wedderburn's Theorem, every finite plane of the form π_D is Pappian. Although this result is fundamentally a geometric theorem, its only known proof relies on algebra.

THEOREM 8.2.5   *Every finite Desarguesian plane is Pappian.*

Proof. The proof follows from Example 6.1.4(ii) and Theorems 7.1.11 and 8.2.2.

Another surprising consequence is that every plane π that satisfies the Second Desargues (L) Theorem for some L satisfies the Second Desargues (L) Theorem for all lines L. Equivalently, if all possible homologies exist with some given axis L, then all possible homologies exist in π.

THEOREM 8.2.6 *The plane π satisfies the Second Desargues (L) Theorem for some L if and only if π satisfies the Desargues Theorem.*

Proof. The proof follows from Theorems 7.3.3 and 8.2.2.

THEOREM 8.2.7 *The following statements are equivalent:*

π *is (p,L) transitive for all $p \notin L$.*
$\mathrm{Hom}(p,L)$ *is transitive for all $p \notin L$.*
$\mathrm{Hom}(p,L)$ *is transitive for all p and L such that $p \notin L$.*
π *is (p,L) transitive for all p and L.*

Proof. The proof is based on Corollary 5.2.2(ii) and (iv) and Theorems 5.2.4 and 8.2.5.

Similarly, if π satisfies the specialized Pappus (L) Theorem for some line L, then π satisfies the Pappus Theorem in general.

THEOREM 8.2.8 *The plane π satisfies the Pappus (L) Theorem for some line L if and only if π satisfies the Pappus Theorem.*

Proof. If π satisfies the Pappus (L) Theorem, then, by Theorem 6.2.1, π satisfies the Second Desargues (L) Theorem and thus $\pi \sim \pi_{D_\pi}$. By Theorem 8.1.5, D_π is a field, so π_{D_π} satisfies the Pappus Theorem. The converse of this theorem is immediate.

Theorems 8.2.6 and 8.2.8 justify the use of a rather specialized version of the Affine Desargues and Affine Pappus theorems. We see from these theorems that α satisfies the Second Affine Desargues Theorem exactly when its projective extension satisfies the Desargues Theorem. A similar statement holds for the Pappus Theorem.

THEOREM 8.2.9 *The affine plane α satisfies the Second Affine Desargues Theorem if and only if $\pi(\alpha^+)$ satisfies the Desargues Theorem; α satisfies the Affine Pappus Theorem if and only if $\pi(\alpha^+)$ satisfies the Pappus Theorem.*

Proof. The proof is based on Theorems 8.2.6 and 8.2.8.

Another consequence of Theorem 8.2.2 concerns the subplanes of a Desarguesian plane. We may completely catalog the different subplanes of a Desarguesian plane π if we know the subdivision rings of D where $D = D_\pi$. This result is stated formally in the following theorem.

THEOREM 8.2.10 ·*Let π be a Desarguesian plane and let π_D be an isomorphic copy of π. Then π' is a subplane of π if and only if $\mathscr{P}' \subseteq \mathscr{P}$ and $\pi' \sim \pi_{D'}$ where D' is a subdivision ring of D.*

Proof. The proof is left as an exercise.

COROLLARY 8.2.11   *The real projective plane contains no finite subplanes.*

The curious student may be interested to know that the subfields of the field $\mathrm{GF}(p^n)$ are of the form $\mathrm{GF}(p^k)$ where k divides n. Thus different possible subplanes of finite Desarguesian planes may easily be found.

Although the consequences of Theorems 8.2.5 through 8.2.10 and Corollary 8.2.10 are important, perhaps the most important aspect of Theorem 8.2.2 is the idea behind it: A geometric plane can be represented in algebraic terms. This idea is not new; the Cartesian coordinate plane was used to represent the Euclidean plane long before projective planes were invented. Even the coordinatization of Desarguesian and Pappian planes is classical—see Veblen and Young (1910). Nevertheless, the idea of representing planes algebraically is still the magic force responsible for the discovery of new planes and the synthesis of new ideas in projective geometry.

EXERCISES

1. Prove that a Desarguesian plane π satisfies FT-II if and only if π is Pappian.

2. Prove that if π is (a, L_o), (b, L_o), and (c, L_o) transitive for three distinct, noncollinear points a, b, and c not on L_o, then π is (p, L) transitive for all points p and lines L.

3. Prove Theorem 8.2.4.

4. Prove Theorem 8.2.10.

5. a. Show that no finite projective plane can be embedded in α_R, the Cartesian plane.
   b. Can any finite affine plane be embedded in α_R? If so, which one(s)?

6. State and prove a theorem for affine planes that serves the same purpose that Theorem 8.2.6 serves for planes satisfying the Second Desargues (L) Theorem.

7. A division ring D is *ordered* if and only if there exists a subset $S \subset D$ such that:

   (1) If $a \in D$, then exactly one of the following three alternatives holds: $a = 0$, $a \in S$, $-a \in S$.
   (2) If $a,b \in S$, then $a + b \in S$.
   (3) If $a,b \in S$, then $ab \in S$ and $ba \in S$.

   Write $a > b$ if and only if $a - b \in S$. The elements of S are called the *positive elements* of D.

   A projective plane is *ordered* if and only if there exists a relation O on ordered pairs of points of π that satisfies the following seven conditions (use the notation $((p,q),(r,s)) \in O$ if and only if $p,q\,|\,r,s$):

   (1) There exists a line with at least four points on it.
   (2) If $p,q\,|\,r,s$, then p,q,r,s are distinct collinear points.
   (3) If $p,q\,|\,r,s$, then $p,q\,|\,s,r$.
   (4) If $p,q\,|\,r,s$, then $p,r\,\slash\!\!\!|\,q,s$.
   (5) If p,q,r,s are distinct collinear points, then at least one of the following must hold: $q,r\,|\,p,s$; $p,r\,|\,q,s$; $p,q\,|\,r,s$.
   (6) If $p,q\,|\,r,s$ and $p,r\,|\,q,t$, then $p,q\,|\,s,t$.
   (7) If $p,q\,|\,r,s$ and f is a perspectivity where $f: p,q,r,s \to p',q', r',s'$, then $p',q'\,|\,r',s'$.

   If $p,q\,|\,r,s$, then p,q *separates* r,s.

   a. Give examples of ordered division rings.
   b. Show that an ordered division ring has characteristic 0.
   c. Let D be an ordered division ring. Define the following relation $/$ on ordered pairs of D: $p,q/r,s$ if and only if $R(p,q; r,s) = [k]$ and $k < 0$.
      (1) Show that $p,q/r,s$ is well defined.
      (2) Show that π_D is an ordered projective plane (under the relation $/$).
   d. Give some examples of ordered projective planes.
   e. Show that if π is an ordered Desarguesian plane, then $\pi \sim \pi_D$ where D is an ordered division ring. Hint: There exists a division ring D such that $\pi \sim \pi_D$. Let $D = L - \{u\}$ and define the set S on D as follows: $S = \{x: o,u\,\slash\!\!\!|\,i,x\}$. Show that D is an ordered division ring. (Exercise 7.4.5 is useful in this regard.)

8. An affine plane α is *ordered* if and only if there exists a ternary relation β on ordered collinear triples of points satisfying B1–B5:

B1. Every line of α contains at least three points.

B2. If $(a,b,c) \in \beta$, then $(c,b,a) \in \beta$.

B3. Exactly one of the following three triples is in β: (a,b,c), (b,c,a), (c,a,b).

B4. If (a,c,b) and $(b,e,c) \in \beta$, then $(a,c,e) \in \beta$.

B5. Let $a,b,c \in L$ and $a',b',c' \in L'$ be such that either aa', bb', and cc' are parallel, or $aa' \cap bb' = aa' \cap cc' = d$ and $(a,d,a'),(b,d,b')$, $(c,d,c') \notin \beta$. Then $(a,b,c) \in \beta$ if and only if $(a',b',c') \in \beta$.

If $(a,b,c) \in \beta$, then point b is *between* points a and c.

a. Show that the affine plane is ordered if and only if the projective plane $\pi(\alpha^+)$ is ordered.

b. Show that the ordered affine plane α satisfies the Second Affine Desargues Theorem if and only if $\alpha \sim \alpha_D$ for some ordered division ring D.

NON-DESARGUESIAN PLANES

In this part several non-Desarguesian planes are
introduced and are analyzed with the algebraic
methods that were developed
primarily in Part 1.

CHAPTER 9

TERNARY RINGS AND PROJECTIVE PLANES

9.1 TERNARY RINGS

The main topic of these last five chapters is the algebraic representation of non-Desarguesian planes. Chapter 8 showed that the division ring is the appropriate algebra for representing Desarguesian planes; naturally a different system is necessary for non-Desarguesian planes. Such a system, the nearfield, has already been mentioned in connection with $\pi_N(9)$ (see Section 2.4). The nearfield is just one of many algebraic systems that will be discussed in Part 3; the most general among these is the ternary ring.

DEFINITION 9.1.1 *A* ternary ring *is a system* $T = (R,t)$ *where R is a set and t is a ternary operation on R; that is,* $t: R \times R \times R \to R$ *such that*

   T1. *There exist elements* $0,1 \in R$ *such that* $0 \neq 1$, $t(0,a,b) = t(a,0,b) = b$, *and* $t(1,a,0) = t(a,1,0) = a$.

T2. *Given $a,b,c,d \in R$ such that $a \neq c$, there exists a unique $x \in R$ such that $t(x,a,b) = t(x,c,d)$.*

T3. *Given $a,b,c \in R$, there exists a unique $x \in R$ such that $t(a,b,x) = c$.*

T4. *Given $a,b,c,d \in R$ such that $a \neq c$, there exists a unique pair $(x,y) \in R \times R$ such that $t(a,x,y) = b$ and $t(c,x,y) = d$.*

DEFINITION 9.1.2   *The ternary rings (R,t) and (R',t') are isomorphic if and only if there exists a one-to-one mapping f from R onto R' such that $f(t(a,b,c)) = t'(f(a),f(b),f(c))$.*

EXAMPLE 9.1.3

i. Let $R = \{0,1\}$ and let t be defined as follows:

$$
\begin{array}{ll}
t(0,0,0) = 0 & t(0,1,1) = 1 \\
t(0,0,1) = 1 & t(1,0,1) = 1 \\
t(0,1,0) = 0 & t(1,1,0) = 1 \\
t(1,0,0) = 0 & t(1,1,1) = 0
\end{array}
$$

ii. Let $R = \{0,1,2\}$, let $(R,+,\cdot)$ be the field $J/3$, and let $t(a,b,c) = ab + c$.

iii. Let $(R,+,\cdot)$ be the right nearfield $N(9) = (N,+,\cdot)$ of Construction 2.4.2. Let

$$
t(a,b,c) = \begin{cases} ba + c \text{ if } b = 0,1,2 \\ b(a + d) + d' \text{ if } b \neq 0,1,2 \text{ and } d \text{ and } d' \text{ are such} \\ \quad \text{that } d,d' = 0,1,2 \text{ and } c = db + d' \end{cases}
$$

That such a unique pair d and d' exists can be checked from Table 10 (p. 61). Note that the a,b,c,d in the preceding notation should not be confused with the $a,b,c,d, \ldots$ of the nearfield.

It is laborious to check that these examples, especially (iii), are indeed ternary rings. Ternary rings are generally clumsy to work with unless there is a rule to follow as there is in (ii). In fact, the rule given in (ii) is a special example of a general rule that may be used to generate a class of ternary rings.

THEOREM 9.1.4   *If $(R,+,\cdot)$ is a division ring, then (R,t) is a ternary ring where t is defined by $t(a,b,c) = a \cdot b + c$.*

Proof.

  i. Clearly, T1 of Definition 9.1.1 is satisfied.

 ii. Let $a,b,c,d \in R$ such that $a \neq c$. Let $x = (d - b)(a - c)^{-1}$; then it is obvious that T2 is satisfied.

iii. Let $a,b,c \in R$. If $x = c - ab$, clearly condition T3 is satisfied.

 iv. For T4 let $a,b,c,d \in R$ be given such that $a \neq c$.

Since $a = c = 0$ is impossible, suppose that either $a \neq 0$ or $c \neq 0$. If $c \neq 0$, we may let $y = b - ax$ and $x = c^{-1}(d - y)$; it is then clear that $ax + y = b$ and $cx + y = d$. Thus (x,y) is a unique pair satisfying these equalities. If $a \neq 0$, then $x = a^{-1}(b - y)$ and $y = d - cx$ satisfy the appropriate conditions.

The ternary rings of Theorem 9.1.4 are called linear ternary rings. The concept of linearity on a ternary ring will now be explored more fully.

DEFINITION 9.1.5 *A* loop *is a set L with a binary operation* $\circ$ *such that if $x,y \in L$, then*

  L1. *There exists a unique $z \in L$ such that $x \circ z = y$.*

  L2. *There exists a unique $w \in L$ such that $w \circ x = y$.*

  L3. *There exists a unique $e \in L$ such that $e \circ x = x \circ e = x$ for all $x \in L$.*

EXAMPLE 9.1.6 Note that any Latin square table will represent a finite loop if the additional condition of an identity is met (see Table 19).

TABLE 19

$\circ$	e	a	b	c	d	f
e	e	a	b	c	d	f
a	a	e	d	b	f	c
b	b	d	e	f	c	a
c	c	b	f	e	a	d
d	d	f	c	a	e	b
f	f	c	a	d	b	e

THEOREM 9.1.7 *Let (R,t) be a ternary ring and define $+$ and $\circ$ as follows: $a + b = t(1,a,b)$; $a \circ b = t(a,b,0)$. Then $(R,+)$ and $(R - \{0\},\circ)$ are loops where 0 is the identity of $(R,+)$ and 1 is the identity of $(R - \{0\},\circ)$.*

Proof. First let us examine $(R,+)$.

   i. Let $x,y \in R$. Then by T3 there exists a unique $z \in R$ such that $t(1,x,z) = y$. Thus there is a unique z such that $x + z = y$ and thus L1 is true.

   ii. Let $x,y \in R$. Then by T4 there exists a unique pair (z,w) such that $t(1,z,w) = y$ and $t(0,z,w) = x$. Since $t(0,z,w) = w$, $w = x$. Thus $t(1,z,x) = y$. We may conclude that there is a unique z such that $z + x = y$, and so L2 is satisfied.

   iii. It is easily checked that 0 is the identity because $t(1,0,x) = t(1,x,0) = x$.

Next let us examine $(R - \{0\},\circ)$. To show that $R - \{0\}$ is closed under $\circ$, assume that $x \circ y = 0$ and that $y \neq 0$. Because $t(0,0,0) = 0$ and $t(x,y,0) = 0$, x must be 0 by T2.

   iv. Let $x,y \in R - \{0\}$. By T4 there exists a unique pair (z,w) such that $t(x,z,w) = y$ and $t(0,z,w) = 0$. Since $t(0,z,w) = 0$ implies that $w = 0$, it follows that $t(x,z,0) = y$. We conclude that $xz = y$ and so L1 is true.

   v. Let $x,y \in R - \{0\}$. By T2 there exists a unique z such that $z \circ x = t(z,x,0) = t(z,0,y) = y$. This establishes L2.

   vi. Clearly, 1 is the identity of $(R - \{0\},\circ)$ because $t(1,x,0) = t(x,1,0) = x$.

Notation. P_T denotes the algebra $(R,+,\circ)$ generated by T where $+$ and $\circ$ are as given in Theorem 9.1.7.

EXAMPLE 9.1.8

   i. Let T be the ternary ring of Example 9.1.3(i); then P_T is easily seen to be $J/2$.

   ii. Let T be a ternary ring (R,t) where $t(a,b,c) = a \cdot b + c$ and $(R,+,\circ)$ is a division ring. Then $P_T = (R,+,\circ)$. Verification of this fact is left to the student.

iii. Let T be the ternary ring of Example 9.1.3(iii). The student will find it a challenging exercise to verify that $P_T = N^d(9)$. ($N(9)$ is given in Construction 2.4.2; $N^d(9) = (R,+,\odot)$ where $a \odot b = b \cdot a$.)

Theorem 9.1.4 showed that for division rings $D = (R,+,\cdot)$, we may form a ternary ring $T = (R,t)$ by defining $t(a,b,c)$ by $a \cdot b + c$. From this ternary ring T we may form an algebra P_T, which, as indicated in Example 9.1.8(ii), turns out to be the original division ring D. This phenomenon occurs for lesser structures than division rings. In fact, if $P = (R,+,\circ)$ and $P = P_T$ for some ternary ring $T = (R,t)$ where $t(a,b,c) = a \circ b + c$, then the same state of affairs holds. Such algebras as P are called planar rings; the generating ternary ring is called a linear ternary ring. These terms are defined more precisely in the definition and theorem that follow.

DEFINITION 9.1.9

i. A *ternary ring* (R,t) *is a* linear ternary ring *if and only if* $t(a,b,c)$ $= t(1,t(a,b,0),c)$; *that is,* $t(a,b,c) = a \circ b + c$ *where* $a \circ b + c$ *is defined in* P_T.

ii. $P = (R,+,\circ)$ *is a* planar ring *if and only if* $P = P_T$ *for some linear ternary ring* T.

THEOREM 9.1.10 *If* $P = (R,+,\cdot)$ *is a planar ring and* $t: R \times R \times R$ $\rightarrow R$ *is defined by* $t(a,b,c) = a \cdot b + c$, *then* $T = (R,t)$ *is a linear ternary ring and* $P_T = P$.

Proof. The proof is left as an exercise.

Notice that Theorem 9.1.10 implies that a planar ring is generated by a unique linear ternary ring T.

Notation. The linear ternary ring that generates a planar ring P is denoted by T_P.

THEOREM 9.1.11 *If* T *is a linear ternary ring, then* $T_{P_T} = T$. *If* P *is a planar ring, then* $P_{T_P} = P$.

Proof. The proof is left as an exercise.

THEOREM 9.1.12   *The system $(R,+,\cdot)$ is a planar ring if and only if the following conditions are satisfied:*

   i. *$(R,+)$ and $(R - \{0\},\cdot)$ are loops.*

  ii. *$0 \cdot a = a \cdot 0 = 0$ for all $a \in R$.*

 iii. *Given $a,b,c,d \in R$ such that $a \neq c$, there exists a unique $x \in R$ such that $xa + b = xc + d$.*

 iv. *Given $a,b,c \in R$, there exists a unique $x \in R$ such that $ab + x = c$.*

  v. *Given $a,b,c,d \in R$ such that $a \neq c$, there exists a unique pair $(x,y) \in R \times R$ such that $ax + y = b$ and $cx + y = d$.*

Proof.   The proof is left as an exercise.

EXAMPLE 9.1.13

   i. A division ring is clearly a planar ring.
  ii. The nearfield $N(9)$ is a planar ring.
 iii. The nearfield $N^d(9)$ is also a planar ring.

The details of (ii) and (iii) are left for the student to verify.

    Notice that a planar ring may be generated by a nonlinear ternary ring as well as by a ternary ring. The planar ring $P = N^d(9)$ may be generated by the linear ternary ring T_P and also by the ternary ring T of Example 9.1.3(iii) (see Example 9.1.8(iii)). Clearly, then, for the ternary ring T of Example 9.1.3(iii), $T \neq T_{P_T}$. It is even possible to generate a field with a nonlinear ternary ring (see Pickert, 1956).

EXERCISES

  1. Show that if T is the ternary ring of Example 9.1.3(iii), then $P_T = N^d(9)$.

  2. Prove Theorem 9.1.10.

  3. Prove Theorem 9.1.11.

  4. Prove Theorem 9.1.12.

  5. Refer back to exercise 2.1.6. Show that a Cartesian group is a planar ring.

*6. If $P = (R,+,\cdot)$ is a planar ring, then $(R,+,\circ)$ is called the *dual of P* where $x \circ y = y \cdot x$. If $T = (R,t)$ is a ternary ring, then (R,t') is called the *dual of T* where $t'(x,y,z) = t(y,x,z)$.

    a. Is the dual of a planar ring also a planar ring?

    b. Is the dual of a ternary ring also a ternary ring?

*7. Verify that (R,t) of Example 9.1.3(iii) is a ternary ring.

9.2 TERNARY RINGS AND PROJECTIVE PLANES

Although a ternary ring is a strange and awkward algebraic system, it is ideal for representing a projective plane. Each axiom is tailored for this role, as will be shown in this section.

CONSTRUCTION 9.2.1   Let $T = (R,t)$ be a given ternary ring. Let

$$\mathscr{P} = \{[x,y],[x],z \colon x,y \in R\}$$
$$\mathscr{L} = \{\langle x,y \rangle, \langle x \rangle, Z \colon x,y \in R\}$$
$$\mathscr{I} = \{([x,y],\langle m,k \rangle)) \colon t(x,m,k) = y; \, x,y,m,k \in R\}$$
$$\quad \cup \; \{([x,y],\langle x \rangle) \colon x,y \in R\} \; \cup \; \{([x],\langle x,y \rangle) \colon x,y \in R\}$$
$$\quad \cup \; \{(z,\langle x \rangle) \colon x \in R\} \; \cup \; \{([x],Z) \colon x \in R\} \; \cup \; \{(z,Z)\}$$

Here we assume that $\mathscr{P} \cap \mathscr{L} = \varnothing$, that z is distinct from points of the form $[x,y]$ and $[x]$, and that Z is distinct from lines of the form $\langle x,y \rangle$ and $\langle x \rangle$.

Notice that if (R,t) is a linear ternary ring, then $[x,y] \in \langle m,k \rangle$ if and only if $y = x \circ m + k$. In analytic geometry this is recognized as the familiar slope intercept form of a straight line.

THEOREM 9.2.2   *The triple* $(\mathscr{P},\mathscr{L},\mathscr{I})$ *as defined in Construction 9.2.1 is a projective plane.*

Proof. First we show that there exists exactly one line passing through two points p and q.

Case 1.   Let $p = [a,b]$ and let $q = [c,d]$.

Subcase A. $a = c$. Then p and q lie on $\langle a \rangle$. To show that this line is unique, observe that p and q certainly do not occupy $\langle k \rangle$ for

$k \neq a$. Also if $p,q \in \langle m,k \rangle$, then $b = t(a,m,k) = d$, and so $p = q$, a contradiction.

Subcase B. $a \neq c$. By T4 of Definition 9.1.1, there exists a unique $(m,k) \in R \times R$ such that $t(a,m,k) = b$ and $t(c,m,k) = d$. Thus $\langle m,k \rangle$ contains both points. Clearly, neither Z nor any line $\langle k \rangle$ may pass through both points, so the line $\langle m,k \rangle$ is unique.

Case 2.   Let $p = [a,b]$ and let $q = [c]$. By T3 there exists a unique $k \in R$ such that $t(a,c,k) = b$. Thus $[a,b] \in \langle c,k \rangle$ and $[c] \in \langle c,k \rangle$. It is easily checked that no other line contains p and q.

Case 3.   Let $p = [a,b]$ and let $q = z$. The line $\langle a \rangle$ is easily seen to be the only line containing p and q.

Case 4.   Let $p = [a]$ and let $q = [b]$. Here Z is the only line containing p and q.

Case 5.   Let $p = [a]$ and let $q = z$. Here also Z is the unique line containing both points.

A similar argument shows that two lines intersect in exactly one point.

The fact that points $[0,0]$, $[0]$, $[1,1]$, and z are a four-point is easily verified.

Notation.   This projective plane is denoted by π_T.

EXAMPLE 9.2.3   Let T be the ternary ring of Example 9.1.3(i). Thus the points are $[0,0]$, $[0,1]$, $[1,0]$, $[1,1]$, $[0]$, $[1]$, and z; the lines are $\langle 0,0 \rangle$, $\langle 0,1 \rangle$, $\langle 1,0 \rangle$, $\langle 1,1 \rangle$, $\langle 0 \rangle$, $\langle 1 \rangle$, and Z. We know that

$$\langle 0,0 \rangle = \{[0],[0,0],[1,0]\}$$
$$\langle 0 \rangle = \{z,[0,0],[0,1]\}$$
$$\langle 1,0 \rangle = \{[0,0],[1,1],[1]\}$$
$$\langle 0,1 \rangle = \{[1,1],[0,1],[0]\}$$
$$\langle 1 \rangle = \{z,[1,1],[1,0]\}$$
$$\langle 1,1 \rangle = \{z,[0],[1]\}$$
$$Z = \{z,[0],[1]\}$$

so π_T is clearly $\pi(2)$.

This process can be reversed to construct a ternary ring T from an arbitrary projective plane π. But before this is done for general projective planes, the construction will be carried out within the familiar confines of the Cartesian plane α_R. Specifically, a geometric method will be generated for constructing the operation $\alpha\beta + \gamma$ on the real numbers α, β, and γ.

Let us begin with Figure 77, consisting of the two axes, the origin $o = (0,0)$, the point $e = (1,1)$, and the three points $a = (\alpha,0)$, $b = (\beta,0)$, and $c = (\gamma,0)$ on the x-axis. (We shall assume that all points are distinct and different from o. Denoting by L_p the line through the point p parallel to the y-axis and by M_p the line through p parallel to the x-axis, we construct the following points:

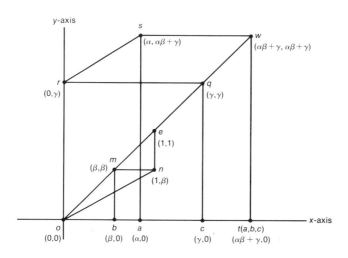

FIGURE 77

$$m = oe \cap L_b, \text{ thus } m = (\beta,\beta)$$
$$n = L_e \cap M_m, \text{ thus } n = (1,\beta)$$
$$q = oe \cap L_c, \text{ thus } q = (\gamma,\gamma)$$
$$r = L_o \cap M_q, \text{ thus } r = (0,\gamma)$$
$$s = L_a \cap N_r \text{ where } N_r \text{ is the line through } r \text{ parallel to the line}$$
$$\text{on; thus } N_r \text{ is the line } y = \beta x + \gamma, \text{ and } s = (\alpha, \alpha\beta + \gamma)$$
$$w = oe \cap M_s, \text{ thus } w = (\alpha\beta + \gamma, \alpha\beta + \gamma)$$
$$t(a,b,c) = M_o \cap L_w, \text{ thus } t(a,b,c) = (\alpha\beta + \gamma, 0)$$

We see that the ternary operation t defined on the points of the x-axis reflects the operation $\alpha\beta + \gamma$ defined on the real numbers α, β, and γ.

With the addition of the line at infinity, this construction may be carried out for general projective planes.

CONSTRUCTION 9.2.4   Let $\pi = (\mathscr{P}, \mathscr{L}, \mathscr{I})$ be a given projective plane
and let u, v, o, e be a given four-point. Define (R, t) as follows: $R = ou - \{u\}$ and $t(a,b,c) = vw \cap ou$ where $a, b, c \in R$ and w is defined
as in Figure 77. Let

$$m = vb \cap oe \qquad n = um \cap ve$$
$$p = on \cap uv \qquad q = cv \cap oe$$
$$r = uq \cap ov \qquad s = pr \cap av$$
$$w = su \cap oe$$

It must be verified that each point exists (see Figure 78).

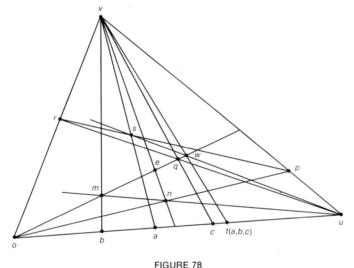

FIGURE 78

THEOREM 9.2.5   *The structure (R, t) of Construction 9.2.4 is a ternary ring.*

Proof.

i. First it must be shown that o and $ve \cap ou = i$ play the roles of
0 and 1, respectively.
   Let $a = o$. Then $s = pr \cap av = r$ and $w = ru \cap oe = q$, so
$t(o,b,c) = vq \cap ou = c$.
   Let $b = o$. Then $m = vo \cap oe = o$, $n = uo \cap ve$, $p = on \cap uv$
$= u$ because $n \in ou$, $s = ur \cap av$, $w = su \cap oe = q$ because
$s \in ru$, and thus $t(a,o,c) = vq \cap ou = c$.

Let $c = o$ and $a = i$. Then $q = ov \cap oe = o$, $r = uo \cap ov = o$, $s = po \cap iv = po \cap ev = n$, and $w = nu \cap oe = m$, so $t(i,b,o) = vm \cap ou = b$.

Let $c = o$ and $b = i$. Then $m = vi \cap oe = ve \cap oe = e$, $n = ue \cap ve = e$, $p = oe \cap uv$, $q = r = o$ (as indicated in the preceding paragraph), $s = po \cap av = oe \cap av$ because $p \in oe$, and $w = su \cap oe = s$, so $t(a,i,o) = vw \cap ou = vs \cap ou = av \cap ou = a$.

ii. Let $b,c,b',c' \in R$ be given such that $b \neq b'$. Let b' generate p' and let c' generate r' in the same manner that b generates p and c generates r in Construction 9.2.4; it is easily checked that $p'r' \neq pr$. Let $s' = pr \cap p'r'$ and let $a = vs' \cap ou$. Define $t(a,b',c')$ with respect to s' in the same manner that $t(a,b,c)$ is defined by s; then $t(a,b,c) = t(a,b',c')$. Furthermore, a is the unique point yielding this equality. Thus T2 of Definition 9.1.1 is true.

iii. Let $a,b,c \in R$ be given. Now w, s, r, q, and c may be defined as follows: $w = cv \cap oe$, $s = wu \cap av$, $r = ps \cap ov$, $q = ru \cap oe$, and $x = vq \cap ou$. Let q,r,s,w assume the same roles as in Construction 9.2.4; then it is obvious that $t(a,b,x) = c$. Furthermore, x is unique, so T3 is satisfied.

iv. Let $a,a',x,x' \in R$ be given such that $a \neq a'$. Also let $w = xv \cap oe$, $w' = x'v \cap oe$, $s = uw \cap av$, and $s' = uw' \cap a'v$. By construction, $s \neq s'$, so we may formulate the following definitions for p,r,q, c,n,m,b:

$$p = ss' \cap vu \qquad r = ss' \cap ov$$
$$q = ru \cap oe \qquad c = vq \cap ou$$
$$n = po \cap ve \qquad m = un \cap oe$$
$$b = vm \cap ou$$

Letting these letters play the same roles as in Construction 9.2.4, we see that $t(a,b,c) = x$ and $t(a',b,c) = x'$. Also (b,c) is the unique pair yielding these equations. Thus T4 is satisfied.

Notation. The ternary ring of Construction 9.2.4 is denoted by $T_{\pi(u,v,o,e)}$ or $T(u,v,o,e)$.

Notice that the notation reflects that this ternary ring is defined in terms of a given four-point u,v,o,e. Presumably another four-point could generate another ternary ring; this possibility is discussed in Section 9.3.

Section 8.1 showed that every Desarguesian plane is associated with a division ring and that every division ring is related to a Desarguesian plane. Theorem 8.2.2 provided the key to these relationships by exhibiting an isomorphism between a given Desarguesian plane π and the plane generated by the division ring associated with π. This established the fundamental theorem that a Desarguesian plane can be represented by a division ring. Theorem 9.2.6 is the analogous theorem for projective planes and ternary rings.

THEOREM 9.2.6 *Let π be an arbitrary projective plane. Then $\pi \sim \pi_{T(u,v,o,e)}$ where u,v,o,e is an arbitrary four-point of π.*

Proof. Let $\mathscr{P}$ denote the points of π, let $\mathscr{P}_T$ denote the points of $\pi_{T(u,v,o,e)}$, and let f be a function from $\mathscr{P}$ onto $\mathscr{P}_T$ defined as follows (see Figure 79):

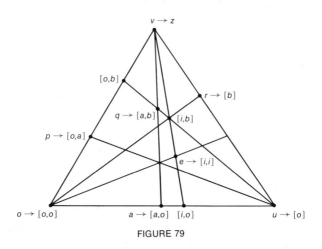

FIGURE 79

For points on ou, f: $a \to [a,o]$ for $a \in R = ou - \{u\}$, $u \to [o]$.

For points on ov, f: $p = ov \cap u(oe \cap av) \to [o,a]$, $v \to z$.

For points not on the triangle ouv, $q = av \cap (f^{-1}[o,b])u \to [a,b]$.

For points on uv, $r = uv \cap (f^{-1}[i,b])o \to [b]$.

Recall that $i = ve \cap ou$.

Proof that f is a one-to-one, onto mapping is left to the student. The present proof can be completed by showing that f preserves incidence. As in Theorem 8.2.2, this is done for the affine plane $\alpha(\pi_{uv})$ only.

Case 1. Suppose that L passes through v. Let $a = L \cap ou$, let p be a third point on L, and let $b = up \cap ov$. Then $f(L) = f(v)f(a)$ $= z[a,o] = \langle a \rangle$. Also $f(p) = [a,c]$ where $c \in R$ such that $f(b)$ $= [o,c]$. Thus $f(p) \in \langle a \rangle$ also, so $f(p) \in f(L)$, and incidence is preserved.

Case 2. Suppose that L does not pass through v. Then let $L \cap uv = p$, let $L \cap ov = r$, and let s be any point on pr. Then $f(p) = [b]$, $f(r) = [o,c]$, and $f(s) = [a,x]$ for some $a,b,c,x \in R$. It is easily verified that $f(p)f(r) = \langle b,c \rangle$, so $f(s) \in f(p)f(r)$ iff $x = t(a,b,c)$. The proof is concluded by showing that $s \in pr$ implies that $x = t(a,b,c)$.

Suppose that $s \in pr$. An examination of $t(a,b,c)$ reveals the following: $f(p) = [b]$, so $p = uv \cap (f^{-1}[1,b])o$; $f(r) = [o,c]$, so $r = ov \cap u(oe \cap cv)$; and $f(s) = [a,x]$, so $s = av \cap (f^{-1}[ox])u$. Also $s \in pr$ and $s \in av$, so $s = av \cap pr$. From this information, it is easily checked that s,p,r may assume the same roles as in Construction 9.2.4.

Let $y = f^{-1}[o,x]$. Then $f(y) = [o,x]$ and so $y = ov \cap u(oe \cap xv)$. Let $w = oe \cap xv$; then $y = ov \cap uw$. It is easily verified that $s \in yu$ and that $w \in yu$, so $w \in su$ and $w = su \cap oe$. But this means that w also denotes the same point as in Construction 9.2.4.

From the preceding information, it is easily seen that $x = ou \cap vw$. But by definition, $t(a,b,c) = ou \cap vw$ also. Thus $t(a,b,c) = x$ and the proof is complete.

This is the key theorem in Part 3. Every plane from now on will consist of points and lines denoted by objects from a ternary ring. For the original paper on projective planes and ternary rings, see Hall (1943).

There is another relationship between the plane and the ternary ring. Theorem 8.2.1 has established this relationship for division rings and Desarguesian planes as follows: There is an isomorphism between an arbitrary division ring D and the division ring D' associated with the plane generated by D. Theorem 9.2.7 states the analogous relationship for ternary rings and projective planes.

THEOREM 9.2.7 *Let T be an arbitrary ternary ring. Then*
$T \sim T_{\pi_T([0],z,[0,0],[1,1])}$.

Proof. The proof is left as an exercise.

If T is a linear ternary ring, it has an associated planar ring P_T. The plane π_T could be represented just as well by P_T as by T. Accordingly, the following natural construction is introduced for a plane represented by a planar ring.

CONSTRUCTION 9.2.8 Let $P = (R,+,\cdot)$ be an arbitrary planar ring. Let

$$\mathcal{P} = \{[x,y],[x],z: x,y \in R\}$$
$$\mathcal{L} = \{\langle x,y \rangle, \langle x \rangle, Z: x,y \in R\}$$
$$\mathcal{I} = \{([x,y],\langle m,k \rangle): y = xm + k; x,y,m,k \in R\}$$
$$\cup \{([x,y],\langle x \rangle): x,y \in R\}$$
$$\cup \{([x],\langle x,y \rangle): x,y \in R\} \cup \{(z, \langle x \rangle): x \in R\}$$
$$\cup \{([x],z): x \in R\} \cup \{(z,Z)\}$$

It is clear from this construction that $(\mathcal{P},\mathcal{L},\mathcal{I})$ is a projective plane isomorphic to the plane defined by the ternary ring T_P. This fact is stated formally in the following theorem.

THEOREM 9.2.9 *The triple $(\mathcal{P},\mathcal{L},\mathcal{I})$ defined in Construction 9.2.8 is a projective plane π, and the mapping*

$$f: [x,y] \rightarrow [x,y]$$
$$[x] \rightarrow [x]$$
$$z \rightarrow z$$

is an isomorphism from π to π_{T_P}.

Proof. The proof is left as an exercise.

Notation. The plane of Construction 9.2.8 is denoted by π_P.

THEOREM 9.2.10 *If T is a linear ternary ring, then the mapping*

$$f: [x,y] \rightarrow [x,y]$$
$$[x] \rightarrow [x]$$
$$z \rightarrow z$$

is an isomorphism from π_T onto π_{P_T}.

Proof.   The proof is left as an exercise.

The linearity condition is necessary in this theorem; for example, if T is the ternary ring of Example 9.1.3(iii), then $P_T = N^d(9)$, but $\pi_{N^d(9)}$ $\neq \pi_{P_T}$. $(\pi_{N^d(9)} = (\pi_N(9))^d$ and $\pi_T = \pi_H(9)$; the former fact is shown in Chapter 12; the latter is shown in Section 9.4.)

This section is concluded with a remark about the different representations of Desarguesian planes. If π is Desarguesian, then presumably there are three ways of representing it: with a linear ternary ring T, with a planar ring P, and with a division ring D.

We know that if $P = P_T$ or if $T = T_P$, then we have the natural isomorphism between π_P and π_T. Since D is a planar ring, it would be convenient if there were a natural correspondence between $\mathscr{P}_P$ and $\mathscr{P}_D$ — in fact, there is, after some minor juggling. Let $D = (S,+,\cdot)$ and define $D^d = (S,+,\circ)$ as follows: $a \circ b = b \cdot a$. Then the following theorem can be stated.

THEOREM 9.2.11   *Let D be a given division ring and let P_D denote the division ring as a planar ring. Then the mappings $f: \mathscr{P}_{P_D} \to \mathscr{P}_{D^d}$ and $F: \mathscr{L}_{P_D} \to \mathscr{L}_{D^d}$ defined by:*

$$f: \quad [x,y] \to [x,y,1]$$
$$[x] \to [1,x,0]$$
$$z \to [0,1,0]$$

and

$$F: \quad \langle m,k \rangle \to \langle m,-1,k \rangle$$
$$\langle k \rangle \to \langle -1,0,k \rangle$$
$$Z \to \langle 0,0,1 \rangle$$

yield an isomorphism from π_{P_D} onto π_{D^d}.

Proof.   The proof is left as an exercise (cf. exercise 7.2.2(b)).

That D as a division ring and D as a planar ring should represent dual planes rather than the same plane (note that $\pi_{D^d} \sim \pi_D{}^d$ as in exercise 7.2.3) is an awkward state of affairs that could have been avoided by adopting different conventions. For example, in Chapter 7 incidence in π_D could have been defined as $[x,y,z] \in \langle a,b,c \rangle$ if and only if $xa + yb + zc = 0$; or in this chapter the operation t for T_P could have been defined by $t(a,b,c) = ba + c$ and incidence in π_{P_T} could have been altered accordingly; or incidence in π_T could have been defined by $[x,y] \in \langle m,k \rangle$ if and only if $y = t(m,x,k)$, a convention used by Pickert (1955). The system adopted here is convenient except in this instance, and since Desarguesian planes are not central to Part 3, such a system will be utilized throughout the remaining text. The important fact still remains that a plane is Desarguesian if and only if it is representable by a planar ring that is a division ring. This fact follows easily from Theorem 9.2.11 and is stated formally in the following theorem.

THEOREM 9.2.12 *A plane π is Desarguesian (Pappian) if and only if $\pi \sim \pi_T$ for some linear ternary ring T whose associated planar ring P_T is a division ring (field).*

Proof. The proof is left as an exercise.

Notice that the linearity condition on T is necessary in this theorem, as it was in Theorem 9.2.10. Pickert's example of a nonlinear ternary ring T, whose associated planar ring P_T is a field, is constructed from a non-Desarguesian plane π; that is, one of the coordinatizing ternary rings $T(u,v,o,e)$ of π has the property that $P_{T(u,v,o,e)}$ is a field.

EXERCISES

1. Show that in Theorem 9.2.2, two lines intersect at exactly one point.

2. Prove Theorem 9.2.7.

3. Prove Theorem 9.2.9.

4. Prove Theorem 9.2.10.

5. Prove Theorem 9.2.11.

6. Prove Theorem 9.2.12.

*7. a. Describe the points and lines of $(\pi_T)^d$. Can you characterize a ternary ring T' such that $\pi_{T'} \sim (\pi_T)^d$?

   b. Let P be a planar ring. Describe the points and lines of $(\pi_P)^d$. Does there exist a planar ring P' such that $\pi_{P'} \sim (\pi_P)^d$?

9.3 ISOTOPISM

There is one major drawback to the method of representation described in Section 9.2: The ternary ring that was constructed to represent π is dependent on a fixed four-point u,v,o,e. In general, if another four-point had been chosen, a nonisomorphic ternary ring would result. This difficulty was not encountered in Chapter 8 because the fact that $C(\pi)$ is transitive on four-points means that one four-point is not geometrically different from any other four-point in a Desarguesian plane π. In fact, this condition of being related by a collineation or, more generally, by an isomorphism, is exactly what ensures that two four-points generate isomorphic ternary rings.

THEOREM 9.3.1 *Let π and π' be given projective planes and let u,v,o,e and u',v',o',e' be two given four-points in π and π', respectively. Then $T(u,v,o,e) \sim T(u',v',o',e')$ if and only if there exists an isomorphism f from π onto π' such that $f: u,v,o,e \rightarrow u',v',o',e'$.*

Proof. Suppose that f is an isomorphism such that $f: u,v,o,e \rightarrow u',v',o',e'$. Let $T(u,v,o,e) = T = (R,t)$ and let $T(u',v',o',e') = T' = (R',t')$. By the definition of R and R', f is a one-to-one map from R onto R'. Furthermore, f preserves the ternary operation because t and t' are defined strictly in terms of points, lines, and incidence in π and π', and these are preserved by f. Thus $T(u,v,o,e) \sim T(u',v',o',e')$.

Suppose that h is an isomorphism from T onto T' and define the mappings f and F as follows: f maps the points of π_T onto the points of $\pi_{T'}$ such that

$$f: \quad [x,y] \rightarrow [h(x),h(y)] \qquad \text{where } x,y \in R$$
$$[x] \rightarrow [h(x)] \qquad \text{where } x \in R$$
$$z \rightarrow z' \qquad \text{where } z' \text{ represents the point } v' \text{ in } \pi_{T'}$$

and F maps the lines of π_T onto the lines of $\pi_{T'}$ such that

$$F: \quad \langle m,k \rangle \rightarrow \langle h(m),h(k) \rangle \qquad \text{where } m,k \in R$$
$$\langle k \rangle \rightarrow \langle h(k) \rangle \qquad \text{where } k \in R$$
$$Z \rightarrow Z' \qquad \text{where } Z' \text{ represents the line } u'v' \text{ in } \pi_{T'}$$

Incidence is preserved because $y = t(x,m,k)$ if and only if $h(y) = t'(h(x),h(m),h(k))$. Thus (f,F) is an isomorphism from π_T onto $\pi_{T'}$. Let g be the isomorphism from π onto π_T and let k be the isomorphism from π' onto $\pi_{T'}$ as described in Theorem 9.2.6. Then j is an isomorphism from π onto π' such that j: $u,v,o,e \rightarrow u',v',o',e'$ where $j = k^{-1} \circ f \circ g$.

In general, the relationship between the ternary rings generated by two four-points in two isomorphic planes is not completely known. However, some understanding of this relationship can be gained from an understanding of the concept of isotopism.

DEFINITION 9.3.2   *Let* $T = (R,t)$ *and* $T' = (R',t')$ *be two ternary rings. Then* T *is* isotopic *to* T' *if and only if there exist three functions* f,g,h, *each one-to-one from* R *onto* R' *such that* $h(0)=0$ *and* $h(t(a,b,c)) = t'(f(a),g(b),h(c))$.

Notation.   The isotopism of Definition 9.3.2 is denoted by (f,g,h).

It is easily checked that an isotopism is an equivalence relation. Also if $f = g = h$, an isomorphism exists between T and T'. Nontrivial examples of isotopic mappings will be given in Chapter 11.

As proved in Theorem 9.3.1, isomorphic ternary rings represent isomorphic projective planes. Isomorphism is a stronger condition than is necessary in this regard. In fact, isotopic ternary rings represent isomorphic planes, as shown in Theorem 9.3.3.

THEOREM 9.3.3   *If* $T = (R,t)$ *and* $T' = (R',t')$ *are isotopic ternary rings, then* π_T *and* $\pi_{T'}$ *are isomorphic.*

Proof.   Define functions j and J as follows:

j: $\mathscr{P}_T \rightarrow \mathscr{P}_{T'}$ where j: $[x,y] \rightarrow [f(x),h(y)]$ where $x,y \in R$

$$[x] \rightarrow [g(x)] \text{ where } x \in R$$
$$z \rightarrow z'$$

J: $\mathscr{L}_T \rightarrow \mathscr{L}_{T'}$ where J: $\langle m,k \rangle \rightarrow \langle g(m),h(k) \rangle$ where $m,k \in R$

$$\langle k \rangle \rightarrow \langle f(k) \rangle \text{ where } k \in R$$
$$Z \rightarrow Z'$$

Clearly, both j and J are one-to-one, onto mappings. The proof is completed by showing that incidence is preserved.

Case 1.   Let $[x,y] \in \langle m,k \rangle$. Therefore, $y = t(x,m,k)$, and so $h(y)$ $= h(t(x,m,k)) = t'(f(x),g(m),h(k))$, that is, $[f(x),h(y)] \in \langle g(m),h(k) \rangle$. Since $j([x,y]) = [f(x),h(y)]$ and $J(\langle m,k \rangle) = \langle g(m),h(k) \rangle$, it follows that $j([x,y]) \in J(\langle m,k \rangle)$.

Case 2.   Let $[x,y] \in \langle k \rangle$. Thus $x = k$. Since $j([x,y]) = [f(x),h(g)]$ and $J(\langle k \rangle) = J(\langle x \rangle) = \langle f(x) \rangle$, it follows that $j([x,y]) \in J(\langle k \rangle)$.

Case 3.   Let $[x] \in \langle m,k \rangle$. Thus $x = m$. Since $j([x]) = [g(x)]$ and $J(\langle m,k \rangle) = J(\langle x,k \rangle) = \langle g(x),h(k) \rangle$, it follows that $j([x]) \in J(\langle m,k \rangle)$.

Case 4.   Clearly, points on Z map into points on Z' and lines through z map into lines through z'.

Thus (j,J) preserves collinearity and therefore π_T and $\pi_{T'}$ are isomorphic.

The converse of this theorem is not true, as will be seen in Chapter 11. There is a useful partial converse to the theorem, however.

THEOREM 9.3.4   *If there exists an isomorphism* (j,J) *from* $\pi_{T(u,v,o,e)}$ *to* $\pi_{T(u',v',o',e')}$ *such that* j: $u,v,o \rightarrow u',v',o'$, *then* $T(u,v,o,e)$ *is isotopic to* $T(u',v',o',e')$.

Proof.   Let $T(u,v,o,e) = T = (R,t)$ and let $T(u',v',o',e') = T' = (R',t')$. Since j maps uv onto $u'v'$, ov onto $o'v'$, and ou onto $o'u'$, and since $u,v,o \rightarrow u',v',o'$, the following one-to-one correspondences may be defined from R to R':

g: $a \rightarrow a'$ where $a \in R$ and $a' \in R'$ such that $j([a]) = [a']$
h: $b \rightarrow b'$ where $b \in R$ and $b' \in R'$ such that $j([0,b] \rightarrow [0,b']$
f: $c \rightarrow c'$ where $c \in R$ and $c' \in R'$ such that $j([c,0]) \rightarrow [c',0]$

First note that $h(0) = 0$ because $j([0,0]) = j(u) = u' = [0,0]$. (Here 0 is used to denote the additive identity in both rings.) Also note that

J: $\langle k \rangle \rightarrow \langle f(k) \rangle$ because j: $z \rightarrow z'$, $[k,0] \rightarrow [f(k),0]$
J: $\langle m,k \rangle \rightarrow \langle g(m),h(k) \rangle$ because j: $[m] \rightarrow [g(m)]$, $[0,k] \rightarrow [0,h(k)]$
J: $Z \rightarrow Z'$ because j: $u,v \rightarrow u',v'$

and

j: $[x,y] \rightarrow [f(x),h(y)]$ because J: $\langle x \rangle \rightarrow \langle f(x) \rangle$, $\langle 0,y \rangle \rightarrow \langle 0,h(y) \rangle$

Since j is an isomorphism, $[x,y] \in \langle m,k \rangle$ iff $j([x,y]) \in J(\langle m,k \rangle)$. Thus $[x,y] \in \langle m,k \rangle$ iff $[f(x),h(y)] \in \langle g(m),h(k) \rangle$. It follows that $y = t(x,m,k)$ iff $h(y) = t'(f(x),g(m),h(k))$, that is, $h(t(x,m,k)) = t'(f(x),g(m),h(k))$. Therefore, (f,g,h) is an isotopism from T to T'.

The following definition of isotopism will be used here for planar rings.

DEFINITION 9.3.5   *The planar rings $P = (R,+,\cdot)$ and $P' = (R',\oplus,\circ)$ are isotopic if and only if there exist functions f, g, and h, all of which are one-to-one correspondences from R onto R' such that $h(x + y) = h(x) \oplus h(y)$ and $h(x \cdot y) = f(x) \circ g(y)$.*

The following two theorems show that this definition is an appropriate analogue to the definition of isotopism for ternary rings (Definition 9.3.2).

THEOREM 9.3.6   *If $P = (R,+,\cdot)$ and $P' = (R',\oplus,\circ)$ are isotopic planar rings, then their associated linear ternary rings, $T = T_P$ and $T' = T_{P'}$, are isotopic.*

Proof.   Let (f,g,h) be an isotopism from P to P'. Then $t(a,b,c) = a \cdot b + c$ and $t'(a',b',c') = a' \circ b' \oplus c'$, so $h(t(a,b,c)) = h(a \cdot b + c) = h(a \cdot b) \oplus h(c) = f(a) \circ g(b) \oplus h(c) = t'(f(a),g(b),h(c))$. Also $h(0) = 0$ because h is an isomorphism from $(R,+)$ to $(R',\oplus)$ and thus preserves the additive identity.

COROLLARY 9.3.7   *If P and P' are isotopic planar rings, then $\pi_P \sim \pi_{P'}$.*

THEOREM 9.3.8   *If $T = (R,t)$ and $T' = (R',t')$ are isotopic linear ternary rings, then their associated planar rings, $P = P_T = (R,+,\cdot)$ and $P' = P_{T'} = (R',\oplus,\circ)$, are isotopic.*

Proof.   Let (f,g,h) be an isotopism from T to T'. Since T and T' are linear, $h(t(a,b,c)) = h(a \cdot b + c) = t'(f(a),g(b),h(c)) = f(a) \circ g(b) \oplus h(c)$. As a result, $h(a + c) = h(1 \cdot a + c) = f(1) \circ g(a) \oplus h(c) = h(1 \cdot a) \oplus h(c) = h(a) \oplus h(c)$, and $h(a \cdot b) = h(a \cdot b + 0) = f(a) \circ g(b) \oplus h(0) = f(a) \circ g(b)$ since $h(0) = 0$. Thus (f,g,h) is an isotopism from P to P'.

As we know, an isomorphic mapping between algebraic structures preserves all the algebraic properties: associativity, commutativity, distributivity, and so on. The question naturally arises as to which algebraic properties are preserved by isotopic maps. This question is partially answered by the following theorem.

THEOREM 9.3.9   *Let $P = (R,+,\cdot)$ and $P' = (R',\oplus,\circ)$ be isotopic planar rings. Then P satisfies the right (left) distributive law if and only if P' does also; P satisfies the associative law of multiplication (addition) if and only if P' does also.*

Proof. Suppose that P satisfies the right distributive law, that is, $(a + b)c = ac + bc$ for $a,b,c \in R$. Thus $f(a + b) \circ g(c) = h((a + b)c) = h(ac + bc) = h(ac) \oplus h(bc) = f(a)\circ g(c) \oplus f(b)\circ g(c)$. In particular, if $g(c) = 1$ (the multiplicative identity of R'), then $f(a + b) = f(a) \oplus f(b)$. Therefore, $(f(a) \oplus f(b))\circ g(c) = f(a + b) \circ g(c) = f(a)\circ g(c) \oplus f(b)\circ g(c)$. Since f and g are one-to-one mappings from R onto R', right distributivity exists in P', that is, $(a' \oplus b')\circ c' = a'\circ c' \oplus b'\circ c'$ for $a', b', c' \in R'$.

The result for left distributivity is proved in a similar way.

For associativity of addition, the result is clear because h is an isomorphism from $(R,+)$ to $(R',\oplus)$. Suppose that P satisfies the associative law of multiplication, that is, that $a(bc) = (ab)c$ for $a,b,c \in R$. Then:

i. $h(a(bc)) = f(a) \circ g(bc) = f(a) \circ (f(1)^{-1} \circ h(bc)) = f(a) \circ (f(1)^{-1} \circ (f(b) \circ g(c)))$. The second equality obtains because $h(a) = f(1) \circ g(a)$, so $g(a) = f(1)^{-1} \circ h(a)$. Also,

ii. $h((ab)c) = f(ab) \circ g(c) = (h(ab) \circ g(1)^{-1}) \circ g(c) = ((f(a) \circ g(b)) \circ g(1)^{-1}) \circ g(c)$. The second equality results because $h(a) = f(a) \circ g(1)$, so $f(a) = h(a) \circ g(1)^{-1}$. Because P satisfies the associative law of multiplication, (i) = (ii). If $f(a) = 1$, then (i) and (ii) result in

iii. $f(1)^{-1} \circ (f(b) \circ g(c)) = (g(b) \circ g(1)^{-1}) \circ g(c)$. Furthermore, if $g(c) = 1$, then

iv. $f(1)^{-1} \circ f(b) = g(b) \circ g(1)^{-1}$. Thus (iii) becomes

v. $f(1)^{-1} \circ (f(b) \circ g(c)) = (f(1)^{-1} \circ f(b)) \circ g(c)$. Similarly, we may obtain the equality

vi. $(f(a) \circ g(b)) \circ g(1)^{-1} = f(a) \circ (g(b) \circ g(1)^{-1})$. Using (v) and (vi) in the equality (i) = (ii), we establish

vii. $f(a) \circ ((f(1)^{-1} \circ f(b)) \circ g(c)) = (f(a) \circ (g(b) \circ g(1)^{-1})) \circ g(c)$. Letting $j(b) = f(1)^{-1} \circ f(b) = g(b) \circ g(1)^{-1}$, we have, finally, $f(a) \circ (j(b) \circ g(c)) = (f(a) \circ j(b)) \circ g(c)$. Now j is a one-to-one map from R onto R' because f is, so f, j, and g are all one-to-one, onto mappings. Thus $a' \circ (b' \circ c') = (a' \circ b') \circ c'$ for $a', b', c' \in R'$, and associativity is preserved.

Commutativity of multiplication is not necessarily preserved under isotopic maps, as will be seen in Chapter 11.

This section is concluded with a rather surprising result about isotopic maps on division rings.

THEOREM 9.3.10 *If D and D' are isotopic division rings, then they are isomorphic.*

Proof. Let $D = (R,+,\cdot)$ and $D' = (R',\oplus,\odot)$ and let (f,g,h) be the isotopism from D onto D'. Since f and g are one-to-one maps from R onto R', there exist elements b and c of R such that $f(b) = g(c) = 1$ (the multiplicative identity of R'). Since $h(by) = f(b) \odot g(y) = g(y)$ and $h(xc) = f(x) \odot g(c) = f(x)$, we know that $h(xy) = h(xc) \odot h(by)$ and that $h((xc^{-1})(b^{-1}y)) = h(x) \odot h(y)$. Since $(xc^{-1})(b^{-1}y) = x(c^{-1}b^{-1})y = xa^{-1}y$ where $a = bc$, we know that $h(x) \odot h(y) = h(xa^{-1}y)$.

Let $j: x \to xa$. The proof is completed by showing that $h \circ j$ is an isomorphism from D to D'. First we have $h \circ j(x + y) = h(j(x + y)) = h((x + y)a) = h(xa + ya) = h(xa) \oplus h(ya) = h\circ j(x) \oplus h\circ j(y)$. Also we have $h \circ j(xy) = h(j(xy)) = h((xy)a) = h(x(aa^{-1})ya) = h((xa)a^{-1}(ya)) = h(xa) \odot h(ya) = h \circ j(x) \odot h \circ j(y)$.

The property that isotopic maps are isomorphisms also holds for structures that are weaker than division rings. This will be of considerable importance later, especially in Chapter 14.

EXERCISES

1. Recall algebras A and C of exercises 2.1.6 and 2.4.4, respectively.
   a. Show that (g,h,j) is an isotopism from A onto C where g is the identity on R, where $h: x \to x \odot f$ (notice that f is not a function but that $f \in R$), and where $j = h$.

b. Show that isotopic mappings do not preserve commutativity.

2. Show that the converse of Theorem 9.3.3 is not true. Hint: Refer back to exercise 2.8.6(b).

9.4   THE PLANE $\pi_H(9)$

This section will describe in detail the representation of a plane by a ternary ring. The plane $\pi_H(9)$ is chosen as an example because it admits a nonlinear ternary ring; subsequent chapters will be concerned only with planes represented by linear ternary rings. One nonlinear ternary ring that is generated by a four-point in $\pi_H(9)$ is the ring of Example 9.1.3(iii). This fact will not be proved here (see Hughes, 1957), but it will become evident from our analysis. The plane $\pi_H(9)$ will be represented first in a manner somewhat different from any heretofore described; this representation will then be related to that of the ternary ring.

CONSTRUCTION 9.4.1   Let $(N,+,\cdot)$ be the right nearfield $N(9)$. Let $\mathscr{P} = \{[x_1,x_2,x_3]: x_i \in N$ and $x_i = 0$ for all i is not true$\}$ and let $[x_1,x_2,x_3] = [y_1,y_2,y_3]$ if and only if $x_i = y_i c$ for some $c \in N$. Thus points are equivalence classes of ordered triples as in Desarguesian and Pappian planes. Let

$$A = \begin{pmatrix} 1 & 1 & 1 \\ 1 & 0 & 0 \\ 0 & 1 & 0 \end{pmatrix}$$

Note that $A^{13} = I$ and $A^i \neq I$ for $1 \leq i < 13$. Let h denote the mapping in $\mathscr{P}$ induced by A and let $h^m = h \circ \ldots \circ h$, m times. Let $M_t = \{[x_1,x_2,x_3]: x_1 + tx_2 + x_3 = 0\}$ where $t \in N$, $h^m M_t = \{[y_1,y_2,y_3]: (y_1,y_2,y_3) = (x_1,x_2,x_3)A^m; [x_1,x_2,x_3] \in M_t\}$, and $\mathscr{L} = \{h^m M_t: m = 0, \ldots, 12; t = 1,a,b,c,d,e,f\}$. Let $\mathscr{I} = \in$.

A definition of $(\mathscr{P},\mathscr{L},\mathscr{I})$ may be formulated in the following manner. Define L_i, $i = 0, \ldots, 6$, and $a_0,b_0,c_0,d_0,e_0,f_0,g_0$ as follows: $L_0 = M_1, L_1 = M_a, L_2 = M_b, L_3 = M_c, L_4 = M_d, L_5 = M_e, L_6 = M_f, a_0 = [a,0,1], b_0 = [a,f,1], c_0 = [b,e,1], d_0 = [c,d,1], e_0 = [d,c,1], f_0 = [e,b,1], g_0 = [f,a,1]$. Define points $a_k,b_k, \ldots, g_k$ by $a_k = h^k(a_0), b_k = h^k(b_0), \ldots, g_k = h^k(g_0)$ where $k = 0, \ldots, 12$. For example:

$$a_1 = [1,0,1] \qquad a_5 = [0,1,1] \qquad a_9 = [0,1,0]$$
$$a_2 = [1,2,1] \qquad a_6 = [1,1,0] \qquad a_{10} = [1,0,0]$$
$$a_3 = [0,2,1] \qquad a_7 = [2,1,1] \qquad a_{11} = [1,1,1]$$
$$a_4 = [2,1,0] \qquad a_8 = [0,0,1] \qquad a_{12} = [2,2,1]$$

Tedious calculation shows that:

$$L_0 = \{a_0, a_3, a_4, a_{11}, b_0, c_0, d_0, e_0, f_0, g_0\}$$
$$L_1 = \{a_0, b_1, b_6, b_{12}, e_4, e_5, f_3, f_7, g_8, g_{11}\}$$
$$L_2 = \{a_0, c_1, c_6, c_{12}, g_4, g_5, e_3, e_7, f_8, f_{11}\}$$
$$L_3 = \{a_0, d_1, d_6, d_{12}, f_4, f_5, g_3, g_7, e_8, e_{11}\}$$
$$L_4 = \{a_0, e_1, e_6, e_{12}, b_4, b_5, c_3, c_7, d_8, d_{11}\}$$
$$L_5 = \{a_0, f_1, f_6, f_{12}, d_4, d_5, b_3, b_7, c_8, c_{11}\}$$
$$L_6 = \{a_0, g_1, g_6, g_{12}, c_4, c_5, d_3, d_7, b_8, b_{11}\}$$

Let $L_{j+7k} = h^k M_t$ where $L_j = M_t$ and $k = 0, \ldots, 12$; then $(\mathscr{P}, \mathscr{L}, \mathscr{I})$ is simply the plane $\pi_H(9)$ of Construction 2.4.1.

Several lemmas are now utilized to prove that the triple $(\mathscr{P}, \mathscr{L}, \mathscr{I})$ is a plane and that $\pi_H(9)$ is therefore a projective plane.

LEMMA 9.4.2 $\mathscr{P}$ has ninety-one members.

Proof.   There are $9^3 - 1$ ordered triples excluding $(0,0,0)$ and there are 8 elements in $N - \{0\}$. Thus there are $(9^3 - 1)/8 = 91$ different points.

LEMMA 9.4.3   *The lines M_t and M_s for $s \neq t$ have exactly one point in common.*

Proof.   It is easily checked that $a_0 = [2,0,1]$ is the only point in common.

LEMMA 9.4.4   *Every two distinct lines have exactly one point in common.*

Proof.   (Hughes, 1957) We wish to show that $h^n M_s \cap h^m M_t$ is a single point. Since h^k is a one-to-one correspondence (for any k) on $\mathscr{P}$, we need only show that $h^m M_t \cap M_s$ is one point. Let

$$B = \begin{pmatrix} a_{11} & a_{12} & a_{13} \\ a_{21} & a_{22} & a_{23} \\ a_{31} & a_{32} & a_{33} \end{pmatrix} = A^{-m}$$

If $[x,y,z] \in h^m M_t$, then $(x,y,z)B = (x',y',z')$ where $x' + ty' + z' = 0$. Thus

i. $xa_{11} + ya_{21} + za_{31} + t(xa_{12} + ya_{22} + za_{32}) + xa_{13} + ya_{23} + za_{33}$
   $= 0$. If $[x,y,z] = M_s$, we have

ii. $x + sy + z = 0$. Solving for x in (ii) and substituting in (i), we obtain

iii. $uy + pz + t(vy + qz) = 0$ where

$$u = a_{21} + a_{23} - s(a_{11} + a_{13})$$
$$v = a_{22} - sa_{12}$$
$$p = a_{31} + a_{33} - (a_{11} + a_{13})$$
$$q = a_{32} - a_{12}$$

Case 1. $q \neq 0$. Thus we may write (iii) in the following form:

iv. $pq^{-1}(vy + qz) + (u - pq^{-1}v)y + t(vy + qz) = 0$ because p, q, and therefore pq^{-1} are in $J/3$ and thus commute with all members of N. Using this reasoning again, (iv) becomes

v. $(pq^{-1} + t)(vy + qz) + (u - pq^{-1}v)y = 0$.

Subcase A. $t = 1$. Hence we have the simultaneous equations

vi. $x + sy + z = 0$ and $(u + v)y + (p + q)z = 0$. The latter equality follows from (iii). This subcase is proved by showing that $u + v = p + q = 0$ is not possible, thus implying that (vi) has a one-dimensional, that is, single-point, solution. If $p + q = 0$, then $a_{11} + a_{12} + a_{13} = a_{31} + a_{32} + a_{33}$. If $u + v = 0$ also, then $a_{21} + a_{22} + a_{23} - s(a_{11} + a_{13}) - sa_{12} = a_{21} + a_{22} + a_{23} - s(a_{11} + a_{12} + a_{13}) = 0$. The first equality holds because $a_{11} + a_{13}$ and a_{12} commute with all elements of N and thus yield the needed left distributivity from the given right distributivity in N. If $s \neq 1$, it follows that $s \notin J/3$, so $a_{11} + a_{12} + a_{13} = a_{21} + a_{22} + a_{23} = 0$. This is a contradiction because B is nonsingular. If $s = 1$, then $a_{21} + a_{22} + a_{23} = a_{11} + a_{12} + a_{13}$. This equality occurs if $B = I$ but not otherwise, which we can see if we let $m_1 = a_{11} + a_{12} + a_{13}$, $m_2 = a_{21} + a_{22} + a_{23}$, and $m_3 = a_{31} + a_{32} + a_{33}$, and list (m_1, m_2, m_3) for powers of A. The result is: A: $(0,1,1)$; A^2: $(2,0,1)$; A^3: $(0,2,0)$; A^4: $(2,0,2)$; A^5: $(1,2,0)$; A^6: $(0,1,2)$; A^7: $(0,0,1)$; A^8: $(1,0,0)$; A^9: $(1,1,0)$; A^{10}: $(2,1,1)$; A^{11}: $(1,2,1)$; A^{12}: $(1,1,2)$; $A^{13} = I$: $(1,1,1)$. Thus $B = I$ and so $q = 0$, a contradiction.

Subcase B. $t \neq 1$. Thus $t \notin J/3$, so if we let $w = pq^{-1} + t$, then $w \neq 0$. Now (v) becomes $w(vy + qz) = -(u - pq^{-1}v)y$; hence the simultaneous equations

vii. $x + sy + z = 0$ and $(v + w^{-1}(u - pq^{-1}v))y + qz = 0$. Since $q \neq 0$, we again have a point solution.

Case 2. $q = 0$, $p \neq 0$. Then (iii) becomes

viii. $(u + tv)y + pz = 0$. Since $p \neq 0$, (ii) and (viii) have a point solution.

Case 3. $p = q = 0$. Then $a_{31} + a_{33} = a_{11} + a_{13}$; $a_{32} = a_{12}$. First it must be shown that B holds $(1,0,-1)$ fixed. Let $d = (1,0,-1)$; then $dB = (1,0,-1)B = (a_{11} - a_{31}, a_{12} - a_{32}, a_{13} - a_{33}) = (a_{11} - a_{31}, 0, a_{31} - a_{11}) = cd$ where $c = a_{11} - a_{31}$. Since A is of order 13, either B must be of order 13 or $B = I$. Letting j be the collineation induced by B, we see that if B is of order 13, j is also of order 13 and thus must move every point of the subplane $\pi(3)$, consisting of points $[x,y,z]$, $x,y,z \in J/3$. But j fixes $[1,0,-1] \in \pi(3)$. Therefore, $B = I$ and it follows that $B = B^{-1} = A^m = I$. Thus $h^m M_t \cap M_s = M_t \cap M_s$, and this is a unique point by Lemma 9.4.3.

LEMMA 9.4.5   *$\mathcal{L}$ has ninety-one members.*

Proof.   There are seven different values for t and thirteen different values for m, and, by Lemma 9.4.4, each m and t yield a distinct line $h^m M_t$. Thus there are $13 \cdot 7 = 91$ lines.

LEMMA 9.4.6   *Each line has ten points.*

Proof.   This fact is easily checked from Construction 9.4.1.

THEOREM 9.4.7   *The triple $(\mathcal{P}, \mathcal{L}, \mathcal{I})$ is a projective plane.*

Proof.   The proof is left as an exercise.

The representation of $\pi_H(9)$ as given in Construction 9.4.1 enables us to choose a specific four-point u,v,o,e and generate a ternary ring. Let $u = [1,0,0] = a_{10}$, $v = [0,1,0] = a_9$, $o = [0,0,1] = a_8$, $e = [1,1,1] = a_{11}$.

Let $T(u,v,o,e) = (R,t)$; then $R = ou - \{u\}$ and $ou - \{u\} = L_{70} - \{a_{10}\}$
$= \{a_8,a_1,a_0,b_{10},c_{10},d_{10},e_{10},f_{10},g_{10}\}$. It is easily checked that

$$
\begin{array}{ll}
a_8 = [0,0,1] & d_{10} = [e,0,1] \\
a_1 = [1,0,1] & e_{10} = [a,0,1] \\
a_0 = [2,0,1] & f_{10} = [c,0,1] \\
b_{10} = [d,0,1] & g_{10} = [b,0,1] \\
c_{10} = [f,0,1]
\end{array}
$$

Also $ov = L_{35} = \{a_9,a_8,a_5,a_3,b_5,c_5,d_5,e_5,f_5,g_5\}$ and

$$
\begin{array}{ll}
a_9 = [0,1,0] & c_5 = [0,f,1] \\
a_8 = [0,0,1] & d_5 = [0,e,1] \\
a_5 = [0,1,1] & e_5 = [0,a,1] \\
a_3 = [0,2,1] & f_5 = [0,c,1] \\
b_5 = [0,d,1] & g_5 = [0,b,1]
\end{array}
$$

Finally, $uv = L_{42} = \{a_{10},a_9,a_6,a_4,b_6,c_6,d_6,e_6,f_6,g_6\}$ and

$$
\begin{array}{ll}
a_9 = [0,1,0] & c_6 = [1,b,0] \\
a_{10} = [1,0,0] & d_6 = [1,c,0] \\
a_6 = [1,1,0] & e_6 = [1,d,0] \\
a_4 = [1,2,0] & f_6 = [1,e,0] \\
b_6 = [1,a,0] & g_6 = [1,f,0]
\end{array}
$$

The points of $\mathscr{P} - L_{42}$ in $\pi_H(9)$ have coordinates of the form $[x,y,1]$:
$x,y \in N$, and the points on L_{42}, except for a_9, have the form $[1,x,0]$:
$x \in N$. Using the notation of Construction 9.4.1 for $\pi_H(9)$ and letting
f denote the isomorphism from $\pi_H(9)$ onto $\pi_{T(u,v,o,e)}$ as defined in
Theorem 9.2.6, we have:

$$
\begin{array}{ll}
f\colon [x,y,1] \to [[x,0,1],[y,0,1]] & \text{where } x,y \in N \\
 [1,x,0] \to [[x,0,1]] & \text{where } x \in N \\
 [0,1,0] \to z
\end{array}
$$

If we simplify our ternary notation in the "natural" way, that is, if we
denote $[[x,0,1],[y,0,1]]$ by $[x,y]$ and $[[x,0,1]]$ by $[x]$, then the isomorphism f becomes:

$$
\begin{array}{l}
f\colon [x,y,1] \to [x,y] \\
 [1,x,0] \to [x] \\
 [0,1,0] \to z
\end{array}
$$

Using this latter isomorphism it is easily checked that, for example,

L_0 and L_1 correspond to $\langle 2,2 \rangle$ and $\langle a,a \rangle$, respectively. Verification that $T(u,v,o,e)$ is isomorphic to the ternary ring of Example 9.1.3(iii) is left as a difficult exercise.

Example 4.3.6 states some facts about the collineation groups of $\pi_H(9)$, among them that (f,F), (g,G), and collineations of the form h^+ generate the entire collineation group $C(\pi_H(9))$. Recalling (f,F) and (g,G) from Example 3.1.5(ii), we can easily verify that they generate a group of six collineations. The collineations of type h^+, defined in Example 4.3.6(ii), were extensions of the collineations h on $\mathscr{P}|A$ where $A = \{a_0, \ldots, a_{12}\}$. It is easily seen from our new representation of $\pi_H(9)$ given in Construction 9.4.1 that the collineations h are the matrix-representable collineations on $\pi(3)$. Thus the collineations h^+ are evidently the mappings induced on all of $\mathscr{P}$ by these matrices. That such mappings are collineations has been shown by Rosati (1958). The student should attempt a proof of this fact himself. For a complete analysis of the collineation group of $\pi_H(9)$, see Room and Kirkpatrick (1971).

This section is concluded with the observation that $\pi_H(9)$ is just one example of a class of planes called *Hughes planes*. Hughes planes of all orders p^2 exist where p is an odd prime. In fact, infinite analogues have now been generated (Rosati, 1960). The construction of $\pi_H(p^2)$ may be based on a nearfield of order p^2 in the same way that $\pi_H(9)$ was generated by $N(9)$. Such nearfields will be defined in Chapter 12. There may also be the "cyclic" representation similar to the one first used to denote $\pi_H(9)$ in Section 2.4. The student is encouraged to generalize such a representation.

EXERCISES

1. Verify that $A^{13} = I$ and that $A^i \neq I$ for $0 \le i < 13$ in Construction 9.4.1 (cf. exercise 7.3.5).

2. Prove Theorem 9.4.7.

3. Find the coordinates of L_2, L_3. Can you establish a rule for the co-ordinates of L_k, $k \le 91$?

*4. Let M be a 3×3 matrix over $J/3$. Also let $h = f_M$, a collineation on $\pi_H(9) \mid \{a_0, \ldots, a_{12}\}$. Show that h^+ is a collineation on $\pi_H(9)$.

*5. Attempt a "cyclic" representation of $\pi_H(25)$ using as a guide the representation of $\pi_H(9)$ given in Section 2.4. Notice that $\pi_H(9)$ has 91 points and that $91 = 13 \cdot 7$; notice also that $\pi_H(25)$ has 651 points and that $651 = 31 \cdot 21$.

*6. Show that $T(u,v,o,e) \sim T$ where T is the ternary ring of Example 9.1.3(iii).

CHAPTER 10

PLANES OVER PLANAR RINGS WITH ASSOCIATIVITY

In the remainder of Part 3, the projective planes examined will be those representable by linear ternary rings, or equivalently, by planar rings. This chapter will explore the geometric consequences of: (1) the linearity of the ternary ring, (2) the associativity of addition and the associativity of multiplication of the planar ring, and (3) the combined laws of associativity and commutativity of addition and multiplication.

10.1 CONSEQUENCES OF LINEARITY

If T is a linear ternary ring, we shall find that the plane π_T must necessarily satisfy a Desarguesian-like constraint.

CONSTRUCTION 10.1.1   Let u,v,o,e be a given four-point in π_T and let a, b, and c be points such that $a \in ou$, $b \in ov$, $c \in uv$, $a \neq u$, and $b,c \neq v$ (see Figure 80). Furthermore, let $p = bc \cap av$, $q = oc \cap av$,

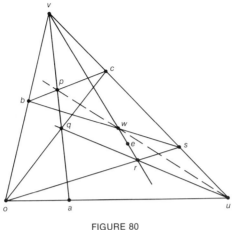

FIGURE 80

$r = uq \cap ev$, $s = or \cap uv$, and $w = bs \cap ev$. Notice that different choices of a, b, and c generate different points p, q, r, s, and w.

THEOREM 10.1.2 *Let* $\pi = \pi_{T(u,v,o,e)}$. *Then* $T(u,v,o,e) = (R,t)$ *is a linear ternary ring if and only if points* p, w, *and* u *in Construction 10.1.1 are collinear.*

Proof. Let $a = [x,0]$, $b = [0,z]$, and $c = [y]$ for some $x,y,z \in R$. The following equalities are easily checked: $p = bc \cap av = \langle y,z \rangle \cap \langle x \rangle = [x,t(x,y,z)]$; $q = oc \cap av = \langle y,0 \rangle \cap \langle x \rangle = [x,xy]$; $r = uq \cap ev = \langle 0,xy \rangle \cap \langle 1 \rangle = [1,xy]$; $s = or \cap uv = \langle xy,0 \rangle \cap Z = [xy]$; and $w = bs \cap ev = \langle xy,z \rangle \cap \langle 1 \rangle = [1,t(1,xy,z)] = [1,xy + z]$. It follows that $wu \cap av = [x,xy + z] = p$ iff $t(x,y,z) = xy + z$. We may conclude that p, w, and u are collinear iff $t(x,y,z) = xy + z$.

The following two corollaries explain how the conclusion $p \in wu$ in the preceding proof is Desarguesian in nature.

COROLLARY 10.1.3 *The ternary ring* $T(u,v,o,e)$ *is a linear ternary ring if and only if all triangles bpw and oqr that are centrally perspective from v are axially perspective from uv.*

COROLLARY 10.1.4 *The ternary ring* $T(u,v,o,e)$ *is linear if and only if all triangles qpc and rws that are axially perspective from ov are centrally perspective from u.*

It should be noted that the proofs of Corollaries 10.1.3 and 10.1.4 require a little thought since there are choices of a, b, and c that do not generate triangles; for example, if $a = ev \cap ou$, then $p = w$, and therefore bpw is not a triangle.

10.2 PLANES OVER CARTESIAN GROUPS

Now that we have represented the linearity of T geometrically, we may proceed to planar rings P. First we supplement the Cartesian group with the property of associative addition.

DEFINITION 10.2.1 *A* Cartesian group *is a planar ring* $P = (R,+,\cdot)$ *such that* $(R,+)$ *is a group.*

EXAMPLE 10.2.2

    i. Clearly, every division ring is a Cartesian group.

    ii. The nearfield $N(9)$ is also clearly a Cartesian group.

    iii. Let $P = (R,+,\circ)$ where R is the set of real numbers, $+$ is usual addition, and $\circ$ is defined as follows:

$$a \circ b = ab \text{ if } ab \geq 0 \text{ (that is, if } a \text{ and } b \text{ have the same sign)}$$
$$a \circ b = a^2b \text{ if } a > 0, b < 0$$
$$a \circ b = ab^2 \text{ if } a < 0, b > 0$$

It can be verified (Spencer, 1960) that this system is a Cartesian group. The system does not, however, satisfy any other standard property, such as associativity of multiplication or left or right distributivity.

    iv. Let $P = (R,\oplus,\circ)$ where $R = \{(x,y): x,y \in J/5\}$; let $(a,b) \oplus (c,d) = (a + c, b + d)$ and let

$$(a,b) \circ (c,d) = \begin{cases} (a,b) \cdot (c,d) \text{ if } b = 0 \text{ or if } (bc - ad)^2 - 2d^2 \\ \qquad\qquad\qquad = 0,1,4; \text{ and} \\ -(a,b) \cdot (c,d) \text{ otherwise, where} \end{cases}$$

$$(a,b) \cdot (c,d) = \begin{cases} (ac,bd) \text{ if } d = 0, \text{ and} \\ (ac - bd^{-1}(c^2 - 2), ad - bc) \\ \text{if } d \neq 0 \end{cases}$$

This is the smallest of the Cartesian groups that fail to satisfy both distributive laws and associativity of multiplication. There are such Cartesian groups of order p^2 where p is a prime ≥ 5 and of order 2^{2k} where k is an odd number ≥ 3. Thus the next largest such Cartesian group has thirty-two members. (For a general method of generating Cartesian groups, see Panella, 1965.)

The following paragraphs explore the geometric consequences of the associative law of addition.

Notation. $P(u,v,o,e)$ denotes $P_{T(u,v,o,e)}$.

THEOREM 10.2.3 *If $P(u,v,o,e) = (R,+,\cdot)$ is a Cartesian group, then π_P is (v,uv) transitive.*

Proof. Define the following mappings f_a, F_a on π_P for $a \in R$:

$$f_a: \quad [x,y] \rightarrow [x,y+a]$$
$$[x] \rightarrow [x]$$
$$z \rightarrow z$$

$$F_a: \quad \langle m,k \rangle \rightarrow \langle m, k+a \rangle$$
$$\langle k \rangle \rightarrow \langle k \rangle$$
$$Z \rightarrow Z$$

The first step is to show that (f_a, F_a) is a collineation on π_T.

Case 1. Suppose that $[x,y] \in \langle m,k \rangle$. Then $y = xm + k$. Also we know that $f_a([x,y]) = [x, y+a]$, $F_a(\langle m,k \rangle) = \langle m, k+a \rangle$, and $y+a = (xm + k) + a = xm + (k + a)$. The last equality holds because addition is associative. Thus $f_a([x,y]) \in F_a(\langle m,k \rangle)$.

Case 2. Suppose that $[x,y] \in \langle k \rangle$. Thus $k = x$, and since $f_a([x,y]) = [x, y+a]$ and $F_a(\langle x \rangle) = \langle x \rangle$, it follows that $f_a([x,y]) \in F_a(\langle k \rangle)$.

Case 3. Suppose that $[x] \in \langle m,k \rangle$. By an argument similar to that in Case 2, $f_a([x]) \in F_a(\langle m,k \rangle)$.

It follows immediately from the definitions of the mappings at the beginning of the proof that (f_a, F_a) leaves Z pointwise fixed and z linewise fixed. Thus (f_a, F_a) preserves incidence for points in Z and lines through z. Therefore, (f_a, F_a) is a collineation and, in fact, an elation in $\text{El}(z, Z)$.

Let p be an arbitrary point on $ov - \{v\}$. Thus $p = [0, c]$ for some $c \in R$. Since (f_a, F_a) exists for all $a \in R$ and since $f_a: [0,0] \to [0,a]$, there exists an element $f_c \in \text{El}(v, uv): o \to p$. We may conclude that π_P is (v, uv) transitive.

THEOREM 10.2.4 *If $T(u, v, o, e) = T = (R, t)$ is a ternary ring and π_T is (v, uv) transitive, then T is a linear ternary ring.*

Proof. The proof follows from Corollary 10.1.3.

THEOREM 10.2.5 *If $P(u, v, o, e) = (R, +, \cdot)$ and π_P is (v, uv) transitive, then $(R, +) \sim \text{El}(v, uv)$.*

Proof. Let f_a be the member of $\text{El}(v, uv)$ such that $f_a: [0,0] \to [0,a]$ and define the mapping γ from R onto $\text{El}(v, uv)$ by $\gamma: a \to f_a$ where $a \in R$. Observe that $\gamma(a + b) = f_{a+b}$ and $\gamma(b) \circ \gamma(a) = f_b \circ f_a$. The proof is completed by showing that γ is an isomorphism, that is, that $f_{a+b} = f_b \circ f_a$.

Since $f_b: [0,0] \to [0,b]$ and since $[a] \in uv$, it follows that $f_b: \langle a, 0 \rangle = [0,0][a] \to [0,b][a] = \langle a, b \rangle$. Similarly, since $v \in \langle 1 \rangle, f_b: \langle 1 \rangle \to \langle 1 \rangle$, and so $f_b: [1, a] = \langle 1 \rangle \cap \langle a, 0 \rangle \to \langle 1 \rangle \cap \langle a, b \rangle = [1, a + b]$. Also $[0] \in uv$, so $f_b: \langle 0, a \rangle = [0][1, a] \to [0][1, a + b] = \langle 0, a + b \rangle$. Since $v \in \langle 0 \rangle, f_b: [0, a] = \langle 0 \rangle \cap \langle 0, a \rangle \to \langle 0 \rangle \cap \langle 0, a+b \rangle = [0, a+b]$. Therefore, $f_b \circ f_a(o) = [0, a + b]$. Clearly, $f_{a+b}(o) = [0, a + b]$, so we may conclude that $f_b \circ f_a = f_{a+b}$.

THEOREM 10.2.6 *If P is a planar ring and π_P is (v, uv) transitive, then P is a Cartesian group.*

Proof. From Theorem 10.2.5, $(R, +) \sim \text{El}(v, uv)$, and since $\text{El}(v, uv)$ is a group, R is also a group. Thus P is a Cartesian group.

The preceding four theorems may be summarized in the following theorem and corollary.

THEOREM 10.2.7   *Let $T = T(u,v,o,\grave{e})$ be a ternary ring associated with π. Then π is (v,uv) transitive if and only if T is a linear ternary ring whose associated planar ring $P_T = (R,+,\circ)$ is a Cartesian group. Furthermore, $(R,+) \sim El(v,uv)$.*

COROLLARY 10.2.8   *π_T satisfies the Desargues (v,uv) Theorem if and only if T is a linear ternary ring and its associated planar ring P_T is a Cartesian group.*

Thus the property of associativity of addition in P is equivalent to the existence of the Desargues (p,L) Theorem for some $p \in L$ in π_P.

EXERCISES

1. Show that $(R,+,\circ)$ of Example 10.2.2(iii) satisfies neither associativity of multiplication nor either distributive law.

2. a. Show that $P = (R,\oplus,\circ)$ of Example 10.2.2(iv) satisfies neither associativity of multiplication nor either distributive law.
   b. Find the multiplicative identity of P.

3. Letting P be as in Example 10.2.2(iv),
   a. Find a subplane $\pi(2)$ in π_P.
   b. Find a subplane $\pi(5)$ in π_P.

4. Show that if π is (p,pq) transitive for some p and q, then there exists a Cartesian group P such that $\pi_P \sim \pi$.

*5. Does there exist a four-point u,v,o,e in $\pi_H(9)$ such that $T(u,v,o,e)$ is a linear ternary ring?

*6. Construct an economical axiom system for a Cartesian group.

10.3   PLANAR RINGS WITH ASSOCIATIVE MULTIPLICATION

This section explores the consequences of associativity of multiplication.

EXAMPLE 10.3.1   Let $P = (R, \oplus, \cdot)$ where R is the set of real numbers, $\cdot$ is the usual multiplication, and $\oplus$ is defined as follows:

$$a \oplus b = \begin{cases} a + b \text{ if } ab \geq 0 \\ a + (\text{sign } b)b^2 \text{ if } ab \leq 0 \text{ and } |a| \geq b^2 \\ b + (\text{sign } a)\sqrt{|a|} \text{ if } ab \leq 0 \text{ and } |a| < b^2 \end{cases}$$

It has been shown (Spencer, 1960) that P is a planar ring. Clearly, multiplication is associative. That addition is not associative and that neither distributive law holds is easily verified.

THEOREM 10.3.2   *If $P(u,v,o,e) = (R,+,\cdot)$ is a planar ring with associative multiplication, then π_P is (u,ov) transitive.*

Proof.   Define the following mappings g_a, G_a on π_P for $a \in R$, $a \neq 0$:

$$g_a: \quad [x,y] \to [xa,y]$$
$$[x] \to [a^{-1}x]$$
$$z \to z$$

$$G_a: \quad \langle m,k \rangle \to \langle a^{-1}m,k \rangle$$
$$\langle k \rangle \to \langle ka \rangle$$
$$Z \to Z$$

The first step is to show that (g_a, G_a) preserves incidence on π_P. Suppose that $[x,y] \in \langle m,k \rangle$. Then $y = xm + k$. Also $g_a([x,y]) = [xa,y]$, $G_a(\langle m,k \rangle) = \langle a^{-1}m,k \rangle$, and $y = x(aa^{-1})m + k = (xa)(a^{-1}m) + k$. The last equality holds because multiplication is associative. Thus $g_a([x,y]) \in G_a(\langle m,k \rangle)$.

For other cases p and L such that $p \in L$, it is easily shown that $g_a(p) \in G_a(L)$. Thus (g_a, G_a) is a collineation.

It is easily checked that $(g_a, G_a) \in \text{Hom}(u,ov)$ and that $g_a: [1,0] \to [a,0]$. Since (g_a, G_a) exists for all $a \in R$, we conclude that π_P is (u,ov) transitive.

THEOREM 10.3.3   *If $T(u,v,o,e) = T = (R,t)$ is a ternary ring and π_T is (u,ov) transitive, then T is a linear ternary ring.*

Proof.   The proof follows from Corollary 10.1.4.

THEOREM 10.3.4   *If $P(u,v,o,e) = (R,+,\cdot)$ and π_P is (u,ov) transitive, then $(R - \{0\},\cdot) \sim \text{Hom}(u,ov)$.*

Proof.   Let h_a be the member of $\text{Hom}(u,ov)$ such that h_a: $[1,0] \to [a,0]$ and define the mapping γ from $R - \{0\}$ onto $\text{Hom}(u,ov)$ by γ: $a \to h_a$ where $a \in R$ and $a \neq 0$. Observe that $\gamma(a \cdot b) = h_{ab}$ and $\gamma(b) \circ \gamma(a) = h_b \circ h_a$. The proof is completed by showing that γ is an isomorphism, that is, that $h_{ab} = h_b \circ h_a$.

Since h_b: $[1,0] \to [b,0]$ and $z \in ov$, it follows that h_b: $\langle 1 \rangle = [1,0]z \to [b,0]z = \langle b \rangle$. Similarly, since $u \in \langle 0,b \rangle$, it follows that h_b: $[1,b] = \langle 1 \rangle \cap \langle 0,b \rangle \to \langle b \rangle \cap \langle 0,b \rangle = [b,b]$. Since $[0,0] \in ov$, h_b: $\langle b,0 \rangle = [0,0] \cap [1,b] \to [0,0] \cap [b,b] = \langle 1,0 \rangle$. Since $u \in \langle 0,ab \rangle$, h_b: $[a,ab] = \langle 0,ab \rangle \cap \langle b,0 \rangle \to \langle 0,ab \rangle \cap \langle 1,0 \rangle = [ab,ab]$. Since $z \in ov$, h_b: $\langle a \rangle = [a,ab]z \to [ab,ab]z = \langle ab \rangle$. Since $u \in \langle 0,0 \rangle$, h_b: $[a,0] = \langle 0,0 \rangle \cap \langle a \rangle \to \langle 0,0 \rangle \cap \langle ab \rangle = [ab,0]$. Therefore, $h_b \circ h_a([1,0]) = [ab,0]$. Clearly, $h_{ab}([1,0]) = [ab,0]$, so we may conclude that $h_b \circ h_a = h_{ab}$.

THEOREM 10.3.5   *If P is a planar ring and π_P is (u,ov) transitive, then P satisfies the associative law of multiplication.*

Proof.   The proof follows easily from Theorem 10.3.4.

The following theorem and corollary summarize the preceding four theorems.

THEOREM 10.3.6   *Let $T = T(u,v,o,e)$ be a ternary ring associated with π. Then π is (u,ov) transitive if and only if T is a linear ternary ring and its associated planar ring P_T satisfies the associative law of multiplication.*

COROLLARY 10.3.7   *The plane π_T satisfies the Desargues (u,ov) Theorem if and only if T is a linear ternary ring and its associated planar ring P_T satisfies the associative law of multiplication.*

Thus the property of associativity of multiplication in P is equivalent to the existence of the Desargues (p,L) Theorem for some $p \notin L$ in π_P.

EXERCISES

1. Show that $(R, \oplus, \cdot)$ of Example 10.3.1 satisfies neither associativity of addition nor either distributive law.

2. Show that if π is (p,qr) transitive for some p, q, and r such that $p \notin qr$, then there exists a planar ring P with associative multiplication such that $\pi \sim \pi_P$.

3. Is there a four-point u,v,o,e in $\pi_H(9)$ such that $T(u,v,o,e)$ is a linear ternary ring and
    a. $P_{T(u,v,o,e)}$ is a Cartesian group?
    b. $P_{T(u,v,o,e)}$ has associative multiplication?

10.4   PLANAR RINGS SATISFYING BOTH ASSOCIATIVE LAWS

If a planar ring P satisfies both associative laws, naturally π_P is (v,uv) and (u,ov) transitive. The question is: Does π_P satisfy even stronger transitivity conditions? Certainly if P is a division ring, the answer is yes. If $P = N(9)$, the answer is also yes. (Recall from Example 4.5.13(ii) that $\pi_N(9)$ is (uv,uv) transitive.) Of course, neither of these answers is surprising since division rings satisfy nearly every algebraic property and since $N(9)$ also satisfies the right distributive law, which is more than necessary. Thus we seek a planar ring that satisfies only the associative laws of addition and multiplication.

EXAMPLE 10.4.1   Let $P = (R,+,\circ)$ where R is the set of real numbers, $+$ is the usual addition, and $\circ$ is defined as follows:

$$a \circ b = \begin{cases} ab & \text{if either } a \text{ or } b \geq 0; \\ kab & \text{if } a,b < 0. \ k \text{ may be any fixed number} \neq 0,1 \end{cases}$$

It is left as an exercise to show that if at least two of a, b, and c are negative, then $a \circ (b \circ c) = (a \circ b) \circ c = kabc$; otherwise, $a \circ (b \circ c) = (a \circ b) \circ c = abc$. This establishes the associative law of multiplication. Proof that neither distributive law holds is also left to the student.

The plane π_P is called a Moulton plane (see exercise 2.1.6(f)). This plane is (v,uv) and (u,ov) transitive because of Theorem 10.2.7 and

Theorem 10.3.6, but it is also (o,ou) transitive. Furthermore, the existence of collineations in El(v,uv) mapping $o \to p$ for any $p \in ov$ $- \{v\}$ implies that π_P is (p,pu) transitive for all $p \in ov$. This additional transitivity in π_P can be accounted for algebraically. It has been shown (Spencer, 1960) that if a planar ring satisfying both associativity laws satisfies the additional property that $(a \odot c) \odot (b \odot c) = a \odot b$ where $a \odot b = a(a - b)^{-1}(-b)$, then the plane is (o,ou) transitive and therefore (p,pu) transitive for all $p \in ov$. The Moulton plane π_P does satisfy this condition.

Thus we have a totally unfamiliar law for P that yields additional transitivity in π_P. Evidently, the familiar laws of algebra cannot alone provide a fine enough scale to distinguish the variations of Desarguesian-like structure possible on projective planes.

To answer the question posed at the beginning of this section, there are planar rings P that satisfy both associative laws and that induce planes π_P satisfying (v,uv) and (u,ov) transitivity and no more (see Dembowski, 1968, footnote p. 126). Thus the laws of associativity taken together do not induce any more transitivity on π_P than does the sum of transitivities induced by each law separately.

EXERCISES

1. Show that $(R,+,\circ)$ of Example 10.4.1 satisfies associativity of multiplication. Hint: Verify that if at least two of a,b,c are negative, then $a \circ (b \circ c) = (a \circ b) \circ c = kabc$; otherwise, $a \circ (b \circ c) = (a \circ b) \circ c = abc$.

2. Show that $(R,+,\circ)$ of Example 10.4.1 does not satisfy either distributive law.

3. Show that $(a \odot c) \odot (b \odot c) = a \odot b$ in a planar ring P of the type described in Example 10.4.1.

4. Are all Moulton planes isomorphic?

5. Describe in geometric terms the following affine planes $\alpha((\pi_P)_z^-)$ where P is the planar ring of:
   a. Example 10.4.1 (see exercise 1.3.8).
   b. Example 10.3.1.
   c. Example 10.2.2(iii).

6. Show that a Moulton plane is (p,pu) transitive for all $p \in ov$.

10.5 PLANAR RINGS WITH ASSOCIATIVITY AND COMMUTATIVITY

This section will supplement P with both associativity and commutativity. Thus planar rings $(R,+,\cdot)$ will be considered such that either $(R,+)$ is an Abelian group, or $(R - \{0\},\cdot)$ is an Abelian group, or both. The property of commutativity occurring alone in planar rings will not be analyzed.

EXAMPLE 10.5.1

   i. In Example 10.2.2, $(R,+)$ is an Abelian group.

   ii. In Example 10.3.1, $(R - \{0\},\cdot)$ is an Abelian group.

   iii. If P is a field, then both $(R,+)$ and $(R - \{0\},\cdot)$ are Abelian groups. This is also true in Example 10.4.1.

Recall that Chapter 7 showed that commutativity of multiplication in a division ring induces the Pappus Theorem in the associated plane. Here it will be shown that commutativity in a planar ring is the basis of a specialized Pappus property in the associated plane. The following discussion will proceed along the lines followed by Burn (1968).

DEFINITION 10.5.2 *Let L and M be distinct lines in π and let p be a point not on L or M. Define the function $f_{L,M,p}$: $L \cup M \to L \cup M$ such that $f_{L,M,p}$: $q \to r$ where $r = pq \cap M$ if $q \in L$ and where $r = pq \cap L$ if $q \in M$.*

Notice that $f_{L,M,p}$ is an involution on $L \cup M$; that is, $(f_{L,M,p})^2$ and not $f_{L,M,p}$ is the identity on $L \cup M$.

LEMMA 10.5.3 *The plane π satisfies the Pappus (L, M, N) Theorem if and only if $f_{L,M,r} \circ f_{L,M,q} \circ f_{L,M,p}$ is an involution on $L \cup M$ for all p, q, r on N but not on L or M.*

Proof. The proof is left as an exercise.

THEOREM 10.5.4 *If $P(u,v,o,e) = (R,+,\cdot)$ and $(R - \{0\},\cdot)$ is an Abelian group, then π_P satisfies the Pappus (ou, ov, uv) Theorem.*

Proof. The points on uv and not on ou or ov may be denoted by $[a]$ where $a \in R$ and $a \neq 0$. Let $p = [a]$ and denote $f_{ou,ov,p}$ by f_a. Then f_a: $[x,0] \rightarrow [0,-xa]$ (because $[x,0] \in ou$), $[0,-xa] \in ov$, and $[x,0]$, $[a]$, and $[0,-xa]$ all lie on $\langle a,-xa \rangle$. Also f_a: $[0,x] \rightarrow [xa^{-1},0]$ (because $[0,x]$ $\in ov$), $[xa^{-1},0] \in ou$, and $[0,x]$, $[a]$, and $[-xa^{-1},0]$ all lie on $\langle a,x \rangle$. (The preceding is based on the fact that $(R - \{0\},\cdot)$ is an Abelian group.) Thus $(f_c \circ f_b \circ f_a)^2$ has the following effect: $[x,0] \rightarrow [0,-xa] \rightarrow [xab^{-1},0]$ $\rightarrow [0,-xab^{-1}c] \rightarrow [xab^{-1}ca^{-1},0] \rightarrow [0,-xab^{-1}ca^{-1}b] \rightarrow [xab^{-1}ca^{-1}bc^{-1},0]$ $= [x,0]$ for all $x \in R$. This last equality holds because $(R - \{0\},\cdot)$ is an Abelian group. Therefore, $(f_c \circ f_b \circ f_a)^2$ is the identity. It follows from Lemma 10.5.3 that π_P satisfies the Pappus (ou,ov,uv) Theorem.

THEOREM 10.5.5 *If P is a planar ring and π_P satisfies the Pappus (ou,ov,uv) Theorem, then $(R - \{0\},\cdot)$ is an Abelian group.*

Proof. The first step is to establish the commutativity of the operation $\cdot$. Using the same notation as in the preceding proof, we may easily check that $f_b \circ f_1 \circ f_a$: $[1,0] \rightarrow [0,-(ab)]$ and that $f_a \circ f_1 \circ f_b$: $[1,0] \rightarrow$ $[0,-(ba)]$. From Lemma 10.5.3, we know that $(f_b \circ f_1 \circ f_a)^2$ is the identity, and since f_a and f_b are involutions, $f_b \circ f_1 \circ f_a = f_a \circ f_1 \circ f_b$. Thus $-(ab) = -(ba)$, and so $ab = ba$ for all $a,b \neq 0$. Clearly, $ab = ba$ if $a = 0$ or if $b = 0$.

To establish associativity, first note that $f_c \circ f_1 \circ f_b \circ f_1 \circ f_a = f_b \circ f_1 \circ f_c$ $\circ f_1 \circ f_a$. Let g represent the function on the left of this equation and let h represent the function on the right. It is then easily checked that g: $[1,0] \rightarrow [0,-(-(ab)c)]$ and that h: $[1,0] \rightarrow [0,-(-(ac)b)]$. Thus $-(ab)c$ $= -(ac)b$. Let $b = 1$; then $-a(c) = -(ac)$ for all $a,c \neq 0$. Therefore, $-(ab)c = -((ab)c)$ and $-(ac)b = -((ac)b)$ for all $a,b,c \neq 0$. Since $-(ab)c = -(ac)b$, we obtain $-((ab)c) = -((ac)b)$, and so $(ab)c = (ac)b$. Since we have already established commutativity in $(R - \{0\},\cdot)$, we may conclude that $c(ab) = (ca)b$ for all $c,a,b \neq 0$. Clearly, $c(ab) = (ca)b$ $= 0$ if one of $c,a,b = 0$. Thus we have established the associativity of the operation.

THEOREM 10.5.6 *If $P(u,v,o,e) = (R,+,\cdot)$ and $(R,+)$ is an Abelian group, then π_P satisfies the Pappus (ov,ev,uv) Theorem.*

Proof. The points on uv but not on ov or ev are denoted by $[a]$ where $a \in R$; $f_{ov,ev,p}$ is denoted by f_a where $p = [a]$. (This notation is similar to that included in the proof of Theorem 10.5.4.) We know that f_{-a}: $[1,x] \rightarrow [0,a + x]$ because $[1,x] \in ev$ and $[0,a + x] \in ov$, and $[1,x]$, $[-a]$, and $[0,a + x]$ all lie on $\langle -a,a + x \rangle$. (Notice that $[1,x] \in \langle -a,a + x \rangle$ because $-a + (a + x) = x$.) Furthermore, f_{-a}: $[0,x] \rightarrow [1,-a + x]$ because $[0,x] \in ov$ and $[1,-a + x] \in ev$, and $[0,x]$, $[-a]$, and $[1,-a + x]$ all lie on $\langle -a,x \rangle$. Thus $(f_{-c} \circ f_{-b} \circ f_{-a})^2$: $[1,x] \rightarrow [1,-c + b - a + c - b + a + x] = [1,x]$. This equality holds because $(R,+)$ is an Abelian group. The proof of the theorem now follows from Lemma 10.5.3.

THEOREM 10.5.7   *If P is a planar ring such that* $-x + (x + y) = y$ *and* π_P *satisfies the Pappus (ov, ev, uv) Theorem, then (R,+) is an Abelian group.*

Proof.   The first step is to show the commutativity of the operation $+$. Using the notation included in the proof of Theorem 10.5.6, we can easily check that $f_{-b} \circ f_0 \circ f_{-a}$: $[1,0] \rightarrow [0,a] \rightarrow [1,a] \rightarrow [0,b + a]$. The last mapping, f_{-b}: $[1,a] \rightarrow [0,b + a]$, holds because $[1,a]$, $[-b]$, and $[0,b + a]$ all lie on the line $\langle -b,b + a \rangle$. ($[1,a]$ lies on $\langle -b, b + a \rangle$ because $-b + (b + a) = a$.) Similarly, $f_{-a} \circ f_0 \circ f_{-b}$: $[1,0] \rightarrow [0,a + b]$. Since $f_{-a} \circ f_0 \circ f_{-b} = f_{-b} \circ f_0 \circ f_{-a}$, we know that $a + b = b + a$.

   To show associativity, first note that $f_{-c} \circ f_0 \circ f_{-b} \circ f_0 \circ f_{-a} = f_{-b} \circ f_0 \circ f_{-c} \circ f_0 \circ f_{-a}$. Let g be the function on the left of this equation and let h be the function on the right. It is then easily checked that g: $[1,0] \rightarrow [0,c + (b + a)]$ and that h: $[1,0] \rightarrow [0,b + (c + a)]$. Thus $c + (b + a) = b + (c + a)$. Using the commutativity of $+$ proved previously, we have $c + (a + b) = (c + a) + b$ for all $a,b,c \in R$. Therefore, the operation of addition is associative.

EXERCISES

1. Prove Lemma 10.5.3.

2. Find all triples (L,M,N) in the Moulton plane of exercise 2.1.6(f) such that the Pappus (L,M,N) Theorem holds.

*3. Formulate and prove as strong a theorem as you can using the following format: If π satisfies both _____ and the Pappus (L,M,N) Theorem, then π satisfies the Desargues $(L \cap M,N)$ Theorem.

*4. Attempt to weaken or delete entirely the awkward condition of $-x + (x + y) = y$ in the hypothesis of Theorem 10.5.7.

*5. State and prove the duals of Theorems 10.5.4 through 10.5.7.

*6. Isolate the geometric property directly related to the commutativity of $+$ and the commutativity of $\cdot$ in a planar ring.

10.6   CLASSIFICATION OF PLANES

Before any further attempt is made to analyze the properties of the planar ring, a classification system will be formalized for projective planes. Section 2.8 explored ways of distinguishing between projective planes: The criteria of cardinality, self-duality, and subplane structure were found to be either superficial or very difficult to apply; the criterion of embeddability of certain configurations was found to be somewhat more useful. Sections 10.1 through 10.4 indicated that the embeddability of pairs of centrally and axially perspective triangles provides important information; Section 10.5 showed that the embeddability of the Pappus configuration may also be helpful. The former property of embeddability, which is Desarguesian in nature, has offered the more fruitful approach to the classification of projective planes.

DEFINITION 10.6.1   *A pair of triangles abc and a'b'c' that are centrally perspective from p and axially perspective from L are called a* nontrivial pair *if and only if the following conditions hold:*
$p \neq a,b,c,a',b',c'$; $a \neq a'$; $b \neq b'$; $c \neq c'$; $L \neq ab,ac,bc,a'b',a'c',b'c'$; $ab \neq a'b'$; $ac \neq a'c'$; $bc \neq b'c'$.

Notice that if abc and $a'b'c'$ are a nontrivial pair of triangles, centrally perspective from p and axially perspective from L, then points $p,a,b,c,a',b',c', ab \cap a'b', ac \cap a'c'$, and $bc \cap b'c'$, as well as lines L, $aa', bb', cc', ab, ac, bc, a'b', a'c'$, and $b'c'$, exist, are distinct, and form a confined configuration. There are many possible configurations of ten points and ten lines that a nontrivial pair of triangles may generate, but only one such configuration is symmetric: $\Sigma(10_3)$, the Desargues configuration.

Every finite plane except $\pi(2)$ contains a nontrivial pair of triangles. That $\pi(3)$ and $\pi(4)$ contain such a pair is easily verified; that higher-order planes also contain such a pair is shown by the following theorem.

THEOREM 10.6.2   (Ostrom, 1957)   *A finite plane of order $n > 4$ contains the Desargues configuration.*

Proof.   Let L_1, L_2, L_3 be three distinct lines concurrent at p; let r and s be two points such that $p \notin rs$ and $r, s \notin L_1, L_2, L_3$; and consider the set T of triangles, each of which has one vertex on L_1, one vertex on L_2, and one vertex on L_3 and has a side containing r and another side containing s (see Figure 81). It is sufficient to show that T contains at least one pair of triangles satisfying the Desargues configuration.

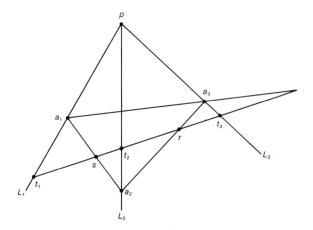

FIGURE 81

Let $t_i = L_i \cap rs$, $i = 1, 2, 3$, and let a_i denote the vertices of a triangle in T such that $a_i \in L_i$. We may assume without loss of generality that $r \in a_2 a_3$ and that $s \in a_1 a_2$. Since a_2 may be any point on L_2 except p and t_2, there exist $n - 1$ triangles in T.

It is easily seen that the points t_1, t_3, r, s are on rs and cannot be on $a_1 a_3$. Furthermore, $t_2 \notin rs$ because, in the Desargues configuration, the axis of perspectivity of two triangles cannot intersect the side of one triangle at a point that is collinear with corresponding vertices. Therefore, there are $n + 1 - 5 = n - 4$ possible points for $a_1 a_3 \cap rs$. Since there are $n - 1$ triangles, there must be at least one pair of triangles

$a_1a_2a_3$, $b_1b_2b_3$ such that $a_1a_3 \cap b_1b_3 \in rs$. Thus $a_1a_2a_3$ and $b_1b_2b_3$ are centrally perspective from p, are axially perspective from rs, and form the Desargues configuration.

Not every infinite plane contains a nontrivial pair of triangles; in fact, for free extension planes if the configuration Σ_o does not contain such a pair, then neither does Σ_o^+ because a nontrivial pair generates a confined configuration and the only confined configurations in Σ_o^+ are in Σ_o.

The frequency of occurrence of nontrivial pairs of triangles in a sense determines the "Desarguesianness" of a plane. If π is a Desarguesian plane, then all centrally perspective triangles are axially perspective and thus all admissible pairs of triangles are nontrivial. If π satisfies the Desargues (p, L) Theorem, then there exist nontrivial pairs of triangles centrally perspective from p and axially perspective from L. The following example describes planes of different Desarguesianness.

EXAMPLE 10.6.3

  i. Let $\pi = \Sigma_o^+$ where Σ_o is a four-point. Then there is no nontrivial pair of triangles.

 ii. Let $\pi = \pi_H(9)$. Then π admits nontrivial pairs of triangles but there is no pair (p, L) in π such that π satisfies the Desargues (p, L) Theorem.

iii. Let π_P be a plane represented by the planar ring P. Then, by Corollaries 10.1.3 and 10.1.4, π_P admits nontrivial pairs of triangles (assuming that $P \neq J/2$).

 iv. Suppose that P is a planar ring satisfying the law of associativity of addition. Then, by Corollary 10.2.8, π_P satisfies the Desargues (v, uv) Theorem.

  v. If P satisfies the law of associativity of multiplication, then, by Corollary 10.3.7, π_P satisfies the Desargues (u, ov) Theorem.

 vi. If P satisfies both laws of associativity, then π_P satisfies the Desargues (p, L) Theorem for pairs (v, uv) and (u, ov).

vii. If P satisfies both laws of associativity, as well as the law of Spencer (that is, $(a \odot c) \odot (b \odot c) = a \odot b$ where $a \odot b =$

$a(a - b)^{-1}(-b)$ (see p. 309), then π_P satisfies the Desarguesian (p,L) Theorem for pairs (u,ov) and (p,pu) where $p \in ov$.

viii. If $\pi = \pi_N(9)$, then, by Example 4.5.13(ii), π satisfies the Desargues (p,L) Theorem for the following pairs (p,L):

$$p \in L \text{ and } L = \langle 0,0,1 \rangle$$
$$p = [0,1,0] \text{ and } L \text{ passes through } [1,0,0]$$
$$p = [1,0,0] \text{ and } L \text{ passes through } [0,1,0]$$
$$p = [1,x,0] \text{ and } L \text{ passes through } [1,-x,0]$$
$$\text{where } x \in N, x \neq 0$$

ix. If π satisfies the Little Desargues Theorem, then π satisfies the Desargues (p,L) Theorem for all p and L such that $p \in L$.

x. If π satisfies the Desargues Theorem, then π satisfies the Desargues (p,L) Theorem for all p and L.

Although we cannot linearly order these ten planes with respect to their Desarguesianness, we can say that in general, the higher the roman numeral, the higher the "Desarguesian quotient." The Desargues (p,L) Theorem applies to certain pairs (p,L) in planes (iv) through (x). Equivalently, we may say that these planes are (p,L) transitive for those appropriate pairs (p,L). These "transitivity pairs" occur in various patterns. Letting $T(\pi) = \{(p,L): \pi \text{ is } (p,L) \text{ transitive}\}$, we have the following categories:

EXAMPLE 10.6.4

i. $T(\pi) = \emptyset$ (see Example 10.6.3(i) through (iii)).

ii. $T(\pi) = \{(p_o,L_o): p_o \in L_o\}$. For the planes π of Example 10.6.3(iv), clearly $T(\pi) \supseteq \{(p_o,L_o): p_o \in L_o\}$. In fact, for Example 10.2.2(iii), we have equality; that is, $T(\pi) = \{(p_o,L_o): p_o \in L_o\}$.

iii. $T(\pi) = \{(p_o,L_o): p_o \notin L_o\}$. For the planes π of Example 10.6.3(v), $T(\pi) \supseteq \{(p_o,L_o): p_o \notin L_o\}$. In fact, for Example 10.3.1, we have equality.

iv. $T(\pi) = \{(p_o,L_o),(q_o,M_o): p_o \neq q_o, L_o \neq M_o, q_o = L_o \cap M_o, p_o \in L_o\}$. For planes π of Example 10.6.3(vi), $T(\pi) \supseteq \{(p_o,L_o),(q_o,M_o)\}$. The last paragraph of Section 10.4 indicates that equality can be achieved.

v. $T(\pi) = \{(p_o,L_o),(p,L): p_o \notin L_o \text{ and } p \in L_o, p_o \in L, p \in L\}$. Here p and L represent variables; p_o and L_o represent fixed elements. For planes π of Example 10.6.3(vii), $T(\pi) \supseteq \{(p_o,L_o),(p,L)\}$. Equality is achieved by Example 10.4.1.

vi. $T(\pi) = \{(p,L): p \in L_o, L \text{ passes through } f(p) \text{ where } f \text{ is an involution that maps } L_o \text{ onto } L_o \text{ but leaves no points of } L_o \text{ fixed}\}$. The plane $\pi_N(9)$ falls into this category.

vii. $T(\pi) = \{(p,L): p \in L\}$. All Little Desarguesian planes that are not Desarguesian fall into this category. Examples of such planes are given in Chapter 14.

viii. $T(\pi) = \{(p,L): p \in \mathscr{P}, L \in \mathscr{L}\}$. All Desarguesian planes belong to this category.

A system of classification of projective planes based on the patterns of transitive pairs (p,L) was developed by Lenz (1954) and Barlotti (1957). Their work establishes existence theorems for certain sets $T(\pi)$. Eight of these sets have been listed here; Dembowski (1968) gives a complete listing of forty-eight different classifications. It is interesting to note that of these forty-eight classes, thirty have no corresponding plane (for example, there is no plane π such that $T(\pi) = \{(p,L): p \in L_o, p_o \in L \text{ for some } p_o \in L_o\}$), seventeen have realizable planes, and one classification remains in question. These classifications will be discussed again and several of the results of Lenz and Barlotti will be proved in subsequent chapters.

This geometric classification is closely related to the algebraic representation of planes that is being developed in Part 3. Thirteen of the seventeen realizable transitivity classes can be characterized algebraically. Also, the algebraic approach distinguishes differences that the Lenz-Barlotti classification does not recognize (for example, Desarguesian and Pappian planes). By combining the geometric criterion of (p,L) transitivity with the algebraic criterion of ternary ring structure, we are able to make a significant analysis of the projective plane. Nevertheless, the projective plane remains for the most part an inscrutable object. The large group of infinite planes in the class $T(\pi) = \varnothing$ has been the most difficult to analyze; even the finite projective planes of small order are difficult to classify. Clearly, an exciting challenge faces the ambitious student of projective geometry.

EXERCISES

1. Show that a nontrivial pair of triangles generates a confined configuration.

2. Find $T(\pi_N(9)^d)$.

3. Show that there are no planes π such that
   a. $T(\pi) = \{(p_o, L_o), (q_o, L_o): p_o \neq q_o\}$.
   b. $T(\pi) = \{(p_o, L_o), (q_o, M_o): \text{either } p_o \neq q_o \text{ or } L_o \neq M_o \text{ or both}\}$.
   c. $T(\pi) = \{(L_o, L_o), (M_o, M_o): L_o \neq M_o\}$.
   d. $T(\pi) = \{(p, L_o), (q_o, L_o): p \in L_o, q_o \notin L_o\}$.
   e. $T(\pi) = \{(p_o, p_o), (L_o, L_o)\}$.
   f. $T(\pi) = \{(p, L), (q_o, M_o): \text{for all } p \text{ and } L \text{ such that } p \in L, q_o \notin M_o\}$.
   (None of these classifications is included in the forty-eight Lenz-Barlotti classes.)

4. The reason that there are no planes π such that $T(\pi)$ equals one of the six sets listed in exercise 3 is that in each instance, additional transitivity is necessarily present in a plane satisfying the given transitivity requirements. In each of the six classifications (a) through (f) and in the accompanying subclassifications (for example, in 3(a), we may have $p_o, q_o \in L$ or $p_o, q_o \notin L$ or $p_o \in L, q_o \notin L$), find the additional transitivity necessitated by the given transitivity.

CHAPTER 11

PLANES OVER QUASIFIELDS

This chapter, like Chapter 10, studies the properties of the class of projective planes defined by planar rings. In Chapter 10 the properties of the planar ring P were supplemented with the property of associativity and the effects of the property of associativity on the geometric structure of π_P were cataloged. In this chapter the properties of the planar ring are supplemented with the property of distributivity and the effects of the property of distributivity on π_P are analyzed.

11.1 QUASIFIELDS

A basic assumption in this section will be that the planar ring P satisfies the associative law of addition; this assumption will be supplemented with a one-sided law of distributivity.

DEFINITION 11.1.1

1. *The planar ring* $P = (R,+,\cdot)$ *is a* right quasifield *if and only if* P *is a Cartesian group such that* $(a + b)c = ac + bc$.

2. P is a left quasifield *if and only if P is a Cartesian group such that*
$$a(b + c) = ab + ac.$$

Such systems are also called right (left) Veblen-Wedderburn systems.

DEFINITION 11.1.2   *Let $P = (R,+,\cdot)$ be a planar ring and define $\circ$ on P as follows: $x \circ y = y \cdot x$ where $x,y \in R$. The system $(R,+,\circ)$ is called the dual of P.*

Notation.   The dual of P is denoted by P^d.

THEOREM 11.1.3   *If $P = (R,+,\cdot)$ is a right quasifield, then P^d is a left quasifield; if P is a left quasifield, then P^d is a right quasifield. The system $(P^d)^d = P$.*

Proof.   First it must be shown that P^d is a planar ring. Recalling Theorem 9.1.12, we must establish five properties.

1. Clearly, $(R,+)$ and $(R - \{0\},\circ)$ are loops.

2. It is also clear that $a \circ 0 = 0 \circ a = 0$.

3. Let $a,b,c,d \in R$ such that $a \neq c$. It must be established that there exists an $x \in R$ such that $x \circ a + b = x \circ c + d$, or, equivalently, that $ax + b = cx + d$. By Theorem 9.1.12(v), there exists a unique pair (x,y) such that $ax + y = -b$ and $cx + y = -d$. Thus $ax + b = -y$ and $cx + d = -y$. Hence there exists a unique x such that $ax + b = cx + d$.

4. Let $a,b,c \in R$ be given. Then it follows easily from Theorem 9.1.12(iv) that there is a unique x such that $a \circ b + x = c$.

5. Let $a,b,c,d \in R$ such that $a \neq c$. It must be established that there exists a unique pair (x,y) such that $a \circ x + y = b$ and $c \circ x + y = d$; or, equivalently, that $xa + y = b$ and $xc + y = d$. By Theorem 9.1.12(iii), there exists a unique $x \in R$ such that $xa + (-b) = xc + (-d)$. Let $y = -xa + b$. Thus $xa + y = b$ and $xc + y = xc + (-xa + b) = xc + (-xc + d) = d$.

Clearly, $(R,+)$ is a group because P is a Cartesian group. Also $c \circ (a + b) = (a + b) \cdot c = a \cdot c + b \cdot c = c \circ a + c \circ b$, so p^d is a left quasifield.

The remainder of the proof is left as an exercise.

EXAMPLE 11.1.4

i. Any division ring is a right (and left) quasifield.

ii. $N(9)$ is a right quasifield.

iii. Let $(R,+,\cdot)$ be defined as follows: $(R,+) = (N,+)$. Multiplication is defined as in Table 20 (cf. Table 12, p. 63). Although (i) and (ii) satisfy additional properties (associativity of multiplication, for example), (iii) satisfies only the bare minimum.

TABLE 20

·	1	2	a	b	c	d	e	f
1	1	2	a	b	c	d	e	f
2	2	1	d	f	e	a	c	b
a	a	e	c	1	d	b	f	2
b	b	d	f	c	2	1	a	e
c	c	f	2	d	b	e	1	a
d	d	c	e	2	a	f	b	1
e	e	b	1	a	f	c	2	d
f	f	a	b	e	1	2	d	c

iv. The *finite Hall systems* are an easily described class of finite right quasifields due to Hall (1943). (Hall does not limit his algebra to the finite case as is done here.) Let F be a Galois Field of order $n > 2$ and let $f(x) = x^2 - rx - s$ be a quadratic polynomial irreducible over F. Let $R = F \times F$. Let $\oplus$ be vector addition on R, that is, $(a,b) \oplus (c,d) = (a + c, b + d)$. Let $\circ$ be defined as follows:

$$(a,b) \circ (c,d) = \begin{cases} (ac,bd) & \text{if } d = 0 \\ (ac - bd^{-1}f(c),\, ad - bc + br) & \text{if } d \neq 0 \end{cases}$$

Then $(R,\oplus,\circ)$ is a finite Hall system.

For example, $N(9)$ is a finite Hall system where $f(x) = x^2 + 1$. The student should display the finite Hall systems for $g(x) = x^2 - x - 1$ and $h(x) = x^2 + x - 1$ as an exercise.

Finite Hall systems are right quasifields. (For a proof of this fact, see Hall, 1959, pp. 364, 365. For other examples of right quasifields, see Dembowski, 1968, p. 228.) Examples of left quasifields can be derived from the duals of the right quasifields (see Theorem 11.1.3).

The nearfield $N(9)$ is the only finite Hall system that satisfies associativity of multiplication; no finite Hall system satisfies left distributivity.

This section is concluded with three elementary results regarding right quasifields that will be used later in the text.

THEOREM 11.1.5 *If $Q = (R,+,\cdot)$ is a right quasifield and $a,b \in R$, then $(-a)b = -(ab)$.*

Proof. By right distributivity, $(-a + a)b = (-a)b + ab = 0$. Thus $(-a)b = -(ab)$.

LEMMA 11.1.6 *If $Q = (R,+,\cdot)$ is a right quasifield, then for any $a,b \in R$ such that $a \neq 1$, the equation $(x \cdot a) + b = x$ has a unique solution in R.*

Proof. By Theorem 9.1.12(iii), there exists a unique $x \in R$ such that $(x \cdot a) + b = x \cdot 1 + 0 = x$.

THEOREM 11.1.7 *If $Q = (R,+,\cdot)$ is a right quasifield, then for any $a,b \in R$, $a + b = b + a$.*

Proof. Let $a,b \in R$. If a or $b = 0$, then clearly $a+b=b+a$. Suppose that $a,b \neq 0$ and let $c = a^{-1}(b + a - b)$. If $c = 1$, we easily see that $a + b = b + a$. Suppose that $c \neq 1$. By Lemma 11.1.6, there exists a unique $x \in R$ such that $xc + b = x$, and thus $-(xc) + x = b$. Using Theorem 11.1.5 and the properties of a right quasifield, we have the following equalities: $-((x + a)c) + (x + a) = (-(x + a))c + x + a = (-a - x)c + x + a = -ac - xc + x + a - ac + b + a = -(b + a - b) + b + a = b - a - b + b + a = b$. Thus $-(xc) + x = b$ and $-((x + a)c) + (x + a) = b$, and since the solution x is unique, $x = x + a$ or $a = 0$, a contradiction. This contradiction implies that $c = 1$, and thus $a + b = b + a$.

EXERCISES

1. Complete the proof of Theorem 11.1.3.

2. Show that the finite Hall system of order 9 generated by
   a. $f(x) = x^2 + 1$ is the nearfield $N(9)$.

b. $g(x) = x^2 - x - 1$ is the algebra A of exercise 2.1.6.

c. $h(x) = x^2 + x - 1$ is the algebra B of exercise 2.4.4.

3. Show that the quasifield of Example 11.1.4(iii) (which is also system C of exercise 2.4.4) is not a finite Hall system.

4. Write out the table for the finite Hall system of order 16.

5. If Q is a right quasifield, prove or disprove that $a(-b) = -(ab)$.

11.2 GEOMETRIC CONSEQUENCES OF DISTRIBUTIVITY

With the addition of distributivity to the properties of the planar ring P, a significant amount of Desarguesian structure is added to the plane π_P. As with the addition of associativity in Chapter 10, this structure is described most clearly in terms of (p, L) transitivity.

THEOREM 11.2.1   *If $Q(u, v, o, e) = Q = (R, +, \cdot)$ is a right quasifield, then π_Q is (uv, uv) transitive.*

Proof.   Define the mappings $f_{r,s}$ and $F_{r,s}$ on π_Q for $r, s \in R$ as follows:

$$f_{r,s}:\ [x, y] \to [x + r, y + s]$$
$$[x] \to [x]$$
$$z \to z$$

$$F_{r,s}:\ \langle m, k \rangle \to \langle m, -rm + k + s \rangle$$
$$\langle k \rangle \to \langle k + r \rangle$$
$$Z \to Z$$

The first step is to show that $(f_{r,s}, F_{r,s})$ is a collineation on π_P.

Case 1. $[x, y] \in \langle m, k \rangle$ iff $y = xm + k$ iff $y + s = xm + k + s$ iff $y + s = xm + rm - rm + k + s$ iff $y + s = (x + r)m - rm + k + s$ iff $[x + r, y + s] \in \langle m, -rm + k + s \rangle$ iff $f_{r,s}([x, y]) \in F_{r,s}(\langle m, k \rangle)$.

Case 2. $[x, y] \in \langle k \rangle$ iff $k = x$ iff $k + r = x + r$ iff $[x + r, y + s] \in \langle k + r \rangle$ iff $f_{r,s}([x, y]) \in F_{r,s}(\langle k \rangle)$.

Case 3. $[x] \in \langle m, k \rangle$ iff $x = m$ iff $[x] \in \langle m, -rm + k + s \rangle$ iff $f_{r,s}([x]) \in F_{r,s}(\langle m, k \rangle)$.

Clearly, incidence is preserved for lines through z and points on Z. Thus $(f_{r,s}, F_{r,s})$ is a collineation on π_P.

Each map $(f_{r,s},F_{r,s})$ is an elation with axis Z. Furthermore, all possible elations with axis Z exist because if $[x,y]$ and $[x',y']$ are any two points not on Z, then clearly $f_{r,s}\colon [x,y] \to [x',y']$ where $r = x - x'$ and $s = y - y'$. Thus we may conclude that π_Q is (Z,Z) transitive or, equivalently, (uv,uv) transitive.

The converse of Theorem 11.2.1 is also true, as Theorem 11.2.2 shows.

THEOREM 11.2.2   *If $P(u,v,o,e) = P = (R,+,\cdot)$ is a planar ring and π_P is (uv,uv) transitive, then P is a right quasifield.*

Proof.   Since π_P is (v,uv) transitive, P is a Cartesian group by Theorem 10.2.6. The following argument shows right distributivity.

Let $b \in R$ and let $f \in \mathrm{El}(u,uv)$ such that $f\colon [0,0] \to [b,0]$ (see Figure 82). Since $[1] \in Z$ and Z is held fixed, $f\colon \langle 1,0 \rangle = [0,0][1] \to [b,0][1] = \langle 1,-b \rangle$. We know that $\langle 0,a \rangle$ is fixed for any $a \in R$ because it is on the center $[0]$ of f, so $f\colon [a,a] = \langle 1,0 \rangle \cap \langle 0,a \rangle \to \langle 1,-b \rangle \cap \langle 0,a \rangle = [a+b,a]$. Thus $f\colon \langle a \rangle = [a,a]z \to [a+b,a]z = \langle a+b \rangle$. Since $\langle 0,ac \rangle$ is also held fixed, $f\colon [a,ac] = \langle a \rangle \cap \langle 0,ac \rangle \to \langle a+b \rangle \cap \langle 0,ac \rangle = [a+b,ac]$. Also, for any $c \in R$, $f\colon \langle c,0 \rangle = [c][0,0] \to [c][b,0] = \langle c,-bc \rangle$. Since $[a,ac] \in \langle c,0 \rangle$ and since f is a collineation, $[a+b,ac] \in \langle c,-bc \rangle$. Thus $ac = (a+b)c - bc$ or $ac + bc = (a+b)c$. This completes the proof that P is a right quasifield.

Proof.   The proof is left as an exercise.

The property of left distributivity on a planar ring P imposes a geometric restriction on π_P that is the dual of the restriction imposed by right distributivity as stated in Theorem 11.2.1. This statement is clarified with the help of the following theorem.

THEOREM 11.2.4   *If $Q = (R,+,\cdot)$ is a right quasifield and $Q^d = (R,+,\circ)$ is the dual left quasifield, then $(\pi_Q)^d \sim \pi_{Q^d}$.*

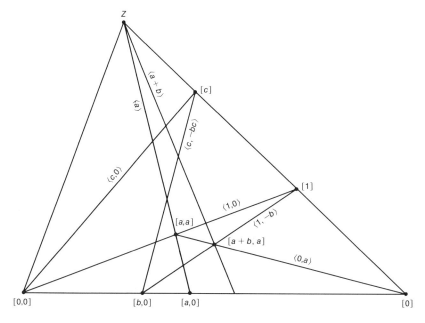

FIGURE 82

Proof. Let $[x,y],[x],z,\langle a,b\rangle,\langle a\rangle,Z$ denote (as usual) the points and lines of π_Q and let $[[x,y]],[[x]],z',\langle\langle a,b\rangle\rangle,\langle\langle a\rangle\rangle,Z'$ denote the points and lines of π_{Q^d}. Consider the following mappings f and F; f maps the lines of π_Q onto the points of π_{Q^d} as follows:

$$f\colon \langle a,b\rangle \to [[a,-b]]$$
$$\langle a\rangle \to [[a]]$$
$$Z \to z'$$

F maps the points of π_Q onto the lines of π_{Q^d} as follows:

$$F\colon [x,y] \to \langle\langle x,-y\rangle\rangle$$
$$[x] \to \langle\langle x\rangle\rangle$$
$$z \to Z'$$

It remains to be shown that (f,F) is an isomorphism from $(\pi_Q)^d$ onto π_{Q^d}. Using I to denote incidence on $(\pi_Q)^d$, we have $\langle a,b\rangle$ I $[x,y]$ iff $[x,y] \in \langle a,b\rangle$ iff $y = xa + b$ iff $y = a\circ x + b$ iff $-b = a\circ x + (-y)$ iff $[[a,-b]]$ $\in \langle\langle x,-y\rangle\rangle$ iff $f(\langle a,b\rangle) \in F([x,y])$. If $L \neq \langle a,b\rangle$ or $p \neq [x,y]$ and p I L, then it is easily checked that $f(L) \in F(p)$.

With this theorem and Theorem 11.2.4, we may formally establish the geometric consequences of left distributivity.

THEOREM 11.2.5   *The plane π is (v,v) transitive if and only if $\pi = \pi_Q$ where $Q = Q(u,v,o,e)$ is a left quasifield. (Points u, o, and e are chosen so that u,v,o,e is a four-point.)*

Proof.   The proof is left as an exercise.

For Cartesian groups P that satisfy both distributive laws, naturally π_P is both (uv,uv) and (v,v) transitive. Such Cartesian groups are called semifields; these algebras and their associated projective planes will be discussed in Chapter 13.

For Cartesian groups P that satisfy the right distributive law and the associative law of multiplication, π_P is both (uv,uv) and (u,ov) transitive. This transitivity necessarily implies further transitivity. The student will find it an easy exercise to show that π_P is (u,v) transitive; proof that π_P is also (v,u) transitive is a more difficult exercise. Such Cartesian groups are called right planar nearfields, and they will be discussed in Chapter 12. Cartesian groups P that satisfy the left distributive law and the associative law of multiplication are called left planar nearfields; their associated planes π_P satisfy the transitivity restrictions dual to the restrictions for right planar nearfields.

It is natural to consider the class $T(\pi)$ of the planes π that are defined over a right quasifield having no further algebraic properties. For example, if Q is the quasifield of Example 11.1.4(iii) or (iv) (excepting $N(9)$), can we conclude that $T(\pi_Q) = \{(p,Z): p \in Z\}$, or might there possibly be more transitivity on π_Q? The answer is that there may or may not be more transitivity. For π_Q, where Q is given as in Example 11.1.4(iii), there is more transitivity; if Q is a finite Hall system other than $N(9)$, then $T(\pi_Q) = \{(p,Z): p \in Z\}$ (see Dembowski, 1968, for further details and references on this subject).

This section is concluded with the introduction of a new notation for planes over quasifields that is analogous to the notation for $\pi_N(9)$ and that will be more convenient for subsequent use than the ternary ring notation. In the remainder of the text, the plane π_Q over the quasifield Q will be represented as in Construction 11.2.6.

CONSTRUCTION 11.2.6 Let $P = (R, +, \cdot)$ be a Cartesian group and define $\mathscr{P}, \mathscr{L}, \mathscr{I}$ as follows:

$$\mathscr{P} = \{[x, y, 1], [1, x, 0], [0, 1, 0]: x, y \in R\}$$
$$\mathscr{L} = \{\langle a, 1, b \rangle, \langle 1, 0, a \rangle, \langle 0, 0, 1 \rangle: a, b \in R\}$$
$$\mathscr{I} = \{([x, y, z], \langle a, b, c \rangle) \text{ if and only if } xa + yb + zc = 0\}$$

THEOREM 11.2.7 *Let $P = (R, +, \cdot)$ be a Cartesian group such that either $-(ab) = (-a)b$ for all $a, b \in R$ or $-(ab) = a(-b)$ for all $a, b \in R$. Then $(\mathscr{P}, \mathscr{L}, \mathscr{I})$ as defined in Construction 11.2.6 is isomorphic to π_P.*

Proof. Suppose that $-(ab) = (-a)b$ for all $a, b \in R$ and let

$$f: \quad [x, y] \to [-x, y, 1]$$
$$[x] \to [1, -x, 0]$$
$$z \to [0, 1, 0]$$

$$F: \quad \langle a, b \rangle \to \langle a, 1, -b \rangle$$
$$\langle a \rangle \to \langle 1, 0, a \rangle$$
$$Z \to \langle 0, 0, 1 \rangle$$

Then it is an easy matter to check that (f, F) is an isomorphism from π_P onto $(\mathscr{P}, \mathscr{L}, \mathscr{I})$. We can check that incidence is preserved in the most difficult case in the following way: $[x, y] \in \langle a, b \rangle$ iff $y = xa + b$ iff $-(xa) + y + (-b) = 0$ iff $(-x)a + y \cdot 1 + (-b) = 0$ iff $([-x, y, 1], \langle a, 1, -b \rangle) \in \mathscr{I}$ iff $(f([x, y], F(\langle a, b \rangle)) \in \mathscr{I}$.

If $-(ab) = a(-b)$, then the desired isomorphism is similar to (f, F). The exact isomorphism is left for the student to find.

COROLLARY 11.2.8 *If Q is a quasifield, then $(\mathscr{P}, \mathscr{L}, \mathscr{I})$ as defined in Construction 11.2.6 is isomorphic to π_Q.*

EXERCISES

1. Prove Theorem 11.2.3.

2. Prove Theorem 11.2.5.

3. Let Q be a right quasifield. Show that if π_Q is self-dual, then π is a Pappian plane.

4. Let Q be a right quasifield. If π_Q is (p, uv) transitive for some $p \notin uv$, then π_Q is a Desarguesian plane.

5. Let Q be a right quasifield and let $T(\pi_Q) = \{(L_o, L_o)\}$. Show that:
   a. Every collineation on π_Q fixes the line L_o.
   b. Q does not satisfy the law of left distributivity or associativity of multiplication.

6. Prove or disprove: If Q is a right quasifield that does not satisfy left distributivity or associativity of multiplication, then $T(\pi_Q) = \{(L_o, L_o)\}$.

7. Complete the proof of Theorem 11.2.7.

8. Suppose that Q is a right quasifield satisfying associativity of multiplication. Show that:
   a. π_Q is (u, v) transitive.
   b. π_Q is (v, u) transitive.

9. a. Let Q be a right quasifield and let α be an affine restriction of π_Q. Can α satisfy the First Affine Desargues Theorem? The Second Affine Desargues Theorem? One theorem and not the other?
   b. Answer the preceding questions for affine restrictions of π_Q if Q is a left quasifield.

11.3 FINITE HALL PLANES

The planes defined by finite Hall systems will be called finite Hall planes. This section examines finite Hall planes in some detail, thus carrying further the analysis of planes π_Q. Although this examination is not completely general, it reveals the clever way in which algebra can be used to analyze the geometric properties of projective planes. It will be proved that a finite Hall plane cannot satisfy Fano's Axiom nor can it be a Fano plane; consequently, in every finite Hall plane, there are subplanes of order 2 and subplanes of order other than 2. It is suspected that every finite non-Desarguesian plane shares this property.

First it will be shown that finite Hall planes of order p^{2a} exist where $p^{2a} \neq 4$. This fact is easily established by a counting procedure. For use later in the text, the theorem is strengthened by the requirement that the irreducible quadratic $x^2 - rx - s$ satisfy the property that $r \neq 1$.

THEOREM 11.3.1 *If $p^{2a} \neq 4$, then there exists an irreducible quadratic $f(x) = x^2 - rx - s$ over the field $F = GF(p^a)$ where $r \neq 1$.*

Proof. (Wagner, 1956) Example 11.1.4(iv) has shown that this theorem is true for $p^{2a} = 9$; thus we can assume that $p^a = q > 3$.

First observe that since b and c may independently assume q values, there are q^2 different polynomials of degree 2 with the leading coefficient of 1, that is, polynomials of the form $x^2 + bx + c$ where $b,c \in F$. Of these q^2 polynomials, $(q/2)(q + 1)$ are reducible. We may verify this by considering all the pairs of roots r_1 and r_2 and noting that if $r_1 \neq r_2$, there exist $q(q - 1)/2$ distinct polynomials of the form $(x - r_1)(x - r_2)$ and that if $r_1 = r_2$, there are q other polynomials of the form $(x - r_1)^2$, for a total of $(q/2)(q - 1) + q = (q/2)(q + 1)$. Thus there are $q^2 - (q/2)(q + 1) = (q/2)(q - 1)$ irreducible polynomials of the form $x^2 + bx + c$ in F. Clearly, there are q polynomials of the form $x^2 - rx - s$ where $r = 1$. But $(q/2)(q - 1) > q$ for $q > 3$; therefore, there are irreducible polynomials of the form $x^2 - rx - s$ where $r \neq 1$.

The answer to the obvious question at this point is that no matter which irreducible quadratic in $GF(p^a)$ we choose, the resulting finite Hall plane is the same.

THEOREM 11.3.2 *All finite Hall planes of the same order are isomorphic.*

Proof. See Hughes (1959).

The following analysis of the subplanes of finite Hall planes is due to Neumann (1955).

CONSTRUCTION 11.3.3 Let $Q = (R,+,\cdot)$ be a finite Hall plane of order p^{2a}, let $F = GF(p^a) = GF(q)$, and let $f(x) = x^2 - rx - s$ be an irreducible quadratic over F. Observe that elements of R are of the form (x,y) where $x,y \in F$. To simplify matters, the element $(t,0) \in R$ will be denoted by t. Observe also that $r + s \neq 1$ because if $r + s = 1$, we would have $1^2 - r \cdot 1 - s = 0$ and f would not be irreducible. Thus we may define $k \in R$ by $k = (r + s - 1)^{-1}$. Let $L_0 = \langle 1,0,0 \rangle$, $L_1 = \langle 0,1,0 \rangle$, $L_2 = \langle 1,0,(-ks,-k) \rangle$, $L_3 = \langle 1,1,0 \rangle$, $L_4 = \langle (1,1),1,(0,-1) \rangle$, $L_5 = \langle (0,1),1,(0,-1) \rangle$, and $L_6 = \langle 1,0,-1 \rangle$; let $p_0 = [0,0,1]$, $p_1 = [0,1,0]$, $p_2 = [1,0,1]$, $p_3 = [0,(0,1),1]$, $p_4 = [1,-1,1]$, $p_5 = [(ks,k),(-ks,-k),1]$, and $p_6 = [(ks,k),0,1]$.

We can show that $\{p_0, \ldots, p_6\}$ forms a subplane $\pi(2)$ in π_Q if and only if Q is of odd order. It is easily checked that $p_0, p_1, p_3 \in L_0$; $p_1, p_2, p_4 \in L_6$; $p_2, p_3 \in L_5$; $p_3, p_4 \in L_4$; $p_4, p_5, p_0 \in L_3$; $p_5, p_6, p_1 \in L_2$; and $p_6, p_0, p_2 \in L_1$. We need only show that $p_5 \in L_5$ and that $p_6 \in L_4$. These facts can be proved with the aid of several lemmas.

LEMMA 11.3.4 $p_5 \in L_5$.

Proof. We must show that $[(ks,k),(-ks,-k),1] \in \langle (0,1),1,(0,-1) \rangle$, or equivalently, that $(ks,k) \cdot (0,1) + (-ks,-k) + (0,-1) = (0,0)$. Keeping in mind that Q is a right quasifield and noting that since $k \in F$, k commutes with all elements of R, we have the following equalities: $(ks,k) \cdot (0,1) = (-k(-s),ks + kr) = (ks,sk + rk) = (ks,(s + r)k)$. Thus $(ks,k) \cdot (0,1) + (-ks,-k) + (0,-1) = (ks,(s + r)k) + (-ks,-k) + (0,-1) = (0,(s + r - 1)k - 1) = (0,k^{-1}k - 1) = (0,0)$.

LEMMA 11.3.5 $p_6 \in L_4$ *if and only if* $r + 2s - 1 = 0$.

Proof.   Clearly, $p_6 \in L_4$ iff $[(ks,k),0,1] \in \langle (1,1),1,(0,-1) \rangle$ iff $(ks,k) \cdot (1,1) + (0,-1) = (0,0)$. Since $(ks,k)\cdot(1,1) = (ks - k(1 - r - s), ks - k + kr) = (ks - k(-k)^{-1},(s + r - 1)k) = (ks + 1,1)$, it follows that $p_6 \in L_4$ iff $ks + 1 = 0$. Obviously, $ks + 1 = 0$ iff $s = -(k^{-1}) = -(r + s - 1)$, that is, $r + 2s - 1 = 0$.

LEMMA 11.3.6   *If* $4s^2 + 1$ *is not a square in* $\mathrm{GF}(q)$, *then* $x^2 - rx - s$ *is irreducible where* $r + 2s - 1 = 0$.

Proof.   By the quadratic formula, $x^2 + (2s - 1)x - s = 0$ has a solution in $\mathrm{GF}(q)$ iff $4s^2 + 1$ is a square. Thus if $4s^2 + 1$ is not a square, then $x^2 - rx - s \neq 0$ for any x where $r = -2s + 1$.

LEMMA 11.3.7   *If* p *is odd, there exists an* $s \in \mathrm{GF}(q)$ *where* $q = p^a$ *such that* $4s^2 + 1$ *is not a square.*

Proof.   Suppose that $4s^2 + 1$ is a square for all s, that is, $(2s)^2 + 1 = t^2$ for all $s \in \mathrm{GF}(q)$. It follows that $(2sc)^2 + c^2 = (ct)^2$ for all s,c in $\mathrm{GF}(q)$. Thus the sum of any two squares is a square and so the squares form a subfield F' of $\mathrm{GF}(q)$. Since $p \neq 2$, the order of F' is $(q - 1)/2 + 1$

$= (q + 1)/2$. This follows because $a^2 = b^2$ iff $(a + b)(a - b) = 0$ iff $a = b$ or $a = -b$, and thus the elements of $GF(q)$ except 0 may be split into doubletons $\{+a,-a\}$, each set uniquely representing one member of F'. Since $(q + 1)/2$ is relatively prime to q, we know that F' has an order that does not divide the order of $GF(q)$. But this is a contradiction, so we may conclude that there exists an $s \in GF(q)$ such that $4s^2 + 1$ is not a square.

THEOREM 11.3.8   *For any finite Hall plane of order p^{2a} where $p \neq 2$, there exists a subplane $\pi(2)$.*

Proof. By Lemma 11.3.7, there exists an s in $GF(p^a)$ such that $4s^2 + 1$ is not a square. By Lemma 11.3.6, if $r + 2s - 1 = 0$, then $f(x) = x^2 - rx - s$ is irreducible in $GF(p^a)$. By Lemma 11.3.5, $p_6 \in L_4$, so $\{p_0, \ldots, p_6\}$ forms a subplane $\pi(2)$ of the finite Hall plane generated by $f(x)$. By Theorem 11.3.2, any finite Hall plane of order p^{2a} contains a subplane $\pi(2)$.

THEOREM 11.3.9   *For any finite Hall plane of order 2^{2a}, there exist subplanes of order other than 2.*

Proof. To show that the four-point p_0,p_1,p_2,p_5 generates a subplane of order unequal to 2, we simply need to show that $p_6 \notin L_4$. By Lemma 11.3.5, this is equivalent to showing that $r + 2s - 1 \neq 0$, which is done as follows.

Since $q = 2^a$, $GF(q)$ is of characteristic 2, and so $r + 2s - 1 = 0$ iff $r - 1 = 0$ iff $r = 1$. By Theorem 11.3.1, there exists an irreducible polynomial $f(x) = x^2 - rx - s$ in $GF(q)$ where $r \neq 1$. Thus there exists a finite Hall plane generated by $f(x)$ such that $p_6 \notin L_4$. Since all finite Hall planes of order 2^{2a} are isomorphic, we may conclude that all finite Hall planes of even order contain subplanes $\pi(m)$ for $m > 2$.

Since a finite Hall plane of order p^{2a} clearly contains a Desarguesian subplane of order p^a coordinatized by the subfield $GF(p^a)$ of Q, the remark made at the beginning of this section can be formally substantiated.

THEOREM 11.3.10   *If π is a finite Hall plane, it does not satisfy Fano's Axiom, nor is it a Fano plane.*

Proof. If π is of even order 2^{2a}, it contains a Desarguesian subplane of order 2^a that is a Fano plane. Therefore, π does not satisfy Fano's Axiom. By Theorem 11.3.9, it also contains a subplane of order $\neq 2$. Thus it is not a Fano plane.

If π is of odd order p^{2a}, it contains both a subplane of odd order (the Desarguesian plane of order p^a) and a subplane of order 2 (by Theorem 11.3.8). Thus we again arrive at the same conclusion.

As mentioned in Section 2.8, Gleason (1956) has established that finite Fano planes must be Desarguesian, so the preceding proof that finite Hall planes are not Fano planes has been generalized. However, the result that all finite Hall planes contain the subplane $\pi(2)$ has not yet been generalized to finite non-Desarguesian planes. It is believed to be true, but a proof (or disproof) has yet to be formulated.

This section is concluded with a cardinality theorem for the subplanes of order p for finite Hall systems of order p^n. In fact, the theorem is proved here also for planes over general right quasifields; it is easily extended to include planes over arbitrary quasifields.

THEOREM 11.3.11 (Killgrove, 1965) *In a plane π_Q over the right quasifield Q, there are at least $n^3(n-1)^2(n+1)/p^2(p^2-1)(p^2-p)$ subplanes of order p where $n = p^k$.*

Proof. By Theorem 11.2.1, π_Q is (L,L) transitive for $L = uv$. For every pair of points (p,q) on L, we may form the algebra $P' = P(u',v',o',e')$ where $u' = p$ and $v' = q$, and where o', e' are an arbitrary pair subject to the restriction that u',v',o',e' be a four-point. Since $\pi_{P'}$ is $(u'v',u'v')$ transitive (because $u'v' = L$), P' is a right quasifield. It is not hard to show that the four-point u',v',o',e' generates a subplane of order p.

Since there are $n(n + 1)$ distinct pairs (u',v') on L and $n^2(n-1)^2$ choices for o' and e' on $\mathscr{P} - L$, there exist $n^3(n-1)^2(n+1)$ four-points u',v',o',e'. Any subplane generated by u',v',o',e' contains $p^3(p-1)^2(p+1)$ four-points. Thus there are $n^3(n-1)^2(n+1)/p^3(p-1)^2(p+1)$ subplanes of order p in π_Q.

For the plane $\pi_N(9)$, there are 1080 subplanes $\pi(3)$, as explained in Section 2.8. It is easily checked that $9^3 \cdot 8^2 \cdot 10/3^3 \cdot 2^2 \cdot 4 = 1080$, so

$\pi_N(9)$ has the minimum number of planes of prime order. For the Desarguesian plane $\pi(9)$, there are 7560 subplanes, not a surprising fact since all the four-points of $\pi(9)$ generate the plane $\pi(3)$. The precise meaning of the cardinality of planes $\pi(p)$ is not at all clear, however; for example, π_H has 1080 of the $\pi(3)$ subplanes and its behavior is considerably different from that of $\pi_N(9)$. The plane $(\pi_N(9))^d$ also has 1080 of these subplanes, but this fact is not surprising since Theorem 11.3.11, as stated previously, can be extended from right quasifields to arbitrary quasifields.

EXERCISES

1. Let Q be a finite Hall system of order 16.
   a. Find a subplane $\pi(2)$ in π_Q.
   b. Find a subplane $\pi(4)$ in π_Q.
   c. Find a four-point in π_Q that does not generate $\pi(2)$.

2. Let Q be a finite Hall system of order 25.
   a. Find a subplane $\pi(2)$ in π_Q.
   b. Find a subplane $\pi(5)$ in π_Q.

3. In the proof of Theorem 11.3.11, it is assumed that u', v', o', e' generates a plane of order p. Prove this assumption.

4. Extend Theorem 11.3.11 to a statement about all quasifields (left as well as right).

11.4 THE COORDINATIZING QUASIFIELDS OF $\pi_N(9)$

This section examines four quasifield representations of $\pi_N(9)$. This concrete study will better acquaint the student with the relationship between a plane and its representative ternary rings. It also will provide an opportunity to reexamine the concept of isotopism. Much of this analysis can be found in Hall (1943).

The four quasifields of order 9 that will be used for this study are the finite three Hall systems and the quasifield of Example 11.1.4(iii) (which is not a finite Hall system). The three finite Hall systems are generated by the three irreducible polynomials $f(x) = x^2 + 1$, $g(x)$

$= x^2 - x - 1$, and $h(x) = x^2 + x - 1$. These quasifields are denoted by $Q(f), Q(g), Q(h)$, respectively; the three planes are denoted by π_f, π_g, π_h, respectively. The fourth quasifield is denoted by Q'; the plane $\pi_{Q'}$ is denoted by π'. The system $Q(f) = N(9)$ is displayed in Table 10 and again in Table 21. Multiplication for Q' is represented by Tables 12 and 20. Tables 3 and 11 are the multiplication tables for $Q(g)$ and $Q(h)$, respectively. From this information it is clear that $Q(g), Q(h), Q'$ are the algebras A, B, C, respectively, that were introduced in the exercises at the end of Chapter 2.

The relationship between the four quasifields and their four associated planes will now be examined.

THEOREM 11.4.1 $Q(g)$ and $Q(h)$ are isotopic but not isomorphic.

Proof.   The proof is left as an exercise.

THEOREM 11.4.2   *The algebras $Q(g)$ and Q' are isotopic but not isomorphic.*

Proof. $Q(g)$ and Q' are not isomorphic because every isomorphism must map $2 \to 2$ and this element behaves differently in $Q(g)$ and Q'. For example, 2 commutes with all members of $Q(g)$, but in Q', $2 \cdot a = d$ and $a \cdot 2 = e$.

We may display the isotopism (g, h, j) from $Q(g) = (R, +, \cdot)$ onto $Q' = (R, +, \circ)$ as follows:

$$g = I \quad \text{(the identity)}$$
$$h: x \to x \circ f \quad \text{(notice that } f \text{ is not a function here; } f \in R)$$
$$j = h$$

Clearly, g, h, j are one-to-one maps from R onto R. For $x, y \in R$, using right distributivity, we have $j(x + y) = (x + y) \circ f = x \circ f + y \circ f = j(x) + j(y)$. By inspecting Tables 3 and 12, we see that $(x \cdot y) \circ f = x \circ (y \circ f)$ for $x, y \in R$. Thus $j(x \cdot y) = (x \cdot y) \circ f = x \circ (y \circ f) = g(x) \circ h(y)$, and therefore (g, h, j) is an isotopism.

COROLLARY 11.4.3   *Isotopism does not preserve commutativity on quasifields.*

Completing this study of the quasifield, we conclude that $Q(f)$ is not isotopic to $Q(g), Q(h), Q'$.

THEOREM 11.4.4 *$Q(f)$ is not isotopic to $Q(g)$.*

Proof. This theorem follows immediately from Theorem 9.3.9, since $Q(f)$ satisfies the law of associativity of multiplication and $Q(g)$ does not.

Now it will be proved that all the associated planes are isomorphic, a result that we could easily have deduced from the listing of the known planes of order 9 given in Chapter 2. By Corollary 9.3.7, π_g, π_h, and π' are isomorphic, since their associated algebras are isotopic. That π_f is isomorphic to π_g remains to be shown.

THEOREM 11.4.5 *The planes π_f and π_g are isomorphic.*

Proof. Let $Q(g) = (R,+,\cdot)$. Choose the four-point $u' = [1,2,0]$, $v' = [0,1,0]$, $o' = [0,0,1]$, $e' = [1,0,1]$ (see Figure 83) and form the ternary ring $T(u',v',o',e') = T = (R,t)$. The remainder of the proof will show that $T(u',v',o',e')$ is a linear ternary ring whose associated planar ring is isomorphic to $Q(f)$.

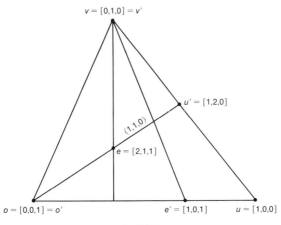

FIGURE 83

The algebras T and $Q(g)$ can be related if we examine the coordinates of the points of π with respect to T and $Q(g)$. Notice that the points of π_g are coordinatized with respect to the four-point $u = [1,0,0]$, $v = [0,1,0]$, $o = [0,0,1]$, $e = [2,1,1]$. (In ternary ring notation, $e = [1,1]$; in π_Q this is $[-1,1,1]$ by Theorem 11.2.7.) Thus points in π_g on $ou - \{u\}$ have coordinates $[x,0,1]$. The line $oe = o'u' = \langle 1,1,0 \rangle$, and so the set $oe - \{oe \cap uv\}$ contains points of the form $[x,0]$ in π_T. Letting j be the perspectivity $oe \overset{v}{\barwedge} ou$, we may define the mapping $F: R \to R$ as follows: $F: x \to x'$ where x' is an element of R such that $j: [x,0] \to [x',0,1]$. It is easily seen that F is a one-to-one map from R onto R; furthermore, $F: 0,1 \to 0,1$.

First consider $t(1,x,y)$. Let b,c,a,m,n,p,q,r,s,w play the same role as in Construction 9.2.4. (Recall, however, that the four-point here is u',v',o',e'.) If $b = [x,0]$, $c = [y,0]$, $a = [1,0]$ in π_T, then $m = [x,x]$, $n = [1,x]$, $p = [x]$, $q = [y,y]$, $r = [0,y]$, $s = [1,t(1,x,y)]$, $w = [t(1,x,y), t(1,x,y)]$ and $wv' \cap o'u' = [t(1,x,y),0]$. Points a, b, and c have the following representations in π_g: $a = [1,2,1]$, $b = [x',2x',1]$, $c = [y',2y',1]$ where $x' = j(x)$ and $y' = j(y)$. This may be seen by observing that the first coordinates of a,b,c in π_g must be the image under j of the first coordinate of a,b,c in π_T and furthermore, that the points must lie on the line $\langle 1,1,0 \rangle$. Therefore, the second coordinates are twice the first (or equivalently, the inverse of the first). It is not difficult to calculate that: $m = [x',0,1]$, $n = [1,x'-1,1]$, $p = [1,x'-1,0]$, $q = [y',0,1]$, $r = [0,y',1]$, $s = [1,x'-1+y',1]$, and $w = [x'+y',0,1]$. Thus $F: t(1,x,y) \to x' + y'$.

Next consider $t(x,y,0)$. Again using Construction 9.2.4 with respect to u',v',o',e', let $b = [y,0]$, $c = [0,0]$, $a = [x,0]$, and thus $m = [y,y]$, $n = [1,y]$, $p = [y]$, $q = [0,0]$, $r = [0,0]$, $s = [x,t(x,y,0)]$, $w = [t(x,y,0), t(x,y,0)]$, $wv' \cap o'u' = [t(x,y,0),0]$. These points have the following coordinates with respect to π_g: $b = [y',2y',1]$, $c = [0,0,1]$, $a = [x',2x',1]$, $m = [y',0,1]$, $n = [1,y'-1,1]$, $p = [1,y'-1,0]$, $q = [0,0,1]$, $r = [0,0,1]$, $s = [x',x'(y'-1),1]$, $w = [x'(y'-1)+x',0,1]$. Thus $F: t(x,y,0) \to x'(y'-1) + x'$.

From these results we may easily conclude that $T(u',v',o',e')$ is a linear ternary ring and that its associated planar ring $P = (R,\oplus,\odot)$ has the following addition and multiplication:

$$x \oplus y = F^{-1}(F(x) + F(y))$$
$$x \odot y = F^{-1}(F(x)(F(y) - 1) + F(x))$$

It is clear that F is an isomorphism from P onto $Q = (R,+,\circ)$ where $x + y = x \oplus y$ and $x \circ y = x(y - 1) + x$. By inspection we may check that $Q = Q(f)$.

COROLLARY 11.4.6   *There exist nonisotopic quasifields that co-ordinatize isomorphic planes.*

This corollary shows that the converse of Corollary 9.3.7 is not true. Thus the relationship between ternary rings, even those ternary rings associated with right quasifields, in a given (L_o, L_o) transitive plane is not as strong as isotopism. We shall see that as more transitivity structure is added to π, the bond between associated ternary rings becomes tighter. As we know from Theorem 9.3.1, if all possible transitivity is available in π, then π is Desarguesian and its associated ternary rings are isomorphic.

EXERCISES

1. Prove Theorem 11.4.2.

2. Let Q be a right quasifield such that $T(\pi_Q) = \{(L_o, L_o)\}$ and let p,q,r,s be a four-point in π_Q. Show that $T(p,q,r,s)$ is a right quasifield if and only if $p,q \in L_o$.

3. In the proof of Theorem 11.4.5, several equalities are assumed without proofs. For example, if $b = [x,0]$, $c = [y,0]$, $a = [1,0]$ in π_T, then $m = [x,x]$, $n = [1,x]$, $p = [x]$, $q = [y,y]$, $r = [0,y]$, $s = [1,t(1,x,y)]$, and $w = [t(1,x,y),t(1,x,y)]$. In π_g these points have another representation, namely, $m = [x',0,1]$, and so on. Verify all these equalities.

4. Consider the plane π_g. By Theorem 11.4.5, we know that $T(u,v,o,e) = Q(g)$ and that $T(u',v',o',e') = Q(f)$ where $u' = [1,2,0]$, $v' = [0,1,0]$, $o' = [0,0,1]$, $e' = [1,0,1]$.
   a. Find other four-points p,q,r,s such that
      (1) $T(p,q,r,s) = Q(g)$.
      (2) $T(p,q,r,s) = Q(f)$.
   b. Also find four-points p,q,r,s such that
      (1) $T(p,q,r,s) = Q(h)$.
      (2) $T(p,q,r,s) = Q'$.
   c. Find four-points p,q,r,s such that
      (1) $T(p,q,r,s)$ is not a quasifield.
      (2) $T(p,q,r,s)$ is not a linear ternary ring.

11.5 PAPPUS THEOREMS AND
PLANES OVER QUASIFIELDS

Section 11.2 discussed the relationship between planes π_Q over quasi-fields and specialized Desargues theorems. This section will show that certain specialized Pappus theorems are also related to such planes. In fact, if we combine the results of this section with those of Section 10.5, we learn the surprising fact that these specialized Pappus theorems are as strong as the Pappus Theorem itself. The results of Burn (1968) are used here, as they were in Section 10.5.

LEMMA 11.5.1 *A projective plane cannot possess an involutory elation and involutory homology with the same axis.*

Proof. The proof is left as an exercise.

An involutory collineation (elation, homology) f is simply a collineation (elation, homology) that is an involution; that is, f^2 is the identity and f is not the identity. Thus an involutory homology is a harmonic homology.

LEMMA 11.5.2 *If $f \in \mathrm{Hom}(p, L)$ and $g \in \mathrm{Hom}(q, L)$ where $p \neq q$ and f and g are involutions, then $g{\circ}f \in \mathrm{El}(pq \cap L, L)$.*

Proof. The proof is left as an exercise.

LEMMA 11.5.3 (Ostrom, 1956) *If f and g are involutions such that $f \in \mathrm{Hom}(p, L)$, $g \in \mathrm{Hom}(q, M)$, and $p \in M$, $q \in L$, then $g \circ f \in \mathrm{Hom}(L \cap M, pq)$ and $g \circ f$ is an involution.*

Proof. First it must be shown that $(g \circ f)^2$ is the identity. Let $x \in L$. Then $(g \circ f)x = g(f(x)) = g(x)$, so $(g \circ f)^2 x = g^2(x) = x$. If $x \in M$, by an analogous argument $(g \circ f)^2 x = x$. Suppose that $x \notin L \cup M \cup pq$ and let $y = px \cap L$, $z = qx \cap M$, $y' = (g \circ f)y$, $z' = (g \circ f)z$, and $x' = py' \cap qz'$. Then $g \circ f(x) = g \circ f(py \cap qz) = py' \cap qz' = x'$. Also $g \circ f(x') = g \circ f(py' \cap qz') = py \cap qz = x$. Thus $(g \circ f)^2 x = x$. Finally, if $x \in pq$, then $(g \circ f)x = x$. The proof of this equality is left as an exercise. Thus $(g \circ f)^2$ is the identity.

It is easily seen that $g \circ f$ is not the identity, so $g \circ f$ must be an involution. Since $g \circ f$ leaves pq pointwise fixed and leaves $r = L \cup M$ fixed, $g \circ f \in \text{Hom}(r, pq)$.

A property in π is now introduced that is nearly equivalent to (L, L) transitivity in π.

PROPERTY $P(p, q)$ For every line L not containing p or q, there exists $f \in \text{CC}(L)$ such that $f: p \to q$ and f is an involution.

Notation.   Such involutions are denoted by f_L.

LEMMA 11.5.4 (Burn, 1968)   *If π satisfies property $P(p, q)$, then all the involutions f_L are either homologies or elations.*

Proof.   Let L be an arbitrary line not containing p or q and let M be a line distinct from L and pq that is held fixed by f_L. First it is shown that L is a fixed line of f_M. Note that $f_L \circ f_M \circ f_L = f_M$. It follows that $f_L = f_M \circ f_L^{-1} \circ f_M^{-1}$, and since $f_L = f_L^{-1}$, we have $f_L = f_M \circ f_L \circ f_M^{-1}$. By Theorem 4.1.4, $f_L \in \text{CC}(f_M(L))$. Thus $f_M(L) = L$.

Suppose that there exist lines L and L' not containing p and q such that f_L is a homology and $f_{L'}$ is an elation. The lemma is proved by deriving a contradiction from this supposition.

Let M be a fixed line of f_L. Since M is fixed by f_L, it follows that $M \cap pq$ is the center of f_L and that M, L, and pq are not concurrent. Thus $f_L \in \text{Hom}(M \cap pq, L)$. Also f_M fixes L and pq, so $f_M \in \text{Hom}(L \cap pq, M)$. Therefore, by Lemma 11.5.3, $f_M \circ f_L \in \text{Hom}(M \cap L, pq)$.

Let M' be a fixed line of $f_{L'}$. Then L', M', and pq are concurrent, so the center of $f_{L'}$ is $M' \cap pq$. Thus $f_{L'} \in \text{El}(pq \cap M', L')$. Since $f_{M'} \circ f_{L'}$ fixes p and q, $f_{M'} \circ f_{L'} \in \text{El}(M' \cap L', pq)$.

Notice that $f_L \circ f_M = f_M \circ f_L$, so $(f_M \circ f_L)^2$ is the identity and thus $f_M \circ f_L$ is an involution. (It is a harmonic homology with axis pq.) Similarly, $f_{M'} \circ f_{L'}$ is an involution. Thus $f_{M'} \circ f_{L'}$ is an involutory elation with axis pq. But pq cannot be the axis for both an elation and a homology because of Lemma 11.5.1. This contradiction establishes the lemma.

THEOREM 11.5.5 (Burn, 1968)   *If π satisfies property $P(p, q)$, then π is (pq, pq) transitive.*

Proof. We may assume that each line of π has at least four points. By Lemma 11.5.4, we may also assume that all collineations f_L are either homologies or elations.

> *Case* 1. All the collineations f_L are homologies. Let f_L be a fixed homology and let M be a line distinct from L and pq held fixed by f_L. As indicated in the proof of Lemma 11.5.4, $f_L \in \text{Hom}(M \cap pq, L)$ and $f_M \in \text{Hom}(L \cap pq, M)$. It must be shown that $\text{El}(L \cap pq, pq)$ is transitive. Consider two arbitrary lines L_1 and L_2 through p distinct from pq and let $N = (f_M(L_1) \cap L_2)(M \cap pq)$. It is easily checked that $f_N \circ f_M(L_1) = L_2$. That $f_N \circ f_M \in \text{El}(L \cap pq, pq)$ remains to be shown. Since $M \cap pq \in N$ and since $M \cap pq$ is the center of f_L, it follows that N is held fixed by f_L. Thus $f_N \in \text{Hom}(L \cap pq, N)$, and from the dual of Lemma 11.5.2, we easily conclude that $f_N \circ f_M \in \text{El}(L \cap pq, pq)$. There exists at least one more line L' distinct from L that may be used in place of L in the preceding argument. Thus $\text{El}(L' \cap pq, pq)$ is also transitive. From Theorem 4.5.15 we conclude that π is (pq, pq) transitive.

> *Case* 2. All the collineations f_L are elations. This proof is similar to that of Case 1 and is left as an exercise.

COROLLARY 11.5.6 *If π satisfies the property $P(u,v)$, then $\pi = \pi_Q$ where $Q = Q(u,v,o,e) = (R,+,\cdot)$ is a right quasifield.*

The converse of Corollary 11.5.6 is not true. In order to satisfy property $P(u,v)$, the plane π must be coordinatizable by a planar ring that is more than a right quasifield. More precisely, π satisfies property $P(u,v)$ if and only if $\pi = \pi_Q$ where $Q = Q(u,v,o,e) = (R,+,\cdot)$ is a right Bol quasifield. A right Bol quasifield is a right quasifield that satisfies the right Bol law, which is a weak associative law stating that for all $a,b,c \in R$, $((ca)b)a = c((ab)a)$.

It will now be shown that if π satisfies the Pappus (p,q) Theorem, then π satisfies property $P(p,q)$. Dually, if π satisfies the Pappus (L,M) Theorem, then π satisfies property $P(L,M)$. The Pappus (p,q) Theorem and property $P(L,M)$ are duals of the more familiar concepts and are stated here for convenience.

PAPPUS (p,q,r) THEOREM *If L,M,N and L',M',N' are triples of concurrent lines through p and q, respectively, such that $r \notin L,M,N,L'$, M',N' and $r \in LM' \cap L'M$, $r \in LN' \cap L'N$, then $r \in MN' \cap M'N$.*

PAPPUS (p,q) THEOREM *π satisfies the Pappus (p,q) Theorem if and only if π satisfies the Pappus (p,q,r) Theorem for all points $r \neq p,q$.*

PROPERTY $P(L,M)$ *For every point $p \notin L \cup M$, there exists $f \in CC(p)$ such that $f: L \to M$ and f is an involution.*

LEMMA 11.5.7 (Burn, 1968) *The plane π satisfies the Pappus (L,M,N) Theorem if and only if $f_{L,M,r} \circ f_{L,M,q} \circ f_{L,M,p} = f_{L,M,s}$ for some $s \in N$ where $p,q,r \in N$ and $p,q,r \notin L \cup M$* ($f_{L,M,p}$ is defined in Definition 10.5.2).

Notation. In the following proof of Lemma 11.5.7 and subsequently, the shorthand notation f_p is used to denote $f_{L,M,p}$. (Notice that this notation differs slightly from that for $f_{L,M,p}$ used in Section 10.5.)

Proof. Suppose that $f_r \circ f_q \circ f_p = f_s$ for some $s \in N$. Since f_s^2 is the identity on $L \cup M$, we see that $(f_r \circ f_q \circ f_p)^2$ is the identity. Thus by Lemma 10.5.3, π satisfies the Pappus (L,M,N) Theorem.

Suppose that π satisfies the Pappus (L,M,N) Theorem. By Lemma 10.5.3, $(f_r \circ f_q \circ f_p)^2$ is the identity on $L \cup M$ for all $p,q,r \in N - (L \cup M)$. Thus $f_r \circ f_q \circ f_p = f_p \circ f_q \circ f_r$. The desired conclusion is established by showing that the group G generated by mappings of the form $f_q \circ f_p$ consists entirely of elements of the form $f_y \circ f_x$ where x is an arbitrary fixed point of $N - (M \cup L)$ and y is a variable point of this set. This is shown as follows. Let $r = x$ and note that $f_q \circ f_p = f_q \circ f_p \circ f_r \circ f_r = f_r \circ f_q \circ f_p \circ f_r = f_s \circ f_r$ for some $s \in N$; thus $f_r \circ f_q \circ f_p = f_s$.

Let x be a fixed point of $N - (L \cup M)$ and let $H = \{f_y \circ f_x : y \in N - (L \cup M)\}$. Clearly, $H \subseteq G$; we must show that $H = G$. Let $a,b \in L - (M \cup N)$ and $y = (f_x(a))(b) \cap N$; then $f_y \circ f_x(a) = b$, and so H is transitive on $L - (M \cup N)$. Also $f_r \circ f_q \circ f_p = f_p \circ f_q \circ f_r$, so $f_s \circ f_r \circ f_q \circ f_p = f_s \circ f_p \circ f_q \circ f_r = f_q \circ f_p \circ f_s \circ f_r$. Thus the generators of G commute and therefore G is Abelian. Since G is Abelian and is transitive on $L - (M \cup N)$, it is sharply transitive (see exercise 4.4.1). Thus there exists one and only one mapping in G that maps a to b. Since H contains that mapping, $H = G$.

LEMMA 11.5.8 *Suppose that π satisfies the Pappus (L, M) Theorem. If $p, q, r \notin L, M$, and $f_r \circ f_q \circ f_p$ is an involution, then p, q, r are collinear.*

Proof. Let s be any point on L and not on M. Consider the following triples of collinear points: s, $f_q \circ f_p(s)$, $f_p \circ f_r \circ f_q \circ f_p(s)$ on L and $f_p(s)$, $f_r \circ f_q \circ f_p(s)$, $f_q \circ f_p \circ f_r \circ f_q \circ f_p(s)$ on M. The hypotheses that $(f_r \circ f_q \circ f_p)^2(s) = s$ and that the Pappus (L, M, pq) Theorem holds in π imply that p, q, r are collinear.

THEOREM 11.5.9 (Burn, 1968) *If π satisfies the Pappus (L, M) Theorem, then f_p extends to an involution $g_p \in CC(p)$.*

Proof. Construct a mapping g_p on the points of π as follows: If $q \in L \cup M$, then $g_p(q) = f_p(q)$; if $a \notin L \cup M$, then $g_p(q) = r$ where $r \in pq$ such that $f_p \circ f_q \circ f_p = f_r$. The existence of such an r is given by Lemma 11.5.7 and the Pappus (L, M, pq) Theorem. Clearly, g_p fixes p and the lines through p, and g_p is an involution because $f_q = (f_p \circ f_p) \circ f_q \circ (f_p \circ f_p) = f_p \circ (f_p \circ f_q \circ f_p) \circ f_p = f_p \circ f_r \circ f_p$; therefore, $g_p^2(q) = g_p(g_p(q)) = g_p(r) = q$. That g_p preserves collinearity remains to be shown.

Let q, r, s be collinear points and let q', r', s' be the images of q, r, s, respectively, under g_p. Three cases are considered here.

Case 1. Suppose that $q, r, s \notin L \cup M$. By Lemma 10.5.3, $f_s \circ f_r \circ f_q$ is an involution because the Pappus (L, M, qr) Theorem is valid. Thus $(f_{s'} \circ f_{r'} \circ f_{q'})^2 = (f_p \circ f_s \circ f_p \circ f_p \circ f_r \circ f_p \circ f_p \circ f_q \circ f_p)^2$ $= (f_p \circ f_s \circ f_r \circ f_q \circ f_p)^2 = f_p \circ f_s \circ f_r \circ f_q \circ f_p \circ f_p \circ f_s \circ f_r \circ f_q \circ f_p = f_p \circ (f_s \circ f_r \circ f_q)^2 \circ f_p = f_p \circ f_p = I$ (the identity on $L \cup M$). Therefore, by Lemma 11.5.8, q', r', s' are collinear.

Case 2. Suppose that $q \notin L \cup M$, $r \in L$, $s \in M$, and $r \neq s$. Since $f_{q'}(r') = f_p \circ f_q \circ f_p(r') = f_p \circ f_q(r) = f_p(s) = s'$, we know that q', r', s' are collinear.

Case 3. Suppose that $q, r \notin L \cup M$, $s \in L \cup M$, $s \neq L \cap M$. Suppose that $s \in L$ and let $t = qr \cap M$. By Case 2, q', s', t' and r', s', t' are collinear; therefore, q', r', s' are also collinear. If $s \in M$, a similar argument holds. Other cases are easily proved.

COROLLARY 11.5.10

i. *If π satisfies the Pappus (L,M) Theorem, then π satisfies property $P(L,M)$.*

ii. *If π satisfies the Pappus (p,q) Theorem, then π satisfies property $P(p,q)$.*

THEOREM 11.5.11

i. *If π satisfies the Pappus (L,M) Theorem, then π is (p,p) transitive where $p = L \cap M$.*

ii. *If π satisfies the Pappus (p,q) Theorem, then π is (pq,pq) transitive.*

Proof.

i. Let π satisfy the Pappus (L,M) Theorem; we know from Corollary 11.5.10(i) that π satisfies property $P(L,M)$. By the dual of Theorem 11.5.5, π is (p,p) transitive.

ii. This proof is the dual of (i).

THEOREM 11.5.12

i. *If π satisfies the Pappus (ov,uv) Theorem, then $\pi = \pi_Q$ where $Q = Q(u,v,o,e) = (R,+,\cdot)$ is a left quasifield.*

ii. *If π satisfies the Pappus (u,v) Theorem, then $\pi = \pi_Q$ where $Q = Q(u,v,o,e) = (R,+,\cdot)$ is a right quasifield.*

THEOREM 11.5.13

i. *If π satisfies the Pappus (L,M) Theorem, then π satisfies the Pappus Theorem.*

ii. *If π satisfies the Pappus (p,q) Theorem, then π satisfies the Pappus Theorem.*

Proof.

i. Let π satisfy the Pappus (L,M) Theorem. Then, by Theorem 11.5.11(i), π is (p,p) transitive. Let $p = v$ and choose points $o \in L$, $u \in M$, and e such that u,v,o,e is a four-point. Then

π is (v,v) transitive. By Theorem 11.2.5, $\pi = \pi_Q$ where Q $= Q(u,v,o,e) = (R,+,\cdot)$ is a left quasifield. Since π satisfies the Pappus (ov,uv,ou) Theorem, π satisfies the Pappus (ou,ov,uv) Theorem (see exercise 6.2.2) and so, by Theorem 10.5.5, $(R - \{0\},\cdot)$ is an Abelian group. Thus Q is a field, and we conclude that π satisfies the Pappus Theorem.

ii. This proof follows by duality from (i).

EXERCISES

1. Prove Lemma 11.5.1.

2. Prove Lemma 11.5.2.

3. In the proof of Lemma 11.5.3, show that $g \circ f(x) = x$ for $x \in pq$.

4. In the proof of Theorem 11.5.5, show that $f_N \circ f_M(L_1) = L_2$.

5. Prove Case 2 in the proof of Theorem 11.5.5.

6. Construct a diagram explaining the proof of Lemma 11.5.8.

7. Show that if the Pappus (ou,ov,uv) Theorem holds in the plane π_Q where Q is a quasifield, then π_Q is a Pappian plane.

8. a. Which of the four quasifields of Section 11.4 are Bol quasifields?
   b. In $\pi_N(9)$ find points p and q such that $\pi_N(9)$ satisfies property $P(p,q)$.

CHAPTER 12

PLANES OVER
PLANAR NEARFIELDS

In a planar nearfield the properties of associativity of multiplication and one-sided distributivity are added to the properties of a Cartesian group P. Since each property has been separately added to P (associativity of multiplication was added in Chapter 10; one-sided distributivity was added in Chapter 11), the student is already acquainted with the minimum transitivity properties of a plane over a planar nearfield. The new content of this chapter will concern the total transitivity $T(\pi)$ of such planes. It will be shown here that $T(\pi)$ is much more easily calculated for planes over planar nearfields than for planes over quasifields. The relationships among the various coordinatizing planar nearfields of a given plane will also be considered.

12.1 PLANAR NEARFIELDS

DEFINITION 12.1.1

1. *The planar ring* $P = (R,+,\cdot)$ *is a* right planar nearfield *if and only if* P *is a Cartesian group such that* $a(bc) = (ab)c$ *and* $(a + b)c = ac + bc$ *for all* $a, b, c \in R$.

2. *P is a* left planar nearfield *if and only if P is a Cartesian group such that $a(bc) = (ab)c$ and $a(b + c) = ab + ac$ for $a, b, c \in R$.*

THEOREM 12.1.2   *$N = (R, +, \cdot)$ is a right planar nearfield if and only if*

   i. *$(R, +)$ is a group.*

   ii. *$(R - \{0\}, \cdot)$ is a group where 0 is the additive identity in R.*

   iii. *$a \cdot 0 = 0 \cdot a = 0$ for $a \in R$.*

   iv. *$(a + b)c = ac + bc$ for $a, b, c \in R$.*

   v. *$-xa + xb = c$ has a unique solution x for a given $a, b, c \in R$, $a \neq b$.*

Proof.   The proof is left as an exercise.

A right (left) nearfield is a structure satisfying the first four properties. (For left nearfield, naturally we assume left distributivity.) In the finite case, a nearfield is a planar nearfield. (Proof of this fact is left as an exercise.) Therefore, we may use "nearfield" and "planar nearfield" interchangeably in that case. This has already been done for $N(9)$. An example of a nearfield that is not planar can be found in Zemmer (1964). The relationship between left and right quasifields shown in Theorem 11.1.3 holds similarly between left and right planar nearfields.

THEOREM 12.1.3   *If $N = (R, +, \cdot)$ is a right planar nearfield, then N^d is a left planar nearfield. If N is a left planar nearfield, then N^d is a right planar nearfield. The system $(N^d)^d = N$.*

Proof.   Since N is a right quasifield, N^d is a left quasifield by Theorem 11.1.3. Since $a \cdot (b \cdot c) = (cb)a = c(ba) = (a \cdot b) \cdot c$, it follows that N satisfies the law of associativity of multiplication, and thus N^d is a left planar nearfield.

EXAMPLE 12.1.4

   i. A division ring is a right (and left) planar nearfield.

   ii. $N(9)$ is a right planar nearfield.

   iii. A class of finite nearfields called regular nearfields can be generated in the following way. Denote such a nearfield by N

$= (R,+,\circ)$ and define R, $+$, and $\circ$ as follows. Let p,r,q,n be integers satisfying the restrictions that p is a prime, that $q = p^r$, that n is such that all prime divisors of n divide $q - 1$, and finally, that if $q = 3 \bmod 4$, then $n \neq 0 \bmod 4$ (for example, $p = 3$, $r = 1$, $q = 3$, $n = 2$ are admissible integers). Let $(R,+) = (F,+)$ where $(F,+,\cdot)$ $= GF(q^n)$. Define $\circ$ in the following complicated manner: Let c be a primitive element of the multiplicative group of $F - \{0\}$ (that is, $c^{q^n-1} = 1$ and $c^k \neq 1$ for $k < q^n - 1$) and consider the subgroup G of $(F - \{0\},\cdot)$ generated by c^n. Each coset of G may be represented by c_i where

$$c_i = c^{\frac{q^i-1}{q-1}}: i = 0, \ldots, n - 1$$

(Proof of this is left as an exercise.) So we may define the mapping f_y for $y \in R$ as follows: $f_y\colon x \to x^{q^i}$ where i is that number such that $y \in [c_i]$. ($[c_i]$ is the coset represented by c_i.) Finally, define $\circ$ as follows: $x \circ y = f_y(x) \cdot y$ for $y \neq 0$ and $x \circ 0 = 0$.

iv. An example of a regular nearfield can be given as follows. If $p = 3$, $r = 1$, $q = 3$, $n = 2$, then $(R,+) = (F,+)$ where $(F,+,\cdot)$ $= GF(9)$ and where c is a of Table 2. The subgroup G generated by a^2 is the set $\{a^0,a^2,a^4,a^6\}$. The two cosets are $\{a^0,a^2,a^4,a^6\}$ and $\{a,a^3,a^5,a^7\}$. Because $a_0 = a^0 = 1$ and because $a_1 = a^1 = a$, we can let 1 and a represent the two cosets. Let $1 = a^0$, $2 = a^4$, $a = a$, $b = a^2$, $c = a^7$, $d = a^5$, $e = a^3$, and $f = a^6$. The multiplication table of this nearfield is represented by Table 21. This table can be checked but the procedure is tedious. For example, $b \circ c \in f_c(b)$ $\cdot c = f_{a^7}(a^2) \cdot a^7 = (a^2)^3 \cdot a^7 = a^6 \cdot a^7 = a^{13} = a^5 = d$. It should be observed that Tables 10 and 21 are identical; therefore, the nearfield of Example 12.1.4 is $N(9)$, the nearfield of Construction 2.4.2.

TABLE 21

$\circ$	1	2	a	b	c	d	e	f
1	1	2	a	b	c	d	e	f
2	2	1	d	f	e	a	c	b
a	a	d	2	e	b	1	f	c
b	b	f	c	2	d	e	a	1
c	c	e	f	a	2	b	1	d
d	d	a	1	c	f	2	b	e
e	e	c	b	d	1	f	2	a
f	f	b	e	1	a	c	d	2

The student is encouraged to construct the table for the regular nearfield of order 25, that is, $p = q = 5, r = 1, n = 2$.

The regular nearfields are all nondistributive right nearfields. There are seven irregular (right) nearfields of order p^2 where $p = 5,7,11,11,$ 23,29,59. For original sources on finite nearfields, see Dickson (1905) and Zassenhaus (1935). Passman (1967) has shown that there are no finite nondistributive right nearfields other than the regular nearfields defined here and the seven irregular nearfields discovered by Dickson. There are, of course, infinite nondistributive nearfields; see Dickson (1905) for such examples.

This section concludes with two lemmas that are followed by two useful theorems concerning planar nearfields.

LEMMA 12.1.5   *In a right quasifield, $(-1) \cdot (-1) = 1$.*

Proof. $(-1 + 1)(-1) = 0$, so by right distributivity, $(-1)(-1) + 1(-1)$ $= (-1)(-1) + (-1) = 0$. Also $1 + (-1) = 0$ and so $(-1)(-1) = 1$.

LEMMA 12.1.6   *In a right nearfield, if $ab = -a$, then $b^2 = 1$.*

Proof.   Suppose that $ab = -a$. Then $ab^2 = (ab)b = (-a)b = -(ab)$ $= -(-a) = a$. Thus $b^2 = 1$.

THEOREM 12.1.7   *Let $N = (R,+,\cdot)$ be a right planar nearfield not of characteristic 2. Then $a(-1) = -a$.*

Proof.   Let $a \in R, a \neq 0$, and let $b \in R$ such that $ab = -a$. From Lemma 12.1.6, we know that $b^2 = 1$; we wish to show that $b = -1$. Let x be the unique member of R such that $-x + xb = -1$. Such an x exists by the fifth property of Theorem 12.1.2, since b clearly is not 1. Thus $(-1)b = (-x + xb)b = -xb + (xb)b = -xb + xb^2 = -xb + x = -1(-x + xb)$ $= (-1)(-1) = 1$. By Lemma 12.1.5, it follows that $b = 1$.

COROLLARY 12.1.8   *In a right planar nearfield not of characteristic 2, $a(-b) = -(ab)$.*

Note that Theorem 12.1.7 is not true for the quasifield Q' of order 9.

THEOREM 12.1.9   *Let N be a right planar nearfield not of characteristic 2. If $b \neq 1$ and $b^2 = 1$, then $b = -1$.*

Proof. Suppose that $b \neq -1$ and $b^2 = 1$. Then $b + 1 \neq 0$ and $(b + 1)b$ $= b^2 + b = 1 + b = b + 1$. Thus $b = 1$. It follows that if $b^2 = 1$, then $b = 1$ or $b = -1$.

EXERCISES

1. Prove Theorem 12.1.2.

2. a. Show that there is a regular nearfield of order p^2 for every odd prime p.
   b. Does there exist a regular nearfield of order p^4 for every odd prime p?
   c. Does there exist a regular nearfield of order 2^k for $k \geq 2$?

3. Prove that, as stated in Example 12.1.4(iii), each coset of G may be represented by c_i where
$$c_i = c^{\frac{q^i - 1}{q - 1}} : i = 0, \ldots, n - 1$$

4. Prove that a finite right nearfield is a right planar nearfield.

5. Are Lemmas 12.1.5 and 12.1.6, Theorems 12.1.7 and 12.1.9, and Corollary 12.1.8 true for left planar nearfields? Explain.

*6. a. Construct a table for a right nearfield of order 25.
   b. Use this table to display the Hughes plane $\pi_H(25)$. (See Section 9.4 for the definition of $\pi_H(9)$ and generalize it to $\pi_H(25)$.)

12.2   THE NEARFIELD $N(9)$

The nearfield $N(9)$ is different from the other regular nearfields in two important respects. One concerns its group of automorphisms and the other concerns the nature of its multiplicative group $(R - \{0\}, \cdot)$. The latter property has geometric consequences and therefore will be studied here in greater detail.

As indicated by Table 21, $x^2 = -1$ for all $x \neq 0, 1, -1$. Among right nearfields, this property is unique to $N(9)$. The result and the proof given here are due to Andre (1955). The geometric consequences of the property are examined in Sections 12.4 and 12.5.

Let $N_o = (R_o, +, \cdot)$ be a nondistributive right nearfield satisfying the property that $a^2 = -1$ for all $a \in R$ such that $a \neq 0, 1, -1$. Then the following three lemmas can be proved.

LEMMA 12.2.1   *If $a \in R_o$, $a \neq 0, 1, -1$, then $a^{-1} = -a$. If $b \in R_o$, $b \neq 0, 1, -1, a, -a$, then $ba = -ab$.*

Proof. First note that $-a = (-1)a = (a^{-1})^2 a = a^{-1}(a^{-1}a) = a^{-1}$. Also $(ab^{-1})(ab^{-1}) = -1$, so $ab^{-1} = -ba^{-1}$. Since $-a = a^{-1}$ and $-b = b^{-1}$, this yields $a(-b) = -b(-a)$ and, by Corollary 12.1.8, $-(ab) = ba$.

LEMMA 12.2.2 *N_o has characteristic 3.*

Proof. Since $N_o \neq \mathrm{GF}(p)$ for $p = 2,3,5$, there exists element a of N_o such that a and $1 + a \neq 0,1$, or -1 and $a + 1 \neq \pm a$. Let c be such an element. Then $-1 = (1 + c)^2 = (1 + c)(1 + c) = 1 + c + c(1 + c) = 1 + c - (1 + c)c = 1 + c - c - c^2 = 1 + 1 = 2$. (The fourth equality follows from Lemma 12.2.1.) Thus $2 = -1$ and N_o has characteristic 3.

LEMMA 12.2.3 *Every element of N_o can be represented by a combination of 1 and c (from Lemma 12.2.2) over GF(3).*

Proof. Suppose that there exists an element of $d \in R_o$ such that $d \neq a + bc$ where $a,b = 0,1,-1$. Then, using Lemma 12.2.1, we may write $-1 = (1 + c + d)^2 = 1 + c + d + c(1 + c + d) + d(1 + c + d) = 1 + c + d - (1 + c + d)c - (1 + c + d)d = 1 + c + d - c - c^2 - dc - d - cd - d^2 = 1 + c + d - c + 1 + cd - d - cd + 1 = 3 = 0$. This contradiction establishes the lemma.

THEOREM 12.2.4 *The right nearfield N_o is isomorphic to $N(9)$.*

Proof. The proof is left as an exercise.

COROLLARY 12.2.5 *If a nondistributive right nearfield satisfies the condition that $a^2 = -1$ for $a \neq 0,1,2$, then it is isomorphic to $N(9)$.*

As mentioned previously, the automorphism group of $N(9)$ is different from the automorphism groups of other nearfields. The automorphisms of $N(9)$ are listed here for future reference. This automorphism group is isomorphic to S_3 (the non-Abelian group of six elements generated by the permutations on three elements), whereas the automorphism groups of other nearfields are cyclic. (For more details and references, see Dembowski, 1968, pp. 35, 225.)

As in Lemma 12.2.3, the elements of $N(9) = (R,+,\cdot)$ may be expressed by linear combinations of 1 and a over GF(3), that is, $x + ya$ where $x,y \in \{0,1,2\}$. Thus $b = 1 + a$, $c = 2 + a$, $d = 2a$, $e = 2a + 1$, $f = 2a + 2$.

DEFINITION 12.2.6 $\alpha_{s,t}\colon R \to R$ *where* $s = 0,1,2$ *and* $t = 1,2$ *is defined by* $\alpha_{s,t}\colon x + ya \to (x + sy) + (ty)a.$

These six mappings $\alpha_{s,t}$ comprise the total set of automorphisms on $N(9)$.

EXAMPLE 12.2.7

i.

$$\alpha_{1,1}\colon \quad 0 \to 0 \qquad a \to b \qquad d \to f$$
$$1 \to 1 \qquad b \to c \qquad e \to d$$
$$2 \to 2 \qquad c \to a \qquad f \to e$$

ii.

$$\alpha_{0,2}\colon \quad 0 \to 0 \qquad a \to d \qquad d \to a$$
$$1 \to 1 \qquad b \to e \qquad e \to b$$
$$2 \to 2 \qquad c \to f \qquad f \to c$$

Note that the two automorphisms of Definition 12.2.6 generate the group of six automorphisms. Observe also that this group is sharply transitive on $R - \{0,1,2\}$.

These automorphisms will be used to define collineations on $\pi_N(9)$ as was done in Example 4.3.4(iii), with which this example should be compared.

EXERCISES

1. Prove Theorem 12.2.4.

2. Formulate and prove a theorem for
   left nearfields analogous to Corollary 12.2.5.

3. Show that the only automorphisms of $N(9)$
   are the six defined in Definition 12.2.6.

12.3   GEOMETRIC CONSEQUENCES OF ASSOCIATIVITY OF MULTIPLICATION AND ONE-SIDED DISTRIBUTIVITY

The geometric consequences of associativity and one-sided distributivity have already been studied separately in previous chapters. Nevertheless, as was mentioned in Section 11.2, the combination of these two properties yields more than the expected transitivity on π. This fact is proved in the following theorem.

THEOREM 12.3.1   *If* $N(u,v,o,e) = N = (R,+,\cdot)$ *is a right planar near-field, then* π_N *is* (u,v), (v,u), *and* (uv,uv) *transitive.*

Proof. Since N satisfies right distributivity, π_N is (uv,uv) transitive; since N satisfies the associative law of multiplication, π_N is (u,ov) transitive. Since π_N is (u,uv) and (u,ov) transitive, by Theorem 4.5.15(ii), it is (u,v) transitive. That π_N is (v,u) transitive remains to be shown.

Consider the following mappings g and G:

$$g: \quad [x,y,1] \rightarrow [y,x,1]$$
$$[1,x,0] \rightarrow [1,x^{-1},0] \text{ where } x \neq 0$$
$$[1,0,0] \rightarrow [0,1,0]$$
$$[0,1,0] \rightarrow [1,0,0]$$

$$G: \quad \langle a,1,b \rangle \rightarrow \langle a^{-1},1,ba^{-1} \rangle \text{ where } a \neq 0$$
$$\langle 0,1,b \rangle \rightarrow \langle 1,0,b \rangle$$
$$\langle 1,0,b \rangle \rightarrow \langle 0,1,b \rangle$$
$$\langle 0,0,1 \rangle \rightarrow \langle 0,0,1 \rangle$$

The following cases show that (g,G) is an involution with center $[1,1,0]$ and axis $\langle -1,1,0 \rangle$.

   Case 1. $[x,y,1] \in \langle a,1,b \rangle$ iff $xa + y + b = 0$ iff $(xa + y + b)a^{-1} = 0a^{-1} = 0$ iff $xaa^{-1} + ya^{-1} + ba^{-1} = 0$ iff $ya^{-1} + x + ba^{-1} = 0$ iff $[y,x,1] \in \langle a^{-1},1,ba^{-1} \rangle$ iff $g([x,y,1]) \in G(\langle a,1,b \rangle)$.

   Case 2. $[1,x,0] \in \langle -x,1,y \rangle$ for $x \neq 0$, and since $(-x)^{-1} = -(x^{-1})$, $[1,x^{-1},0] \in \langle (-x)^{-1},1,y((-x)^{-1}) \rangle$ for $x \neq 0$. Thus $g([1,x,0]) \in G(\langle -x,1,y \rangle)$.

   Case 3. $[x,y,1] \in \langle 1,0,-x \rangle$ and $[y,x,1] \in \langle 0,1,-x \rangle$. Therefore, $g([x,y,1]) \in G(\langle 1,0,-x \rangle)$.

It is easily checked that (g,G) also preserves other incidences. It is also easily seen that $g \in \text{El}([1,1,0],\langle -1,1,0 \rangle)$. Since $g: u \rightarrow v$, $v \rightarrow u$, and since π_N is (u,v) transitive, it follows from Theorem 4.5.14 that π is (v,u) transitive.

THEOREM 12.3.2   *If $P(u,v,o,e) = P = (R,+,\cdot)$ is a planar ring and π_P is (u,v), (v,u), and (uv,uv) transitive, then P is a right planar near-field.*

Proof.   Since π_P is (uv,uv) transitive, it follows from Theorem 11.2.2 that P is a right quasifield. Since π_P is (u,v) transitive, it must be (u,ov) transitive, and therefore, by Theorem 10.3.5, P satisfies the associative law of multiplication. Thus P is a right planar nearfield.

As we can see, the conditions of (uv,uv), (u,v), and (v,u) transitivity on π are redundant. Our parsimonious instincts will be satisfied when we find one condition of transitivity that is equivalent to these three. Our search begins with a lemma about transitivity in planar rings.

LEMMA 12.3.3   *If $P(u,v,o,e) = P = (R,+,\cdot)$ is a planar ring and π_P is (u,uv) transitive, then π_P is (v,uv) transitive.*

Proof.   We know that π_P will be (v,uv) transitive if P is a Cartesian group. Thus we must show that addition is associative in P.
Let $f_a \in \text{El}(u,uv)$ such that $f_a\colon [0,0] \to [a,0]$ (see Figure 84). Thus $f_a\colon \langle 0 \rangle = [0,0]v \to [a,0]v = \langle a \rangle$, and so $f_a\colon [0,a] = \langle 0 \rangle \cap \langle 0,a \rangle \to \langle a \rangle \cap \langle 0,a \rangle = [a,a]$. (Note that $\langle 0,a \rangle \to \langle 0,a \rangle$ since lines through u, that is, lines of the form $\langle 0,k \rangle$, are held fixed.) Also $f_a\colon \langle 1,a \rangle = [0,a][1] \to [a,a][1] = \langle 1,0 \rangle$; therefore, $f_a\colon [x,x + a] = \langle 0,x + a \rangle \cap \langle 1,a \rangle \to \langle 0,x + a \rangle \cap \langle 1,0 \rangle = [x + a,x + a]$. Thus $f_a\colon \langle x \rangle = [x,x + a]v \to [x + a,x + a]v = \langle x + a \rangle$. By a similar argument, $f_a\colon [0,a + b] = \langle 0 \rangle \cap \langle 0,a + b \rangle \to \langle a \rangle \cap \langle 0,a + b \rangle = [a,a + b]$, and so $f_a\colon \langle 1,a + b \rangle = [0,a + b][1] \to [a,a + b][1] = \langle 1,b \rangle$. Therefore, $f_a\colon [x,x + (a + b)] = \langle x \rangle \cap \langle 1,a + b \rangle \to \langle x + a \rangle \cap \langle 1,b \rangle = [x + a, (x + a) + b]$. Since u is the center of f_a, the two points $[x,x + (a + b)]$ and $[x + a, (x + a) + b]$ must have the same second coordinates. Thus $x + (a + b) = (x + a) + b$. Since x, a, and b were arbitrary, we have shown that P is a Cartesian group.

THEOREM 12.3.4   *If $P(u,v,o,e) = P$ is a planar ring and π_P is (u,v) transitive, then P is a right planar nearfield.*

Proof.   Since π_P is (u,v) transitive, it is (u,ov) transitive, and so P has associative multiplication. Also π_P must be (u,uv) transitive and thus,

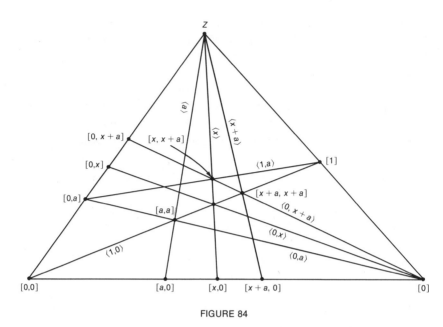

FIGURE 84

by Lemma 12.3.3, it is (v, uv) transitive. But a (u, uv) and (v, uv) transitive plane is (uv, uv) transitive by Theorem 4.5.15(i), and therefore P is a right quasifield. We may then conclude that P is a right planar nearfield.

The following three important theorems follow easily from Theorem 12.3.4.

THEOREM 12.3.5 *The plane π is (u, v) transitive if and only if $\pi = \pi_N$ where $N = N(u, v, o, e)$ is a right planar nearfield. (The points o and e are chosen so that u, v, o, e is a four-point.)*

Proof. The proof is left as an exercise.

THEOREM 12.3.6 *The plane π is (p, q) transitive if and only if it is (q, p) transitive.*

THEOREM 12.3.7 *The plane π is (L, M) transitive if and only if it is (M, L) transitive.*

The situation for left planar nearfields is the dual of that for right planar nearfields. This statement may be verified with the help of the following theorem.

THEOREM 12.3.8 *If $N = (R,+,\cdot)$ is a right planar nearfield, then the plane π_N is isomorphic to $(\pi_{N^d})^d$.*

Proof. The proof is based on Theorem 11.2.4.

THEOREM 12.3.9 *The plane π_P is (uv,ov), (ov,uv), and (v,v) transitive if and only if $P(u,v,o,e)$ is a left planar nearfield.*

Proof. The proof is based on Theorems 11.2.4, 12.3.1, and 12.3.2.

THEOREM 12.3.10 *The plane π is (ov,uv) transitive if and only if $\pi = \pi_N$ where $N = N(u,v,o,e)$ is a left planar nearfield. (The point e is chosen so that u,v,o,e is a four-point.)*

Proof. The proof is based on Theorems 11.2.4 and 12.3.5.

EXERCISES

1. Show that in a right planar nearfield, $(-x)^{-1} = -(x^{-1})$ where $x \neq 0$.

2. Prove Theorem 12.3.5.

3. Let N be a right planar nearfield. Show that if π_N is self-dual, then π_N is Desarguesian.

4. Show that if π is (p,q) and (pr,qr) transitive for three distinct non-collinear points p, q, and r, then π is Desarguesian.

5. Show that if π is (L,M) and (N,N) transitive where $L \neq M$, then π is Desarguesian.

12.4   ADDITIONAL TRANSITIVITY ON π_N

In this section the transitivity class $T(\pi_N)$ will be completely determined where N is any nondistributive right planar nearfield except $N(9)$. The set $T(\pi_N(9))$ is determined in Section 12.5. This is a significant achievement and one that has not been duplicated for such lesser algebraic structures as quasifields. In fact, we will find that all π_N except $\pi_N(9)$ belong to the same transitivity class. From Theorem 12.3.1 we know that π_N is (u,v), (v,u), and (uv,uv) transitive. This section will show that no more (p,L) transitivity is possible (except in $\pi_N(9)$).

The following lemma represents an attempt to impose further transitivity on a plane π_N.

LEMMA 12.4.1   *If π_N is (p,p) transitive for some point p, then π_N is Desarguesian.*

Proof.   This result is shown by constructing a map that sends $p \to v$. Thus by Theorem 4.5.14, π_N is (v,v) transitive and consequently N satisfies left distributivity. This implies that N is a division ring and therefore that π_N is Desarguesian.

Case 1. $p \notin uv$. Since π_N is (p,p) transitive and $p \notin uv$, there exists a collineation f such that $f: v \to q$ for some $q \notin uv$. Since π_N is (uv,uv) transitive, there exists $g \in \text{El}(uv)$ such that $g: q \to p$. Thus $f^{-1} \circ g^{-1}: p \to v$.

Case 2. $p \in uv$. If $p = v$, we have nothing more to show; if $p = u$, we may map $u \to v$ by the involution g of Theorem 12.3.1. If $p \neq u,v$, since π_N is (p,p) transitive, there exists a collineation f such that $f: v \to q$ where $q \in uv$ and $q \neq u,v$. Then there exists $h \in \text{Hom}(u,ov)$ such that $h: q \to p$. Thus $f^{-1} \circ h^{-1}: p \to v$.

THEOREM 12.4.2   *Let $p \in L$, $p \neq u,v$, and $L \neq uv$. Then if π_N is (p,L) transitive, π_N is Desarguesian.*

Proof.

Case 1. $p \notin uv$. Since π_N is (uv,uv) transitive, there exists $f \in \text{El}(uv)$ such that $f: p \to o$. Thus by Theorem 4.5.14, π_N is (o,M) transitive where $M = f(L)$. Now $M \neq uv$ because $o \in M$. Thus we may map $M \to M' \neq M$ by a collineation h where $h = g$ (the involution of Theorem 12.3.1) if $M = ov$ or ou; otherwise, h may be any nonidentity member of $\text{Hom}(u,ov)$. Therefore, π_N is (o,M) and (o,M') transitive and, by Theorem 4.5.15(ii), it is also (o,o) transitive. By Lemma 12.4.1, π_N is Desarguesian.

Case 2. $p \in uv$. Since $L \neq uv$, we may map $L \rightarrow M \neq L$ by an appropriate elation $f \in \mathrm{El}(uv)$. The elation f fixes p, so by Theorem 4.5.14, π_N is (p,L) and (p,M) transitive. Thus π_N is (p,p) transitive, and so by Lemma 12.4.1, π_N must be Desarguesian.

THEOREM 12.4.3 *Let $p \notin L$ and $p \notin uv$. If π_N is (p,L) transitive, π_N is Desarguesian.*

Proof.

Case 1. $L = uv$. With an elation $f \in \mathrm{El}(uv)$, we may map $p \rightarrow q$ for any $q \notin uv$. Thus π_N satisfies the Second Desargues (uv) Theorem, and by Theorem 8.2.6, π_N is Desarguesian.

Case 2. $L \neq uv$. Since π_N is (p,L) transitive, $f \in \mathrm{CC}(p,L)$ exists such that $f\colon uv \rightarrow M \neq uv$. Since π_N is (uv,uv) transitive, it follows by Theorem 4.5.14 that π_N is also (M,M) transitive. Let $q = uv \cap M$. Then π_N is (q,M) and (q,uv) transitive. Thus π_N is (q,q) transitive and, by Lemma 12.4.1, is also Desarguesian.

THEOREM 12.4.4 *Suppose that $p \notin L$, that $p \in uv$, and that one of the following four cases is true:*

$$p = v \text{ and } u \notin L$$
$$p = u \text{ and } v \notin L$$
$$p \neq v \text{ and } u \in L$$
$$p \neq u \text{ and } v \in L$$

Then if π_N is (p,L) transitive, π_N is Desarguesian.

Proof.

Case 1. Suppose that the first case is true. Since π_N is (v,L) and (v,uv) transitive, by Theorem 4.5.15(ii) it is also (v,q) transitive where $q = uv \cap L$. Thus π_N is (v,uo), (v,qo), and therefore (v,o) transitive. By Theorem 12.3.6, π_N is (o,v) transitive and thus (o,uv) transitive. Since π_N is (uv,uv) transitive, there exists $f \in \mathrm{El}(uv)$ such that $f\colon o \rightarrow p$ for any point $p \notin uv$. Thus π_N is (p,uv) transitive for all $p \notin uv$, and it follows that π_N satisfies the Second Desargues (uv) Theorem and therefore the Desargues Theorem.

Proof of the last three cases is left as an exercise.

The transitivity situation for π_N may be summarized as follows. By Theorems 12.4.2 through 12.4.4, $(p,L) \notin T(\pi_N)$ if

$$p \in L, \quad L \neq uv$$
$$p \notin L, \quad p \notin uv$$
$$p \notin L, \quad p \in uv$$

and

$$p = v, \quad u \notin L$$
$$p = u, \quad v \notin L$$
$$p \neq v, \quad u \in L$$
$$p \neq u, \quad v \in L$$

Since π_N is (uv,uv), (u,v), and (v,u) transitive, $(p,L) \in T(\pi_N)$ if

$$p \in L, \quad L = uv$$
$$p \notin L, \quad p \in uv$$

and

$$p = v, \quad u \in L$$
$$p = u, \quad v \in L$$

This leaves only the following condition:

$$p \notin L, \quad p \in uv$$

and

$$p \neq u,v, \quad u,v \notin L$$

This condition will now be examined.

LEMMA 12.4.5   *If π_N is (p,q) and (p,r) transitive for $q \neq r$, then it is Desarguesian.*

Proof.   The proof is left as an exercise.

THEOREM 12.4.6 (Andre, 1955)   *Let $N = (R,+,\cdot)$ be a right planar nearfield and suppose that the previously stated condition is true (i.e., $p \in L, p \in u,v, p \neq u,v,$ and $u,v \notin L$). Then if π_N is (p,L) transitive, either it is $\pi_N(9)$ or it is Desarguesian.*

Proof.   Suppose that π_N is not Desarguesian. First observe that π_N is (p,q) transitive where $q = L \cap uv$. We know this is true because we may map $L \to M \neq L$ with an $f \in El(uv)$; therefore, π_N is (p,L) and (p,M) transitive and thus (p,q) transitive.

Let $p = [1,a,0]$ and $q = [1,b,0]$ and let c be such that $ac = 1$. Define (h_n, H_n) to be the following collineation where $n \neq 0$:

$$h_n: \quad [x,y,1] \rightarrow [x,yn,1]$$
$$[1,x,0] \rightarrow [1,xn,0]$$
$$[0,1,0] \rightarrow [0,1,0]$$

$$H_n: \quad \langle m,1,k \rangle \rightarrow \langle mn,1,kn \rangle$$
$$\langle 1,0,k \rangle \rightarrow \langle 1,0,k \rangle$$
$$\langle 0,0,1 \rangle \rightarrow \langle 0,0,1 \rangle$$

This is easily seen to be a collineation. Thus $h_c: p \rightarrow [1,1,0];\ q \rightarrow [1,bc,0]$. Denoting $[1,1,0]$ by r and $[1,bc,0]$ by s, we see that π_N is (r,s) transitive and also (s,r) transitive. Since $h_{bc}: r \rightarrow s$, we know that $s \rightarrow [1,d^2,0] = t$ where $d = bc$, so π_N is also (s,t) transitive. But by Lemma 12.4.5, (s,r) and (s,t) transitivity implies that $r = t$ since π_N is not Desarguesian. Thus we have $d^2 = 1$. Theorem 12.1.9 then tells us that $d = -1$ since, if $d = 1$, $bc = ac = 1$ and so $p = q$, a contradiction. Therefore, $s = [1,-1,0]$.

The next step is to show that $x^2 = -1$ for all $x \neq 0,1,-1$. This will complete the proof since such a nearfield must be isomorphic to $N(9)$ by Corollary 12.2.5. Let $m = [1,x,0]$, $x \neq 0,1,-1$. Since π_N is (r,s) transitive, there exists $f \in \mathrm{Hom}(r,os)$ such that $f: v \rightarrow m$. Let $n = f(u) = [1,y,0]$. Recall that the involution (g,G) of Theorem 12.3.1 is also in $\mathrm{Hom}(r,os)$. Also recall from Theorem 10.3.4 that $\mathrm{Hom}(r,os) \sim (R' - \{0\},\odot)$ where $(R',\oplus,\odot) = P(r,s,o,e')$ for an appropriate e'. Since π_N is (r,s) transitive, $(R',\oplus,\odot)$ is a right planar nearfield. But $g^2 = I$, so by Theorems 12.1.7 and 12.1.9, g commutes with all members of $\mathrm{Hom}(r,os)$. In particular, $g \circ f = f \circ g$. Thus $g(m) = f \circ f^{-1} \circ g(m) = f \circ g \circ f^{-1}(m) = f \circ g(v) = f(u) = n$. It follows that $[1,x^{-1},0] = [1,y,0]$ or $y = x^{-1}$.

Since π_N is (v,u) transitive, it is also (m,n) transitive because $f: v,u \rightarrow m,n$. Also, π_N is $(h_x(r),h_x(s))$ transitive. Clearly, $h_x(r) = m$, so $h_x(s)$ must equal n, for if $h_x(s) = n' \neq n$, π_N would be (m,n) and (m,n') transitive, and Lemma 12.4.5 tells us that π_N would then be Desarguesian. Thus $[1,-x,0] = [1,y,0]$ and so $y = -x$.

We conclude that $-x = y = x^{-1}$, so $-x^2 = -x(x) = x^{-1}x = 1$, or $x^2 = -1$.

It follows directly from the theorems in this section that for non-distributive right planar nearfields $N(u,v,o,e)$ other than $N(9)$, the only transitivity in π_N is (uv,uv), (u,v), and (v,u) transitivity. This conclusion may be formally stated as follows.

THEOREM 12.4.7 *The set* $T(\pi) = \{(p_o,q_o),(q_o,p_o),(p_oq_o,p_oq_o)\colon p_o \neq q_o\}$ *if and only if* π *can be represented by a nondistributive right planar nearfield* $N \neq N(9)$ *where* $p_o = [1,0,0]$ *and* $q_o = [0,1,0]$.

Proof of a similar statement about left planar nearfields is left as an exercise.

EXERCISES

1. Prove the second, third, and fourth cases of Theorem 12.4.4.

2. Prove Lemma 12.4.5.

3. Formulate and prove a theorem for left planar nearfields analogous to Theorem 12.4.6. (Naturally you may use Theorem 12.4.7 in your proof.)

12.5   THE COLLINEATION GROUP OF $\pi_N(9)$

Section 12.4 established the transitivity properties of π_N for $N \neq N(9)$. This section will examine the transitivity properties of the plane $\pi_N(9)$. This plane has served as an example throughout Part 1, and several statements about its collineation group were made in Theorems 4.3.2 and 4.3.3 and Example 4.3.4. A proof of those statements is provided here. The section begins with a list of five different types of collineations on π_N. The first three are collineations for a general plane π_N; the last two apply specifically to $\pi_N(9)$.

CONSTRUCTION 12.5.1   Let $N = (R,+,\cdot)$ be given and construct mappings: $f_{s,t}, F_{s,t},\ h_{s,t}, H_{s,t},\ g, G,\ j, J,$ and $r_{s,t}, R_{s,t}$ as follows:

$$f_{s,t}\colon\quad [x,y,1] \to [x+s,y+t,1] \text{ where } s,t \in R$$
$$[1,x,0] \to [1,x,0]$$
$$[0,1,0] \to [0,1,0]$$

$$F_{s,t}\colon\quad \langle m,1,k \rangle \to \langle m,1,k-sm-t \rangle$$
$$\langle 1,0,k \rangle \to \langle 1,0,k-s \rangle$$
$$\langle 0,0,1 \rangle \to \langle 0,0,1 \rangle$$

$h_{s,t}$: $[x,y,1] \rightarrow [xs,yt,1]$ where $s,t \in R-\{0\}$
$[1,x,0] \rightarrow [1,s^{-1}xt,0]$
$[0,1,0] \rightarrow [0,1,0]$

$H_{s,t}$: $\langle m,1,k \rangle \rightarrow \langle s^{-1}mt,1,kt \rangle$
$\langle 1,0,k \rangle \rightarrow \langle 1,0,ks \rangle$
$\langle 0,0,1 \rangle \rightarrow \langle 0,0,1 \rangle$

g: $[x,y,1] \rightarrow [-x,-y,1]$
$[1,x,0] \rightarrow [1,x,0]$
$[0,1,0] \rightarrow [0,1,0]$

G: $\langle m,1,k \rangle \rightarrow \langle m,1,-k \rangle$
$\langle 1,0,k \rangle \rightarrow \langle 1,0,-k \rangle$
$\langle 0,0,1 \rangle \rightarrow \langle 0,0,1 \rangle$

(Notice that this is not the involution (g,G) of Theorem 12.3.1.)

Let $N = N(9)$ for the following four mappings:

j: $[x,y,1] \rightarrow [x + y,-x + y,1]$
$[1,x,0] \rightarrow [1,x,0]$ where $x = 0,1,-1$
$[0,1,0] \rightarrow [1,1,0]$
$[1,1,0] \rightarrow [1,0,0]$
$[1,0,0] \rightarrow [1,-1,0]$
$[1,-1,0] \rightarrow [0,1,0]$

J: $\langle m,1,k \rangle \rightarrow \langle m,1,km + k \rangle$ where $m \neq 0,1,-1$
$\langle 1,0,k \rangle \rightarrow \langle -1,1,k \rangle$
$\langle 1,1,k \rangle \rightarrow \langle 1,0,k \rangle$
$\langle -1,1,k \rangle \rightarrow \langle 1,1,k \rangle$
$\langle 0,1,k \rangle \rightarrow \langle 1,1,k \rangle$
$\langle 0,0,1 \rangle \rightarrow \langle 0,0,1 \rangle$

$r_{s,t}$: $[x,y,z] \rightarrow [\alpha_{s,t}(x),\alpha_{s,t}(y),\alpha_{s,t}(z)]$
$R_{s,t}$: $\langle m,n,k \rangle \rightarrow \langle \alpha_{s,t}(m),\alpha_{s,t}(n),\alpha_{s,t}(k) \rangle$

where $\alpha_{s,t}$ is the automorphism defined in Corollary 12.2.5.

These types of mappings have been presented previously in the text but were labeled differently. Compare $f_{s,t}$ with f_r of Theorem 4.3.2;

compare $h_{s,t}$ with g_r of Theorem 4.3.3 and h_n of Theorem 12.4.6; compare g with j of Example 4.3.4; and compare j with h of Corollary 3.4.7.

THEOREM 12.5.2   *The mappings $(f_{s,t}, F_{s,t})$ are elations with axis uv. The mappings $(h_{s,t}, H_{s,t})$ are homologies. If $s = 1$, the center is v; if $t = 1$, the center is u. The mapping (g,G) is a harmonic homology in $Hom(o,uv)$. The mappings (j,J) and $(r_{s,t}, R_{s,t})$ are collineations on $\pi_N(9)$.*

Proof.   The only difficult part of this proof is showing that (j,J) is a collineation on $\pi_N(9)$. Furthermore, there is only one difficult case involved in showing that (j,J) preserves incidence, namely, if $p = [x,y,1]$ and $L = \langle m,1,k \rangle$ where $m \neq 0,1,-1$. Then $p \in L$ implies that $j(p) \in J(L)$. This case is established as follows: $[x,y,1] \in \langle m,1,k \rangle$ iff $xm + y + k = 0$ iff $(xm + y + k) + (xm + y + k)m = 0$ iff $xm + y + k + xm^2 + ym + km = 0$ (by right distributivity) iff $xm + y + k - x + ym + km = 0$ (since $m^2 = -1$) iff $(x + y)m + (-x + y) + km + k = 0$ (by right distributivity and commutativity of addition) iff $[x + y, -x + y, 1] \in \langle m,1,km + k \rangle$ iff $j([x,y,1]) \in J(\langle m,1,k \rangle)$. Completion of the proof that (j,J) is a collineation and proof of the other parts of the theorem are left as an exercise.

Theorem 12.4.6 implies that $\pi_N(9)$ may contain more transitivity than π_N for arbitrary planar nearfields N. This (p,L) transitivity would take the following form: $p \notin L$, $p \in uv$, and $p \neq u,v$; $u,v \notin L$. We know from Lemma 12.4.1 that $\pi_N(9)$ is not (p,p) transitive, and we know from Lemma 12.4.5 that $\pi_N(9)$ is not (p,q) and (p,r) transitive for $q \neq r$. Thus, at best, $\pi_N(9)$ has additional transitivity of the form $(p,q),(q,p)$ for four pairs of distinct points on $uv - \{u,v\}$. This, in fact, is true, as is shown by the following theorem.

THEOREM 12.5.3   *Let $N(9) = (R,+,\cdot)$. The plane $\pi_N(9)$ is (p,q) transitive where $p = [1,x,0]$, $q = [1,-x,0]$, and $x \in R$, $x \neq 0$.*

Proof.   Since $\pi_N(9)$ is (u,v) transitive, we need only show that there exists a collineation mapping $u \to [1,x,0]$ and $v \to [1,-x,0]$. The result will then follow from Theorem 4.5.14. Clearly, $j: u \to [1,-1,0]$, $v \to [1,1,0]$, and $h_{1,-x}: [1,-1,0] \to [1,x,0], [1,1,0] \to [1,-x,0]$, so $h_{1,-x} \circ j$ is the desired collineation.

The following two theorems now follow easily. Their proofs are left as exercises.

THEOREM 12.5.4 *The set* $T(\pi_N(9)) = \{(u,v),(v,u),(uv,uv),(p,q)$ *where* $p = [1,x,0]$, $q = [1,-x,0]$, *and* $x \neq 0\}$.

THEOREM 12.5.5 *If* $T(\pi) = \{(p,L): p \in L_o$ *for some* L_o *and* L *passes through* $f(p)$ *where* f *is an involution that fixes* L_o *but leaves no points of* L_o *fixed*$\}$, *then* π *may be coordinatized by the nearfield* $N(9)$.

Now that the transitivity of $\pi_N(9)$ has been established, an attempt will be made to find the group of all collineations on $\pi_N(9)$ by dividing the possible collineations into two groups: those that fix uv pointwise, that is, those that are elements of $CC(uv)$, and those that permute the points on uv. Such a division assumes that all collineations on $\pi_N(9)$ fix uv — an assumption that is, in fact, true for π_N generally.

THEOREM 12.5.6 *If* f *is a collineation on* π_N, *then* f *fixes uv.*

Proof. The proof is left as an exercise.

First the central collineations with axis uv will be displayed. We are already acquainted with the 81 members of $El(uv)$ of the form $f_{s,t}$: $s,t \in R$ and the harmonic homology g, which, when composed with each elation, yields 81 new members of $CC(uv)$. These 162 central collineations are, in fact, all there are, as is now shown.

LEMMA 12.5.7 *In* π_N, *if* $f \in Hom(o,uv)$ *such that* $f: [1,0,1] \rightarrow [t,0,1]$, *then* $f: [x,y,1] \rightarrow [tx,ty,1]$.

Proof. $f: [1,0,1] \rightarrow [t,0,1]$ and $[0,1,0] \rightarrow [0,1,0]$; therefore, $f: \langle 1,0,-1 \rangle = [1,0,1][0,1,0] \rightarrow [t,0,1][1,0,1] = \langle 1,0,-t \rangle$. Also f fixes $\langle y,1,0 \rangle$ for any y, so $f: [1,y,1] = \langle 1,0,-1 \rangle \cap \langle -y,1,0 \rangle \rightarrow \langle 1,0,-t \rangle \cap \langle -y,1,0 \rangle = [t,ty,1]$. It follows that $f: \langle 0,1,-y \rangle = [1,y,1][1,0,0] \rightarrow [t,ty,1][1,0,0] = \langle 0,1,-ty \rangle$. Therefore, $f: [x,y,1] = \langle 0,1,-y \rangle \cap \langle -x^{-1}y,1,0 \rangle \rightarrow \langle 0,1,-ty \rangle \cap \langle -x^{-1}y,1,0 \rangle = [tx,ty,1]$.

LEMMA 12.5.8 *In* $\pi_N(9)$, *if* $f \in Hom(o,uv)$ *and* $f: [1,0,1] \rightarrow [t,0,1]$, *then either* $t = 1$ *or* $t = 2$.

Proof. Clearly, $t \neq 0$ because o is held fixed. Suppose that $t \in \{a, b, \ldots, f\}$. By Lemma 12.5.7, $f: [t, t + 1, 1] \rightarrow [t^2, t(t + 1), 1] = [-1, -t + 1, 1]$. This equality follows because $t^2 = -1$ and $t(t + 1) = -(t + 1)t = -(t^2 + t) = -(-1 + t) = -t + 1$. Similarly, $f: [0, t - 1, 1] \rightarrow [0, t(t - 1), 1] = [0, t + 1, 1]$. Thus $f: \langle -t, 1, -t + 1 \rangle = [t, t + 1, 1][0, t - 1, 1] \rightarrow [-1, -t + 1, 1][0, t + 1, 1] = \langle t, 1, -t - 1 \rangle$ and $f: [-t, 1, 0] = \langle -t, 1, -t + 1 \rangle \cap \langle 0, 0, 1 \rangle \rightarrow \langle t, 1, -t - 1 \rangle \cap \langle 0, 0, 1 \rangle = [t, 1, 0]$. But $[-t, 1, 0]$ must be held fixed by f, hence a contradiction. Thus $t \notin \{a, \ldots, f\}$.

THEOREM 12.5.9 *In $\pi_N(9)$, $CC(uv) = \{f_{s,t}: s, t \in R\} \cup \{f_{s,t}^{-1} \circ g \circ f_{s,t}: s, t \in R\}$.*

Proof. The proof is left as an exercise.

COROLLARY 12.5.10 *In $\pi_N(9)$ there are 162 central collineations with axis uv.*

Let us consider the collineations on $\pi_N(9)$ that move the points of uv. Observe first that all the mappings $h_{s,t}, j$, and $r_{s,t}$ permute points of uv and that they also move the points in pairs. We may describe this phenomenon in the following way. Label the points of uv as follows: $u = [1, 0, 0]$, $m = [1, 1, 0]$, $n = [1, a, 0]$, $q = [1, b, 0]$, $r = [1, c, 0]$; now define a unary operation of "$-$" on points by $-u = [0, 1, 0] = v$, $-m = [1, -1, 0]$, $-n = [1, -a, 0]$, $-q = [1, -b, 0]$, $-r = [1, -c, 0]$, and $-(-p) = p$ for all $p \in uv$. Then observe that the maps described here preserve the operation of $-$. All such maps must behave this way, as shown by Theorem 12.5.11.

THEOREM 12.5.11 *If f is a collineation on $\pi_N(9)$ that permutes points $p \in uv$, then $f: p \rightarrow p'$ if and only if $f: -p \rightarrow -p'$.*

Proof. We know by Theorem 12.5.4 that $\pi_N(9)$ is $(p, -p)$ transitive and $(p', -p')$ transitive. If $f: p \rightarrow p'$, then $\pi_N(9)$ is also $(p', f(-p))$ transitive. Since $\pi_N(9)$ is not Desarguesian, it follows from Lemma 12.4.5 that $f(-p) = -p'$.

There are $10 \cdot 8 \cdot 6 \cdot 4 \cdot 2 = 3840$ ways that five pairs of points can be permuted; thus there are at most 3840 possible collineations that move points on uv. In actuality, one-half of this number or 1920 collineations exist. One-half of these permutations are not realizable because if a collineation pointwise fixes four of the five pairs $(p,-p)$, it must necessarily fix the fifth pair. This is now shown for the particular case where $p = m,n,q,r$.

THEOREM 12.5.12 (Stevenson, 1970) *If a collineation f on $\pi_N(9)$ fixes $+p$ for $p = m,n,q,r$, then f fixes $\pm u$ (that is, u and v).*

Proof. Suppose that f: $[1,x,0] \to [1,x,0]$, $x = \pm1,\pm a,\pm c$, and also that f: $u \to v$. Clearly, f: $[0,0,1] \to [s,t,1]$ for some s,t. Let $h = f_{s,t}^{-1} \circ f$; then h: $o \to o$, $p \to p$ for $p = m,n,q,r$ and $u \to v$. Let y be a fixed nonzero element of R. Thus h: $\langle 1,0,0 \rangle = [0,1,0][0,0,1] \to [1,0,0][0,0,1] = \langle 0,1,0 \rangle$, so h: $[0,y,1] \to [x(y),0,1]$. Hereafter $x(y)$ will be denoted by x. Notice that $x \neq 0$. As a result h: $\langle 0,1,-y \rangle = [1,0,0][0,y,1] \to [0,1,0][x,0,1] = \langle 1,0,-x \rangle$. Also h: $\langle t,1,0 \rangle \to \langle t,1,0 \rangle$ because h holds fixed $[0,0,1]$ and $[1,t^{-1},0]$. Thus $[x,y,1] = \langle 0,1,-y \rangle \cap \langle -x^{-1}y,1,0 \rangle \to \langle 1,0,-x \rangle \cap \langle -x^{-1}y,1,0 \rangle = [x,y,1]$, so $[x,y,1]$ is held fixed by h.

Let $z \in R$ such that $z \neq 0, yx^{-1}$. Since $\langle -z,1,xz-y \rangle = [x,y,1][1,z,0]$ is held fixed, h: $[0,y-xz,1] = \langle -z,1,xy-y \rangle \cap \langle 1,0,0 \rangle \to \langle -z,1,xz-y \rangle \cap \langle 0,1,0 \rangle = [(xz-y)z^{-1},0,1]$. By the preceding argument, h fixes $[(xz-y)z^{-1},y-xz,1]$. This point will be denoted by p_z.

Let $L_z = \langle z(xz-y)^{-1}xz,1,-y \rangle = [0,y,1]p_z$. Then $L_z \cap uv = [1,-z(xz-y)^{-1}xz,0]$, and this point is fixed since $-z(xz-y)^{-1}xz \neq 0$. Therefore, L_z is a fixed line since it contains two fixed points (o and $L_z \cap uv$). Since $[0,y,1] \in L_z$ and h: $[0,y,1] \to [x,0,1]$, it follows that $[x,0,1] \in L_z$. However, this is not necessarily true, as the following example shows. Suppose that $y \neq \pm1$ and let $z = x^{-1}$. Then $L_z = \langle x^{-1}(1-y)^{-1},1,-y \rangle$. If $[x,0,1] \in L_z$, we would have $xx^{-1}(1-y)^{-1}-y = (1-y)^{-1}-y = (y-1-y) = -1 = 0$, a contradiction. Thus h and therefore f do not exist.

As the following four theorems show, there exists a collineation that induces any permutation that permutes the points pairwise and does not fix exactly four pairs of points.

THEOREM 12.5.13 *There exists a collineation f on $\pi_N(9)$ that maps $m \to p$ for any $p \in uv$.*

Proof. We may map $u \to m$ by j and $u \to -u$ by j^2. Also we may map $m \to p = [1,x,0]$ by $h_{1,x}$ for any $x = \pm 1, \pm a, \pm b, \pm c$. Thus with compositions of the mappings j and $h_{1,x}$, we may map $u \to p$ for any $p \in uv$.

THEOREM 12.5.14 *There exists a collineation f on $\pi_N(9)$ such that f fixes u and v, and $f: m \to p$ for any $p \neq u, v$.*

Proof. As mentioned in Theorem 12.5.13, $h_{1,x}: m \to [1,x,0]$ for $x = \pm 1, \pm a, \pm b, \pm c$. Also $h_{1,x}: u, v \to u, v$.

THEOREM 12.5.15 *There exists a collineation f on $\pi_N(9)$ such that f fixes $u, v, m, -m$, and $f: n \to p$ for $p \neq u, v, \pm m$.*

Proof. It may be easily observed that $h_{b,b}$, $r_{1,1}$, and $h_{a,a}$ hold u and m fixed and that $h_{b,b}: n \to -n$, $r_{1,1}: n \to q$, $r_{1,1}^2: n \to r$, $h_{a,a} \circ r_{1,1}: n \to -q$, $h_{a,a} \circ r_{1,1}^2: n \to -r$. The identity maps $n \to n$.

THEOREM 12.5.16 *There exists a collineation f on $\pi_N(9)$ such that f fixes $u, v, \pm m, \pm n$, and $f: q \to p$ where $p = \pm q, \pm r$.*

Proof. Notice that $r_{0,2}$, $h_{a,a}$, and $h_{b,b}$ hold u and m fixed. Clearly, $r_{0,2}: n \to -n$, $-n \to n$, $q \to -r$, so $h_{b,b} \circ r_{0,2}: n \to n$, $q \to r$. Also $h_{a,a}: n \to n$, $q \to -q$, so $h_{a,a} \circ h_{b,b} \circ r_{0,2}: n \to n$, $q \to -r$. The identity maps $q \to q$.

THEOREM 12.5.17 *There exist 1920 collineations that permute points on uv.*

Proof. There exist 10 collineations that map $u \to p$ where $p \in uv$; 8 collineations that fix $\pm u$ and map $m \to p$, $p \neq \pm u$; 6 collineations that fix $\pm u, \pm m$ and map $n \to p$, $p \neq \pm u, \pm m$; and 4 collineations that fix $\pm u, \pm m, \pm n$ and map $q \to p$ where $p = \pm q, \pm r$. If f fixes $\pm u, \pm m, \pm n$ and $\pm q$, it follows easily from Theorem 12.5.12 that f fixes $\pm r$ and therefore is the identity. Thus there are a total of $10 \cdot 8 \cdot 6 \cdot 4 = 1920$ collineations.

This completes the analysis of $C(\pi_N(9))$. Since $CC(uv)$ is a normal subgroup of $C(\pi_N(9))$, there are $1{,}920 \cdot 162 = 311{,}040$ members of $C(\pi_N(9))$. Clearly, the collineations $f_{s,t}$, g, $h_{s,t}$, j, and $r_{s,t}$ generate this group. Reduction of this set of generators to a minimal set is left for the student.

EXERCISES

1. Complete the proof of Theorem 12.5.2.

2. Let N be a right planar nearfield, let $\pi_N = (\mathscr{P}_N, \mathscr{L}_N, \mathscr{I}_N)$, and let $f_t: \mathscr{P}_N \rightarrow \mathscr{P}_N$ and $F_t: \mathscr{L}_N \rightarrow \mathscr{L}_N$ where $t \in N$, $t \neq 0$, and where

$$f_t: \quad [x,y,1] \rightarrow [yt^{-1}, xt, 1]$$
$$[1,x,0] \rightarrow [1, tx^{-1}t, 0] \text{ where } x \neq 0$$
$$[1,0,0] \rightarrow [0,1,0]$$
$$[0,1,0] \rightarrow [1,0,0]$$

$$F_t: \quad \langle a,1,b \rangle \rightarrow \langle ta^{-1}t, 1, ba^{-1}t \rangle \text{ where } a \neq 0$$
$$\langle 0,1,b \rangle \rightarrow \langle 1,0,bt^{-1} \rangle$$
$$\langle 1,0,b \rangle \rightarrow \langle 0,1,bt \rangle$$
$$\langle 0,0,1 \rangle \rightarrow \langle 0,0,1 \rangle$$

   Prove that (f_t, F_t) is a collineation in π_N.

3. Prove Theorem 12.5.4.

4. Prove Theorem 12.5.5.

5. Refer back to Section 11.4. Find all the transitivity in the following planes: π_g, π_h, π'.

6. Prove Theorem 12.5.6.

7. Prove Theorem 12.5.9.

8. Find a minimal set of generators for $C(\pi_N(9))$.

9. As mentioned in Corollary 3.4.7, there are projective collineations in $\pi_N(9)$ that are not products of central collineations.
   a. Exhibit a set of generators for $CC(\pi_N(9))$.
   b. Exhibit a set of generators for $PC(\pi_N(9))$.
   c. Is (f_t, F_t) of exercise 2 in $PC(\pi_N(9))$? in $CC(\pi_N(9))$?

12.6   THE COORDINATIZING PLANAR NEARFIELDS OF π_N

This section will show that if two nearfields coordinatize isomorphic planes, then the nearfields themselves are isomorphic. This strong bond between π and its representative nearfields $N(u,v,o,e)$ may be unexpected, since, by Corollary 11.4.6, the bond between representative quasifields of a given plane is not necessarily as strong as isotopism. Nevertheless, some thought reveals that the tight restriction on the transitivity of π_N dictates a strong bond on representative nearfields $N(u,v,o,e)$.

THEOREM 12.6.1   *Let π be represented by the right planar nearfield $N(u,v,o,e)$ and let u',v',o',e' be a four-point in π such that $\{u',v'\} = \{u,v\}$, that is, $u' = u$ or v and $v' = v$ or u. Then there exists a collineation f on π such that $f: u,v,o,e \rightarrow u',v',o',e'$.*

Proof.

   Case 1.   Let $u=u'$ and $v=v'$. Also let $o' = [a,b,1]$ and $e' = [c,d,1]$. Then $c \neq a$ because o',e',v' are not collinear and $d \neq b$ because o',e',u' are not collinear. Therefore, $f_{a,b} \circ h_{a-c,d-b}$: $o = [0,0,1] \rightarrow [a,b,1]$, $e = [-1,1,1] \rightarrow [c,d,1]$, so if $f = f_{a,b} \circ h_{a-c,d-b}$, then f: $u,v,o,e \rightarrow u',v',o',e'$.

   Case 2.   Let $u' = v$ and $v' = u$. Let g be the involution in Theorem 12.3.1; it is then easily verified that $f_{a,b} \circ h_{c-a,b-d} \circ g$: $v \rightarrow u$, $u \rightarrow v$, $o \rightarrow [a,b,1]$ and that $e \rightarrow [c,d,1]$. Thus if $f = f_{a,b} \circ h_{a-c,b-d} \circ g$, then f: $u,v,o,e \rightarrow u',v',o',e'$.

THEOREM 12.6.2   *If $N = N(u,v,o,e)$ and $N' = N(u',v',o',e')$ are coordinatizing right planar nearfields of π, then N is isomorphic to N'.*

Proof.   If $\pi = \pi_N(9)$, then $N = N' = N(9)$ because there is only one nearfield of order 9. Suppose that $\pi \neq \pi_N(9)$. Then π is (uv) and $(u'v')$ transitive. By Theorems 12.3.5 and 12.4.6, either $\{u,v\} = \{u',v'\}$ or π is Desarguesian. If π is Desarguesian, we know that there exists a collineation f on π such that f: $u,v,o,e \rightarrow u',v',o',e'$. If π is non-Desarguesian, Theorem 12.6.1 assures us that such an f exists. In either instance we may apply Theorem 9.3.1 to arrive at the desired conclusion that $N \sim N'$.

COROLLARY 12.6.3   *If N and N' are right planar nearfields and π_N is isomorphic to $\pi_{N'}$, then N is isomorphic to N'.*

The statements and proofs of appropriate theorems for left planar nearfields are left as an exercise.

An interesting algebraic result follows directly from Corollary 12.6.3, the left nearfield analogue of that result, and Theorem 9.3.4; namely, that isotopic planar nearfields are isomorphic. We have already proved algebraically that isotopic division rings are isomorphic (Theorem 9.3.10) and that isotopic quasifields are not necessarily isomorphic (Theorem 11.4.2).

THEOREM 12.6.4 *If planar nearfields N and N' are isotopic, then they are isomorphic.*

The student is encouraged to attempt an algebraic proof of this theorem.

EXERCISES

1. Formulate and prove the analogue to Corollary 12.6.3 for left planar nearfields.

2. From what you have learned in Chapter 12, how many planes π are there of order n such that $T(\pi) = \{(p_o,q_o),(q_o,p_o),(p_oq_o,p_oq_o); p_o \neq q_o\}$ where $n = 16?$ 25? 27? 32? 49?

3. Let N be a nondistributive right planar nearfield and let p,q,r,s denote a four-point in π_N. For each of the following algebras explain why restrictions can or cannot be placed on p,q,r,s so that:

   a. $P(p,q,r,s)$ is right quasifield but not a right nearfield.
   b. $P(p,q,r,s)$ is a left quasifield.
   c. $P(p,q,r,s)$ is a Cartesian group that is not a quasifield.
   d. $P(p,q,r,s)$ is a planar ring with associative multiplication but not associative addition.
   e. $P(p,q,r,s)$ is a planar ring with no associativity.
   f. $T(p,q,r,s)$ is a nonlinear ternary ring.

*4. Prove algebraically that isotopic planar nearfields are isomorphic.

CHAPTER 13

PLANES OVER
SEMIFIELDS

This section examines planes π defined over Cartesian groups with two-sided distributivity. We have already explored the geometric consequences of right distributivity and left distributivity separately, so we know that π has a minimum of (p_o, p_o) and (L_o, L_o) transitivity for some $p_o \in L_o$. It will be shown here that for semifields without additional algebraic conditions, there is no other transitivity on π. The relationships among the various coordinatizing semifields of π will also be analyzed.

13.1 SEMIFIELDS

DEFINITION 13.1.1   *A semifield $S = (R, +, \cdot)$ is a Cartesian group such that for $a, b, c \in R$, $a(b + c) = ab + ac$; and $(a + b)c = ac + bc$.*

EXAMPLE 13.1.2   The finite semifield of least order has 16 elements. There are 23 known nonisomorphic semifields of order 16; 18 are in one isotopy class and 5 are in the other. One example from each class is given here (see Knuth, 1965).

i. Recall that GF(4) contains the elements $0,1,a,a+1$ with addition and multiplication as displayed in Table 1 (p. 39). Then define $S(16) = (R,\oplus,\odot)$ as follows:

$$R = \{(x,y): x,y \text{ in GF(4)}\}$$
$$\oplus : (x,y) \oplus (w,z) = (x+w, \; y+z)$$
$$\odot : (x,y) \odot (w,z) = (xw + y^2z, \; yw + x^2z + y^2z^2)$$

ii. Define $S'(16) = (R,\oplus,\circ)$ as follows:

$(R,\oplus)$: same definition as in (i)
$$\circ : (x,y) \circ (w,z) = (xw + ay^2z, \; x^2z + yw)$$

One way to see that S and S' are not isotopic is to verify that their partial associative laws are of different strengths: For one, $a(bc) = (ab)c$ if any two of $a,b,c \in$ GF(4); for the other, $a(bc) = (ab)c$ if any one of $a,b,c \in$ GF(4). It is left to the student to show which is which.

iii. One of several methods for generating finite semifields may be displayed as follows (see Dembowski, 1968, p. 236 ff. for other methods). Let $(R,\oplus,\circ)$ be defined as follows:

$$R = \{(x,y): x,y \text{ are in GF}(p^n), \; n \geq 2\}$$
$$\oplus : (x,y) \oplus (w,z) = (x+w, \; y+z)$$
$$\circ : (x,y) \circ (w,z) = (xw + ty^pz, \; x^pz + yw)$$

where t is not a $(p+1)$th power of any element of GF(p^n) (for a proof that $(R,\oplus,\circ)$ is a semifield, see Albert and Sandler, 1968, p. 86 ff.).

iv. For an example of an infinite semifield, let $(R,\oplus,\circ)$ be defined by:

$$R = \{(x,y,z): x,y,z \in Q, \text{ the field of rationals}\}$$
$$\oplus : (x,y,z) \oplus (x',y',z') = (x+x', \; y+y', \; z+z')$$
$$\circ : (x,y,z) \circ (x',y',z') = (xx' + 2(yz'+zy'), \; xy'$$
$$+ \; yx' - 16zz', \; xz' + zx' + yy')$$

EXERCISES

1. Is either $S(16)$ or $S'(16)$ generated by the method displayed in Example 13.1.2(iii)?

2. Describe a semifield of order n where n is 25; 27; 32; 49.

3. Construct tables for $S(16)$ and $S'(16)$. Which of these two semifields satisfies the associative law $a(bc) = (ab)c$ if any two of a, b, or $c \in \mathrm{GF}(4)$?

4. Define f_c on R as follows: $f_c(x,y) \to (x, cy)$ where $c = 1, -1, a, b$. For which c is f_c an automorphism on $S(16)$? on $S'(16)$?

5. The semifield S satisfies the left (right) alternative law if and only if $a^2b = a(ab)$ $(ab^2 = (ab)b)$. Show that the semifields of Example 13.1.2 do not satisfy the left or right alternative laws.

13.2 TRANSITIVITY IN PLANES OVER SEMIFIELDS

This section attempts to find $T(\pi_S)$ for any plane represented by a semifield. A similar attempt succeeded for planes π_N (Theorem 12.4.7) but failed for planes π_Q (see p. 326). The first theorem states a result that was established in Chapter 11.

THEOREM 13.2.1 *The plane π is (v,v) and (uv, uv) transitive if and only if $\pi = \pi_S$ where $S = S(u, v, o, e)$ is a semifield. (The points o and e are chosen so that u, v, o, e is a four-point.)*

Proof. This follows from Theorems 11.2.3 and 11.2.5.

The following seven theorems explore the feasibility of adding transitivity to π_S. The first question to be answered is whether π_S can be (p, L) transitive for some p and L such that $p \notin L$.

THEOREM 13.2.2 *Suppose that S is a semifield and that π_S is (p, L) transitive where $p \notin L$ and $p \notin uv$. Then π_S is Desarguesian.*

Proof.

Case 1. $L = uv$. Since π_S is (uv, uv) transitive, there exists a collineation $f_q \in \mathrm{El}(uv)$ such that $f_q: p \to q$ for any $q \notin uv$. Thus π_S is (q, uv) transitive for all $q \notin uv$ and it follows that π_S satisfies the Second Desargues (uv) Theorem. We conclude that π_S is Desarguesian.

Case 2. $L \neq uv$. Let $t = L \cap uv$ and let q be a point satisfying the condition that $q \notin L$, $q \notin uv$, $q \notin pt$. Furthermore, let $r = pq$

$\cap$ uv and let $s = pq \cap L$. The (p,L) transitivity in π_S implies the existence of $f \in \text{Hom}(p,L)$ such that $f: q \to r$; the (uv,uv) transitivity implies the existence of $g \in \text{El}(uv)$ such that $g: s \to q$. Thus $f \circ g: L = st \to rt = uv$ and $p \to f(g(p))$, so π_S is $(f(g(p)),uv)$ transitive. It follows from Case 1 that π_S is Desarguesian.

THEOREM 13.2.3   *If π_S is (p,L) transitive where $p \notin L$, $p \in uv$, and $v \in L$, then π_S is Desarguesian.*

Proof.   Consider a four-point p,v,o',e' where $o' \in L$ and e' is any appropriate fourth point. Since π_S is (uv,uv) transitive and since $uv = pv$, it follows that π_S is (pv,pv) transitive. Also π_S is (v,v) and $(p,o'v)$ transitive (since $L = o'v$). Thus $S' = S(p,v,o',e')$ is a division ring (since, according to the given transitivity properties, it must satisfy both of the distributive laws and associativity of multiplication), and therefore $\pi_{S'} = \pi_S$ is Desarguesian.

THEOREM 13.2.4   *If π_S is (p,L) transitive where $p \notin L$, $v \notin L$, and $p \in uv$, then π_S is Desarguesian.*

Proof.   This theorem follows by duality from Theorem 13.2.2.

The remaining theorems explore the possibility of adding (p,L) transitivity where $p \in L$.

THEOREM 13.2.5   *If there exists a collineation f on π_S such that f: $v \to p \notin uv$, then π_S satisfies the Little Desargues Theorem.*

Proof.   Since π_S is (uv,uv) transitive, we may map $v \to q$ for any q $\notin uv$ with a composition of mappings f and $f_{s,t} \in \text{El}(uv)$. Thus π_S is (q,q) transitive for all $q \notin uv$. It is an easy matter to extend the (q,q) transitivity to the line uv. Therefore, π_S is (p,p) transitive for all p and the Little Desargues Theorem easily follows.

THEOREM 13.2.6   *If there exists a collineation f on π_S such that f: $v \to p \in uv$, then π_S is (p,p) transitive for all $p \in uv$.*

Proof.   This theorem follows directly from Theorem 4.5.16(ii), since π_S must be (v,v) and (p,p) transitive.

THEOREM 13.2.7 *If there exists a collineation f on π_S such that f: $uv \rightarrow L$ and $v \in L$, then π_S is (M,M) transitive for all lines M through v.*

Proof. Since π_S must necessarily be (uv,uv) and (L,L) transitive, it follows from Theorem 4.5.16(i) that π_S is (M,M) transitive for all M incident with v.

From these theorems we may derive the following facts about additional (p,L) transitivity in π_S where $p \in L$.

THEOREM 13.2.8 *Suppose that π_S is (p,L) transitive where $p \in L$, $p \neq v$, and $L \neq uv$. It follows that:*

  i. *If $p \notin uv$ and $v \notin L$, then π_S satisfies the Little Desargues Theorem.*

  ii. *If $p \in uv$, then π_S is (p,p) transitive for all $p \in uv$.*

  iii. *If $v \in L$, then π_S is (M,M) transitive for all M through v.*

Proof. Clearly, the first statement follows from Theorem 13.2.5, the second follows from Theorem 13.2.6, and the third follows from Theorem 13.2.7.

Thus non-Desarguesian planes π_S appear to have four different transitivity classifications:

$$T_1(\pi) = \{(p_o,p_o),(L_o,L_o): p_o \in L_o\}$$
$$T_2(\pi) = \{(p,p): p \in L_o\}$$
$$T_3(\pi) = \{(L,L): L \text{ is on } p_o\}$$
$$T_4(\pi) = \{(L,L): L \text{ is an arbitrary line}\}$$

This is not true. As we shall find in Section 14.2, $T_2(\pi)$ and $T_3(\pi)$ cannot be realized. If π has either the transitivity of T_2 or the transitivity of T_3, it must have the transitivity of T_4 and thus it must be a Little Desarguesian plane. Both T_1 and T_4 can be realized. All the planes π_S defined over the semifields of Example 13.1.2 are members of transitivity class T_1 because they do not satisfy the additional algebraic properties requisite for Little Desarguesian planes π_S. This reason will be explained fully in Chapter 14, but it can be stated here that additional properties pertain to a specialized associative law. If π_S is a member of the class T_4, it is necessary that $a(ab) = a^2b$ and that $(ab)b = ab^2$ for all a,b in the semifield. These two laws, called the left and right alternative laws, respectively, do not hold in Example 13.1.2.

EXERCISES

1. Show that the class of projective planes π_S satisfies the principle of duality.

2. Explain why Theorem 13.2.4 follows from Theorem 13.2.2 by duality.

3. Prove that, as stated in the proof of Theorem 13.2.5, the (q,q) transitivity may be extended to the line uv.

4. Show that if there exists a collineation f on π_S mapping $uv \to L \neq uv$, then π_S is a Little Desarguesian plane.

13.3 THE COORDINATIZING SEMIFIELDS OF π_S

We have found that if $\pi = \pi_{N(u,v,o,e)} = \pi_{N(u',v',o',e')}$ for planar nearfields N, then the nearfields must be isomorphic. The same holds true for division rings (that is, if two division rings coordinatize π, they are isomorphic) but not for quasifields. In fact, two coordinatizing quasifields need not be isotopic, as Theorem 11.4.2 shows. For semifields, the situation is not as bad as it is for quasifields but not as good as it is for planar nearfields and division rings.

The following theorem assumes that π_S and $\pi_{S'}$ are not Little Desarguesian planes and thus that S and S' are "proper" semifields; in other words, they do not satisfy the alternative laws.

THEOREM 13.3.1 (Albert, 1960) *If $S = (R,+,\cdot)$ and $S' = (R',+,\circ)$ are semifields such that π_S and $\pi_{S'}$ are isomorphic, then S and S' are isotopic.*

Proof. (Knuth, 1965) Let $h: \pi_S \to \pi_{S'}$ denote an isomorphism between the two planes. Denote the points and lines of π_S by [ ] and $\langle\ \rangle$, respectively, and the points and lines of $\pi_{S'}$ by [[ ]] and $\langle\langle\ \rangle\rangle$, respectively. Since both planes are in the transitivity class $T(\pi)$ = $\{(p_o,p_o),(L_o,L_o): p_o \in L_o\}$, it follows that

$$h: [0,1,0] \to [[0,1,0]]$$
$$\langle 0,0,1 \rangle \to \langle\langle 0,0,1 \rangle\rangle$$

Thus h: $[1,0,0] \to [[1,a,0]]$ for some $a \in R'$ and h: $[0,0,1] \to [[b,c,1]]$ for some $b,c \in R'$. Consider the mappings $g_{s,t}, G_{s,t}$ defined on plane π_S as follows:

$$g_{s,t}: \quad [x,y,1] \to [x,y+xs+t,1]$$
$$[1,x,0] \to [1,x+s,0]$$
$$[0,1,0] \to [0,1,0]$$

$$G_{s,t}: \quad \langle a,1,b \rangle \to \langle a-s,1,b-t \rangle$$
$$\langle 1,0,a \rangle \to \langle 1,0,a \rangle$$
$$\langle 0,0,1 \rangle \to \langle 0,0,1 \rangle$$

This is easily seen to be a collineation. Let $j = f_{(-b,ba-c)} \circ g_{(-a,0)} \circ h$ where $f_{s,t} \in \text{El}(uv)$ as defined in Construction 12.5.1. Clearly, $f_{s,t}$ is a collineation in π_S. Then the components of j yield the following mappings:

$$j: [0,1,0] \to [[0,1,0]]$$
$$j: [1,0,0] \to [[1,a,0]] \to [[1,0,0]] \to [[1,0,0]]$$
$$j: [0,0,1] \to [[b,c,1]] \to [[b,c+b(-a),1]] \to [[0,0,1]]$$

Let $o = [0,0,1]$, $u = [1,0,0]$, $v = [0,1,0]$, and $o' = [[0,0,1]]$, $u' = [[1,0,0]]$, and $v' = [[0,1,0]]$; then j: $u,v,o \to u',v',o'$. By Theorem 9.3.4, S and S' are isotopic.

COROLLARY 13.3.2 *Let S and S' be the semifields of order 16 in Example 13.1.2. Then π_S and $\pi_{S'}$ are not isomorphic.*

Since π_S and $\pi_{S'}$ are in the same transitivity class, the student is encouraged to analyze and formalize the geometric difference between them. It should also be noted that isotopic semifields are not necessarily isomorphic. This can be shown for semifields of order 16 by appropriately choosing an e' in $\pi_{S(16)}$ and comparing $S(u,v,o,e)$ and $S(u,v,o,e')$.

EXERCISES

1. Show that $(g_{s,t}, G_{s,t})$ of Theorem 13.3.1 is a collineation.

2. How many different planes of order n were described in Part 3 where $n = 16$? Where $n = 25$?

3. Let S be a nonassociative semifield and let p,q,r,s denote a four-point in π_S. For each of the following algebras explain why restrictions can or cannot be placed on p,q,r,s so that:

   a. $P(p,q,r,s)$ is a right planar nearfield.
   b. $P(p,q,r,s)$ is a left planar nearfield.
   c. $P(p,q,r,s)$ is a right quasifield but not a right planar nearfield or a semifield.
   d. $P(p,q,r,s)$ is a left quasifield but not a left planar nearfield or a semifield.
   e. $P(p,q,r,s)$ is a Cartesian group but not a quasifield.
   f. $P(p,q,r,s)$ is a planar ring with associative multiplication but not associative addition.
   g. $P(p,q,r,s)$ is a planar ring with no associativity.
   h. $T(p,q,r,s)$ is a nonlinear ternary ring.

*4. Distinguish geometrically between $\pi_{S(16)}$ and $\pi_{S'(16)}$.

*5. Display isotopic but nonisomorphic semifields of order 16 by comparing $S(u,v,o,e)$ and $S(u,v,o,e')$ for an appropriately chosen e'.

CHAPTER 14

PLANES OVER
ALTERNATIVE RINGS

An alternative ring is a semifield with a weak associativity property. This section will show that with the addition of this property, the associated projective planes satisfy the Little Desargues Theorem. In terms of transitivity, no more structure can be achieved in a non-Desarguesian plane.

14.1 ALTERNATIVE RINGS

DEFINITION 14.1.1 *A planar ring $A = (R, +, \cdot)$ is an* alternative ring *if and only if A is a semifield such that*

1. $a(ab) = (aa)b$
2. $(ab)b = a(bb)$ *for all $a, b \in R$*

These two properties are called the *left alternative law* and the *right alternative law*, respectively.

EXAMPLE 14.1.2

  i. Define $A = (R, \oplus, \circ)$ as follows: $R = \{(a_0 + a_1 i_1 + a_2 i_2 + \ldots + a_7 i_7)$: a_k are real numbers, $k = 0, \ldots, 7\}$. Define equality as identity; that is, $a_0 + a_1 i_1 + \ldots + a_7 i_7 = b_0 + b_1 i_1 + \ldots + b_7 i_7$ if and only if $a_i = b_i$, $i = 0, \ldots, 7$. Define $\oplus$ as in a vector space, that is, $(a_0 + a_1 i_1 + \ldots + a_7 i_7) \oplus (b_0, b_1 i_1 + \ldots + b_7 i_7) = (a_0 + b_0) + (a_1 + b_1)i_1 + \ldots + (a_7 + b_7)i_7$. Multiplication is illustrated by Table 22, with the following specifications:

    1. $i_j \circ i_k = -i_k \circ i_j$.
    2. Both distributive laws hold.
    3. $ai_k = a \circ i_k = i_k \circ a$ for real numbers a.
    4. $a \circ (bi_k) = (ab) \circ i_k$ for real numbers a and b.

TABLE 22

$\circ$	i_1	i_2	i_3	i_4	i_5	i_6	i_7
i_1	-1	i_3	$-i_2$	i_5	$-i_4$	$-i_7$	i_6
i_2		-1	i_1	i_6	i_7	$-i_4$	$-i_5$
i_3			-1	$-i_7$	$-i_6$	i_5	$-i_4$
i_4				-1	i_1	i_2	i_3
i_5					-1	$-i_3$	i_2
i_6						-1	$-i_1$
i_7							-1

The octuples in R are called *Cayley numbers*. Notice that $\{(a_0 + a_1 i_1 + a_2 i_2 + a_3 i_3 + 0i_4 + 0i_5 + 0i_6 + 0i_7)$ where a_k are reals$\}$ is isomorphic to the ring of quaternions.

    It can be checked that R may be represented by elements of the form $a + bi$ where a, b are quaternions and addition and multiplication are defined as follows:

$$(a + bi) \oplus (c + di) = (a + c) \oplus (b + d)i$$
$$(a + bi) \circ (c + di) = (ac - \bar{d}b) \oplus (da + b\bar{c})i$$

The element $\bar{x}$ has the following definition: $\bar{x} = s + r - si - tj - uk$ where $x = r + si + tj + uk$; $\bar{x}$ is called the *conjugate* of d in the quaternion algebra.

  ii. A general alternative ring $A = (R, \oplus, \circ)$, called a *Cayley-Dickson algebra over a field F*, is defined as follows:

$$R = (a + bi) \text{ where } a,b \text{ are in a}$$
$$\text{quaternion algebra over } F$$
$$(a + bi) \oplus (c + di) = (a + c) \oplus (b + d)i$$
$$(a + bi) \circ (c + di) = (ac + j\bar{d}b) \oplus (da + b\bar{c})i \text{ where } j = i^2,$$
$$j \in F, \text{ and } j \neq x\bar{x} \text{ for any } x \in Q$$
$$\text{(the quaternion ring)}$$

Since there are quaternion rings of any permissible characteristic (that is, either a prime number or 0 — see Definition 7.1.10), there are also alternative rings of any permissible characteristic.

Following are three main theorems concerning alternative rings; their proofs are beyond the scope of this book. The first two theorems are due to Artin.

THEOREM 14.1.3 *The subalgebra generated by any two elements of an alternative ring is an associative algebra.*

THEOREM 14.1.4 *A finite alternative ring is a field.*

THEOREM 14.1.5 (Bruck and Kleinfeld, 1951; Skornyakov, 1950) *If $A = (R,+,\cdot)$ is an alternative ring, then either A is associative or A is a Cayley-Dickson algebra over some field F.*

This last theorem is a significant completeness theorem for alternative rings. A proof can be found in Pickert (1955).

Every Cayley-Dickson algebra satisfies two partial associative laws called inversive laws that are somewhat stronger than the alternative laws.

The left inversive law: If $aa' = 1$, then $a(a'b) = b$.
The right inversive law: If $a'a = 1$, then $(ba')a = b$.

The left inversive law also implies that the left and right inverse of a are unique and the same. Thus we may state the left inversive law by $a'(ab) = b$. A similar statement is true with regard to the right inversive law.

The inversive laws are said to be stronger than the alternative laws because in semifields, the left (right) alternative law can be derived from the left (right) inversive law (see Moufang, 1935, or Hall, 1959).

For semifields of characteristic other than 2, the validity of the left alternative law implies the validity of the right alternative law (see Kleinfeld, 1953; Skornyakov, 1951a); for semifields of characteristic 2, the validity of the left inversive law implies the validity of the right alternative law (San Soucie, 1955). Combining these remarks results in the following theorem.

THEOREM 14.1.6   *If S is a semifield satisfying the left inversive law, then S also satisfies the right inversive law.*

Proof.   Since S satisfies the left inversive law, we may conclude from the preceding discussion that S satisfies both alternative laws and thus is an alternative ring. By Theorem 14.1.5, S must be a Cayley-Dickson algebra, and therefore it satisfies the right inversive law.

14.2   GEOMETRIC CONSEQUENCES OF THE INVERSIVE LAWS

If A is an alternative ring, then π_A is called an *alternative plane* or a *Moufang plane* (named after Ruth Moufang, who first studied these planes). This section will show that alternative planes satisfy the Little Desargues Theorem.

THEOREM 14.2.1   *If S is a semifield satisfying the left inversive law, then π_S is (L,L) transitive for all L passing through v.*

Proof.   Consider the mapping (h,H) on π_A defined as follows:

$$h: \quad [x,y,1] \to [(x^{-1}+1)^{-1},(x^{-1}+1)^{-1}(x^{-1}y),1] \text{ where } x \neq 0,-1$$
$$[0,y,1] \to [0,y,1]$$
$$[-1,y,1] \to [1,-y,1]$$
$$[1,x,0] \to [1,x,1]$$
$$[0,1,0] \to [0,1,0]$$

$$H: \quad \langle a,1,b \rangle \to \langle a-b,1,b \rangle$$
$$\langle 1,0,a \rangle \to \langle 1,0,(a^{-1}+1)^{-1} \rangle \text{ where } a \neq 0,-1$$
$$\langle 1,0,0 \rangle \to \langle 1,0,0 \rangle$$
$$\langle 1,0,1 \rangle \to \langle 0,0,1 \rangle$$
$$\langle 0,0,1 \rangle \to \langle 1,0,-1 \rangle$$

The mapping (h, H) is a collineation. Incidence is shown here to be preserved in the most difficult case; completion of the proof is left as an exercise.

$[(x^{-1} + 1)^{-1}, (x^{-1} + 1)^{-1}(x^{-1}y), 1] \in \langle a - b, 1, b \rangle$ iff $(x^{-1} + 1)^{-1}(a - b)$ $+ (x^{-1} + 1)^{-1}(x^{-1}y) + b = 0$ iff $(x^{-1} + 1)^{-1}(a - b + x^{-1}y + (x^{-1} + 1)b) = 0$ iff $a - b + x^{-1}y + x^{-1}b + b = 0$ iff $x^{-1}(xa) + x^{-1}y + x^{-1}b = 0$ iff $x^{-1}(xa + y + b) = 0$ iff $xa + y + b = 0$. (Notice that the second and fourth "iffs" are based on the left inversive law.) Thus $[x, y, 1] \in \langle a, 1, b \rangle$ iff $h([x, y, 1])$ $\in H(\langle a, 1, b \rangle)$.

Since (h, H) maps $\langle 0, 0, 1 \rangle \to \langle 1, 0, -1 \rangle$, π_A is both $(\langle 0, 0, 1 \rangle, \langle 0, 0, 1 \rangle)$ transitive and $(\langle 1, 0, -1 \rangle, \langle 1, 0, -1 \rangle)$ transitive. By Theorem 4.5.16, π_S is (L, L) transitive for all L passing through v.

THEOREM 14.2.2   *If π_S is (L, L) transitive for all L passing through v, then S is a semifield having the left inversive property.*

Proof.   Let $f \in El(o, ov)$ such that $f: u \to [-1-a, 0, 1]$ for $a \neq 0$. Clearly, f fixes $[0, 1 + a, 1]$ because ov is its axis, so $f: \langle 0, 1, -1 - a \rangle$ $= [1, 0, 0][0, 1 + a, 1] \to [-1 - a, 0, 1][0, 1 + a, 1] = \langle -1, 1, -1-a \rangle$. Also f fixes $\langle 1 - a, 1, 0 \rangle$ because o is its center, so $f: [1, 1 + a, 1] = \langle -1 - a, 1, 0 \rangle$ $\cap \langle 0, 1, -1 - a \rangle \to \langle -1 - a, 1, 0 \rangle \cap \langle -1, 1, -1 - a \rangle = [(1 + a)a^{-1},$ $((1 + a)a^{-1})(1 + a), 1] = [a^{-1} + 1, a^{-1} + 1 + a + 1, 1]$. The last equality holds because $((1 + a)a^{-1})(1 + a) = (a^{-1} + 1)(1 + a) = a^{-1} + a^{-1}a + 1 + a$ $= a + a^{-1} + 1 + 1$. Let $d = a^{-1} + 1$; then $f: [1, 1 + a, 1] \to [d, d + a + 1, 1]$. Thus $f: \langle 1, 0, -1 \rangle = [0, 1, 0][1, 1 + a, 1] \to [0, 1, 0][d, d + a + 1, 1] =$ $\langle 1, 0, -d \rangle$. For $b \in R$, we have $\langle -b - ab, 1, 0 \rangle \to \langle -b - ab, 1, 0 \rangle$ because o is the center of f, so it follows that $f: [1, b + ab, 1] = \langle 1, 0, -1 \rangle \cap$ $\langle -b - ab, 1, 0 \rangle \to \langle 1, 0, -d \rangle \cap \langle -b - ab, 1, 0 \rangle = [d, d(b + ab), 1]$. Also $f: \langle 0, 1, -b - ab \rangle = [1, 0, 0][0, b + ab, 1] \to [-1 - a, 0, 1][0, b + ab, 1]$ $= \langle -b, 1, -b - ab \rangle$. Since $[1, b + ab, 1] \in \langle 0, 1, -b - ab \rangle$, it follows that $[d, d(b + ab), 1] \in \langle -b, 1, -b - ab \rangle$. Thus $d(-b) + d(b + ab) + (-b - ab)$ $= -db + db + d(ab) - b - ab = (a^{-1} + 1)ab - b - ab = a^{-1}(ab) + ab - b$ $- ab = a^{-1}(ab) - b = 0$. Therefore, $a^{-1}(ab) = b$, and so the left inversive law is satisfied.

THEOREM 14.2.3   *The semifield S satisfies the right inversive law if and only if π_S is (p, p) transitive for all $p \in uv$.*

Proof. Suppose that $S = (R,+,\cdot)$ satisfies the right inversive law. Then $S^d = (R,+,\circ)$ where $a \circ b = ba$ is a semifield satisfying the left inversive law. Thus π_{S^d} is (L,L) transitive for all L passing through v. It follows from Theorem 11.2.4 that $(\pi_{S^d})^d \sim \pi_S$, and so π_S is (p,p) transitive for all $p \in uv$.

Assuming that S satisfies both inversive laws, we conclude that π_S satisfies the Little Desargues Theorem.

THEOREM 14.2.4   *If $P = (R,+,\cdot)$ is a planar ring, then P is an alternative ring if and only if π_P is (L,L) transitive for all lines L in π_P.*

Proof. Suppose that P is an alternative ring. It follows from Theorem 14.1.5 and the remarks thereafter that P satisfies both inversive laws. Since P satisfies the right inversive law, we know that π_P is (p,p) transitive for some $p \in uv$, $p \neq u,v$. Thus π_P is (p,L) transitive for $L = op$, and so there exists $f \in \mathrm{El}(p,L)$ such that $f\colon u \to v$. Since P satisfies the left inversive law, π_P is (ov,ov) transitive, and therefore, by Theorem 4.5.14, π_P is (ou,ou) transitive. But π_P is also (uv,uv) transitive, and so, by Theorem 4.5.17(i), we conclude that π_P is (L,L) transitive for all L.

Suppose that π_P is (L,L) transitive for all L. It is easily shown that π_P is therefore (p,p) transitive for all points p. It follows from Theorems 13.2.1, 14.2.2, and 14.2.3 that P is an alternative ring.

COROLLARY 14.2.5 π *is a Little Desarguesian plane if and only if π is an alternative plane.*

COROLLARY 14.2.6   *A finite Little Desarguesian plane is Pappian.*

Proof. This corollary follows from Theorem 14.1.4 and Corollary 14.2.5.

As indicated in Section 5.7, the Little Desargues Theorem is equivalent to the Axiom of the Fourth Harmonic Point in π if π satisfies Fano's Axiom. Thus it is possible to assume the validity of the Axiom of the Fourth Harmonic Point in π and to develop a representative alternative ring of characteristic $n \neq 2$. This approach has been carried out by Mendelsohn (1956).

The remainder of the section will examine the additional transitivity that the inversive laws on S may induce on π_S.

THEOREM 14.2.7 *If A is an alternative ring and π_A is not Desarguesian, then $T(\pi_A) = \{(L,L): L$ is any line in $\pi_A\}$.*

Proof. We know that $T(\pi_A) \supseteq \{(L,L): L$ is any line in $\pi_A\}$. Suppose that we do not have equality; that is, suppose that π_A is (p,M) transitive for some p and M such that $p \notin M$. Then, since π_A is (L,L) transitive for all L, there exists $f_q \in \text{El}(M)$ such that $f_q: p \rightarrow q$ for any point $q \notin M$. Thus π_A satisfies the Second Desargues (M) Theorem and therefore it is Desarguesian. This contradiction establishes Theorem 14.2.7.

THEOREM 14.2.8 *If S is a semifield satisfying the left inversive law and π_S is not Desarguesian, then $T(\pi_S) = \{(L,L): L$ is any line$\}$.*

Proof. The proof follows from Theorems 14.1.6 and 14.2.7.

THEOREM 14.2.9 *If S is a semifield satisfying the right inversive law and π_S is not Desarguesian, then $T(\pi_S) = \{(L,L): L$ is any line$\}$.*

Proof. The proof is left as an exercise.

The relevant theorems in this section verify that the transitivity classes for non-Desarguesian planes π_S are of the form $T_1(\pi) = \{(p_o,p_o), (L_o,L_o): p_o \in L_o\}$ and $T_2(\pi) = \{(L,L): L$ is any line$\}$, as indicated in Section 13.2.

EXERCISES

1. Establish that incidence is preserved in all cases in the proof of Theorem 14.2.1.

2. Prove Theorem 14.2.9.

3. Show that if π is (p_o, L_o) transitive and $C(\pi)$ is doubly transitive, then π is an alternative plane.

4. This exercise indicates a method of coordinatizing a Little Desarguesian plane π as suggested in Section 5.6. Let π be a Little Desarguesian plane with characteristic $\neq 2$, let L be an arbitrary line, and let o, i, and u be three distinct points on L.

   a. Show that addition may be defined on π as in the first part of Definition 5.6.1.

   b. Show that multiplication cannot be defined on π as in the first part of Definition 5.6.1.

   c. Let M be a line through u distinct from L and let a and b be distinct points on M other than u. Define $m_{a,b}$ on $L - \{u\} \times I - \{u\}$ as follows:

   $$m_{a,b}(o,p) = m_{a,b}(p,o) = o \quad \text{for } p \in L - \{u\}$$
   $$m_{a,b}(i,p) = m_{a,b}(p,i) = p \quad \text{for } p \in L - \{u\}$$
   $$m_{a,b}(p,q) = s \text{ where } s \text{ is generated as in Figure 85}$$
   $$\text{(which is identical to Figure 59)}$$

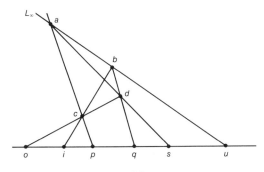

FIGURE 85

Let $p_{(a,b)}^{-1}$ be a point such that $m_{a,b}(p,p_{(a,b)}^{-1}) = i$. Show that $p_{(c,d)}^{-1} = p_{(a,b)}^{-1}$ where $p_{(c,d)}^{-1}$ is a point such that $m_{c,d}(p,p_{(c,d)}^{-1}) = i$ and $m_{c,d}$ is a mapping on $L - \{u\} \times L - \{u\}$ for a permissible $c,d \in M'$.

   d. Letting $p \circ q$ denote $m_{a,b}(p,q)$, show that $p^{-1} \circ (p \circ q) = q$. Hint: Consider Figure 86. Show that i, d, and a' are collinear using 4(c) and the fact that $m_{a',b}(p^{-1} \circ p) = i$; then show that o, w, and t are collinear by considering triangles $ap^{-1}i$ and bcd and referring back to exercise 5.7.8(b).

   e. Show that $p \circ (q + r) = p \circ q + p \circ r$ and that $(p + q) \circ r = p \circ r + q \circ r$. Conclude that $(L - \{u\}, +, \circ)$ is an alternative ring.

*5. Prove geometrically that if π is (p,p) transitive for all $p \in L_o$, then π is a Little Desarguesian plane.

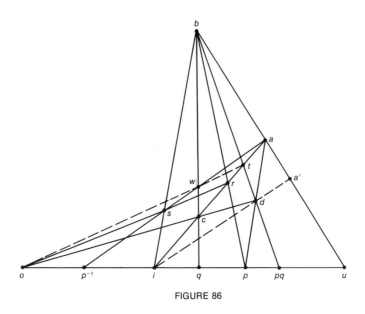

FIGURE 86

14.3   THE COORDINATIZING TERNARY RINGS OF π_A

In this section it will be shown that every ternary ring $T = T(u',v',o',e')$ generated from π_A is isomorphic. Thus every T is linear and P_T is isomorphic with A. This important result should be compared with the lesser results attainable in π_Q, π_N, and π_S. For planes π_Q nothing can be guaranteed about the relationship between two associated ternary rings $T = T(u,v,o,e)$ and $T' = T(u',v',o',e')$, even under the additional restrictions that T and T' are linear and that P_T and $P_{T'}$ are quasifields. With regard to nearfields, once again T and T' do not necessarily share any relationship. However, if T and T' are linear and P_T and $P_{T'}$ are nearfields, then P_T and $P_{T'}$ are isomorphic. For planes π_S, the situation is similar to that for planes π_N: If T and T' are linear and P_T and $P_{T'}$ are semifields, then P_T and $P_{T'}$ are isotopic (but not necessarily isomorphic). The result for planes π_A is as important as that for planes defined over division rings. A first step is verifying this statement is to show that $C(\pi_A)$ is transitive on triangles.

CONSTRUCTION 14.3.1   Let $A = (R,+,\cdot)$ be an alternative ring and construct the mappings $(f,F),(g_s,G_s),(h,H)$ on π_A:

f: $[x,y,1] \rightarrow [y,x,1]$
     $[1,x,0] \rightarrow [1,x^{-1},0]$ where $x \neq 0$
     $[1,0,0] \rightarrow [0,1,0]$
     $[0,1,0] \rightarrow [1,0,0]$

F: $\langle a,1,b \rangle \rightarrow \langle a^{-1},1,ba^{-1} \rangle$ where $a \neq 0$
     $\langle 0,1,b \rangle \rightarrow \langle 1,0,b \rangle$
     $\langle 1,0,b \rangle \rightarrow \langle 0,1,b \rangle$
     $\langle 0,0,1 \rangle \rightarrow \langle 0,0,1 \rangle$

g_s: $[x,y,1] \rightarrow [x,y + xs,1]$
     $[1,x,0] \rightarrow [1,x + s,0]$
     $[0,1,0] \rightarrow [0,1,0]$

G_s: $\langle a,1,b \rangle \rightarrow \langle a - s,1,b \rangle$
     $\langle 1,0,b \rangle \rightarrow \langle 1,0,b \rangle$
     $\langle 0,0,1 \rangle \rightarrow \langle 0,0,1 \rangle$

h: $[x,y,1] \rightarrow [x^{-1},x^{-1}y,1]$ where $x \neq 0$
     $[0,y,1] \rightarrow [1,y,0]$
     $[1,y,0] \rightarrow [0,y,1]$
     $[0,1,0] \rightarrow [0,1,0]$

H: $\langle a,1,b \rangle \rightarrow \langle b,1,a \rangle$
     $\langle 1,0,b \rangle \rightarrow \langle 1,0,b^{-1} \rangle$ where $b \neq 0$
     $\langle 1,0,0 \rangle \rightarrow \langle 0,0,1 \rangle$
     $\langle 0,0,1 \rangle \rightarrow \langle 1,0,0 \rangle$

LEMMA 14.3.2   *The mappings* $(f,F),(g_s,G_s),(h,H)$ *are collineations on* π_A.

Proof.   The proof is left as an exercise.

THEOREM 14.3.3   *If* π_A *is an alternative plane, then* $C(\pi)$ *is transitive on triangles.*

Proof.   (Hall, 1943) Let p,q,r be three noncollinear points. The theorem is proved by showing that there exists a collineation that maps $p,q,r \rightarrow o,u,v$. First observe that we may map $qr \rightarrow uv$ by a combination of maps $f_{s,t}, f, g_r, h$ where $f_{s,t}$ are the members of El(uv) defined in Construction 12.5.1. Such a mapping is explained as follows. If $qr = \langle a,1,b \rangle$, then $h \circ f_{b,0} \circ f \circ g_a$: $\langle a,1,b \rangle \to \langle 0,1,b \rangle \to \langle 1,0,b \rangle \to \langle 1,0,0 \rangle \to \langle 0,0,1 \rangle = uv$; if $qr = \langle 1,0,b \rangle$, then $h \circ f_{b,0}$: $\langle 1,0,b \rangle \to \langle 0,0,1 \rangle = uv$. This collineation is denoted by t.

We may map $t(r) \to v$ by a combination of mappings g_s and f as follows: If $t(r) = [1,x,0]$, then $f \circ g_{-x}$: $t(r) \to u \to v$; if $t(r) = v$, we may use the identity map. This collineation is denoted by w.

We may also map $w \circ t(q) \to u$ by a map g_s. This may be seen by noting that $w \circ t(q) = [1,x,0]$ for some x, so g_{-x}: $w \circ t(q) \to u$. This map is denoted by j.

Finally, we may map $j \circ w \circ t(p) \to o$ by a map $f_{s,t}$. This may be seen by noting that $j \circ w \circ t(p) \notin uv$ because p,q,r are noncollinear, so $j \circ w \circ t(p) = [x,y,1]$. Thus $f_{-x,-y}$ is the desired map. Calling this map k, we conclude that $k \circ j \circ w \circ t$: $p,q,r \to o,u,v$.

THEOREM 14.3.4   *Let π_A be an alternative plane. If T is a ternary ring generated by u', v', o', e' on π_A, then T is isotopic to $T(u,v,o,e)$.*

Proof. This theorem follows from Theorem 9.3.4 and Construction 14.3.1.

The next task will be to prove the difficult result that isotopic alternative rings are isomorphic. This fact was not hard to prove for division rings (Theorem 9.3.10) and nearfields (Theorem 12.6.4), but proving it for alternative rings is a major operation.

CONSTRUCTION 14.3.5   Suppose that $A = (R,+,\cdot)$ and $A' = (R',\oplus,\circ)$ are isotopic Cayley-Dickson algebras and let (f,g,h) be the isotopism. Let $b,c \in R$ such that $f(b) = 1$ and $g(c) = 1$ (where 1 is the multiplicative identity of R') and let Q be a quaternion ring that contains $b,c \in R$. (Such a quaternion ring exists by Theorem 14.1.3.) Denote the elements of R by $x + yi$ where $x,y \in Q$ and define the mapping t on Q by t: $x \to h(xa)$ where $a = bc$. Also let $t(i) = h(ia)$. Finally, define F: $R \to R'$ by $x + yi \to t(x) \oplus t(y) \circ t(i)$. By Theorem 9.3.10, F is an isomorphism from Q to $F(Q)$.

The following five lemmas will help to show that F is an isomorphism from A to A'.

LEMMA 14.3.6 *If A is a Cayley-Dickson algebra and elements are of the form $x + yi$ where $x, y \in Q$, then the following three equalities are true:*

$$(xi)y = \bar{y}(xi)$$
$$(xy)i = y(xi)$$
$$(xi)(yi) = k(\bar{y}x) \text{ where } k = i^2$$

Proof. The proof is left as an exercise.

LEMMA 14.3.7 *If (f, g, h) is an isotopism from $A = (R, +, \cdot)$ to $A' = (R', \oplus, \circ)$ and if $f(b) = g(c) = 1$, then $h(r) \circ h(s) = h((rc^{-1})(b^{-1}s))$ for $r, s \in A$.*

Proof. The proof is left as an exercise.

LEMMA 14.3.8 *Let t be the mapping defined in Construction 14.3.5 and let $x, y \in Q$. Then $(t(x) \circ t(1)) \circ t(y) = \overline{t(y)} \circ (t(x) \circ t(i))$.*

Proof.

$$
\begin{aligned}
(t(x) \circ t(i)) \circ t(y) &= (h(xa) \circ h(ia)) \circ h(ya) \\
&= h((xac^{-1})(b^{-1}(ia))) \circ h(ya) \\
&= h((xb)(b^{-1}(\bar{a}i))) \circ h(ya) \\
&= h((xb)((\bar{a}b^{-1})i)) \circ h(ya) \\
&= h((\bar{a}b^{-1}xb)i) \circ h(ya) \\
&= h((((\bar{a}b^{-1}xb)i)c^{-1})(b^{-1}ya)) \\
&= h((\overline{c^{-1}}((\bar{a}b^{-1}xb)i))(b^{-1}ya)) \\
&= h((((\bar{a}b^{-1}xb\overline{c^{-1}})i)b^{-1}ya) \\
&= h((\overline{b^{-1}ya})((\bar{a}b^{-1}xb\overline{c^{-1}})i)) \\
&= h((\bar{a}b^{-1}xb\overline{c^{-1}}\overline{b^{-1}ya})i) \text{ because } \overline{xy} = \overline{yx} \text{ for } x, y \in Q \\
&= h((\bar{a}b^{-1}xb\overline{c^{-1}}\bar{a}\bar{y}\overline{b^{-1}})i) \\
&= h((\bar{a}b^{-1}xb\overline{ac^{-1}}\bar{y}\overline{b^{-1}})i) \\
&= h((\bar{a}b^{-1}x(b\bar{b})\bar{y}\overline{b^{-1}})i) \\
&= h((\bar{a}b^{-1}x\bar{y}(b\bar{b})\overline{b^{-1}})i) \text{ because } (b\bar{b}) \in F \text{ and thus com-} \\
&\qquad \text{mutes with all elements of } Q \\
&= h((\bar{a}b^{-1}x\bar{y}b)i)
\end{aligned}
$$

Also,

$\overline{t(y)} \circ (t(x) \circ t(i)) = h(\bar{y}a) \circ (h(xa) \circ h(ia))$. (Note that $\overline{t(y)} = t(\bar{y})$ because t is an isomorphism from Q to $Q'(Q' = t(Q))$.)

$$= h(\bar{y}a) \circ h((xac^{-1})(b^{-1}(ia)))$$
$$= h(\bar{y}a) \circ h((\bar{a}b^{-1}xb)i)$$
$$= h((\bar{y}ac^{-1})(b^{-1}((ab^{-1}xb)i)))$$
$$= h((\bar{y}b)((\bar{a}b^{-1}xbb^{-1})i))$$
$$= h((\bar{a}b^{-1}x\bar{y}b)i)$$

Thus we have the desired equality.

LEMMA 14.3.9 *Let t be the mapping in Construction 14.3.5 and let $x,y \in Q$. Then $(t(x) \circ t(y)) \circ t(i) = t(y) \circ (t(x) \circ t(i))$.*

Proof.

$$\begin{aligned}
(t(x) \circ t(y)) \circ t(i) &= (h(xa) \circ h(ya)) \circ h(ia) \\
&= h(xac^{-1}b^{-1}ya) \circ h(ia) \\
&= h(xaa^{-1}ya) \circ h(ia) \\
&= h(xya) \circ h(ia) \\
&= h((xyac^{-1})(b^{-1}(\bar{a}i))) \\
&= h((xyb)((\bar{a}b^{-1})i)) \\
&= h((\bar{a}b^{-1}x\bar{y}b)i)
\end{aligned}$$

Also,

$$\begin{aligned}
t(y) \circ (t(x) \circ t(i)) &= h(ya) \circ (h(xa) \circ h(ia)) \\
&= h(ya) \circ h((xac^{-1})(b^{-1}(ia))) \\
&= h(ya) \circ h((xb)((\bar{a}b^{-1})i)) \\
&= h(ya) \circ h((\bar{a}b^{-1}xb)i) \\
&= h((yac^{-1})(b^{-1}((\bar{a}b^{-1}xb)i))) \\
&= h((yb)((\bar{a}b^{-1}xbb^{-1})i)) \\
&= h((\bar{a}b^{-1}xyb)i)
\end{aligned}$$

Thus we have the desired equality.

LEMMA 14.3.10 *Let t be the mapping in Construction 14.3.5 and let $x,y \in Q$. Then $(t(x) \circ t(i)) \circ (t(y) \circ t(i)) = t(k) \circ (\overline{t(y)} \circ t(x))$ where $k = i^2$.*

Proof.

$$\begin{aligned}
(t(x) \circ t(i)) \circ (t(y) \circ t(i)) &= (h(xa) \circ h(ia)) \circ (h(ya) \circ h(ia)) \\
&= h((xac^{-1})(b^{-1}(ia))) \circ h((yac^{-1})(b^{-1}(ia))) \\
&= h((xac^{-1})((\bar{a}b^{-1})i)) \circ h((yac^{-1})((\bar{a}b^{-1})i)) \\
&= h((((\bar{a}b^{-1}xb)i)c^{-1})(b^{-1}((\bar{a}b^{-1}yb)i))) \\
&= h((\overline{c^{-1}}((ab^{-1}xb)i))((\bar{a}b^{-1}ybb^{-1})i))
\end{aligned}$$

$$= h(((\bar{a}b^{-1}xb\overline{c^{-1}})i)((\bar{a}b^{-1}y)i))$$
$$= h(k\overline{\bar{a}b^{-1}y}\ \bar{a}b^{-1}xb\overline{c^{-1}})$$
$$= h(k\bar{y}\overline{b^{-1}}(a\bar{a})b^{-1}xb\overline{c^{-1}})$$
$$= h(k\bar{y}\overline{b^{-1}}b^{-1}xb\overline{c^{-1}}a\bar{a})$$
$$= h(k\bar{y}xb\overline{^{-1}}b^{-1}b\overline{c^{-1}}a\bar{a})$$
$$= h(k\bar{y}x\overline{c^{-1}}\overline{b^{-1}}\bar{a}a) = h(k\bar{y}x\overline{a^{-1}}\bar{a}a) = h(k\bar{y}xa).$$

Also,

$$t(k) \circ (\overline{t(y)} \circ t(x)) = h(ka) \circ (h(\bar{y}a) \circ h(xa))$$
$$= h(ka) \circ h(\bar{y}ac^{-1}b^{-1}xa)$$
$$= h(kac^{-1}b^{-1}\bar{y}ac^{-1}b^{-1}xa)$$
$$= h(kbb^{-1}\bar{y}bb^{-1}xa) = h(k\bar{y}xa)$$

Thus we have the desired equality.

THEOREM 14.3.11 (Schafer, 1943) *If $A = (R,+,\cdot)$ and $A' = (R',\oplus,\circ)$ are isotopic alternative rings, then they are isomorphic.*

Proof. (see Pickert, 1955, p. 168 ff.) Since A and A' are alternative rings, by Theorem 14.1.5 we may assume that they are Cayley-Dickson algebras. Furthermore, we shall assume that A is a Cayley-Dickson algebra as in Construction 14.3.5. It will be shown that F is an isomorphism where F is as in Construction 14.3.5, $F: x + yi \rightarrow t(x) \oplus t(y)\circ t(i)$, and also that F is an isomorphism from A to A'.

$$F((x + yi) + (x' + y'i)) = F((x+x')+(y+y')i) = t(x+x') \oplus t(y+y')\circ t(i)$$
$$= t(x) \oplus t(x') \oplus (t(y) \oplus t(y'))\circ t(i)$$
$$= t(x) \oplus t(x') \oplus (t(y)\circ t(i)) \oplus (t(y')\circ t(i))$$
$$= t(x) \oplus t(y)\circ t(i) \oplus t(x') \oplus t(y')\circ t(i)$$
$$= F(x + yi) \oplus F(x' + y'i)$$

Thus F preserves addition. Also,

$$F((x + yi)(x' + y'i)) = F((xx' + k\bar{y}'y) \oplus (y'x + y\bar{x}')i)$$
$$= t(xx' + k\bar{y}'y) \oplus t(y'x + y\bar{x}')\circ t(i)$$
$$= t(x)\circ t(x') \ \oplus \ t(k)\circ\overline{t(y')}\circ t(y) \ \oplus \ (t(y'))\circ t(x)$$
$$\oplus \ t(y)\circ\overline{t(x')})\circ t(i) \text{ because } t \text{ is an isomor-}$$
$$\text{phism on } Q$$
$$= t(x)\circ t(x') \ \oplus \ t(k)\circ\overline{t(y')}\circ t(y) \ \oplus \ (t(y')\circ t(x))\circ t(i)$$
$$\oplus \ (t(y)\circ\overline{t(x')})\circ t(i)$$
$$= t(x)\circ t(x') \ \oplus \ t(x)\circ(t(y')\circ t(i)) \ \oplus \ (t(y)\circ t(i))\circ t(x')$$

$$\oplus \ (t(y) \circ t(i)) \circ (t(y') \circ t(i)) \text{(by Lemmas 14.3.7,}$$
$$\text{14.3.8, and 14.3.9)}$$
$$= (t(x) \ \oplus \ (t(y) \circ t(i))) \circ (t(x') \ \oplus \ (t(y') \circ t(i)))$$
$$= F(x + yi) \circ F(x' + y'i)$$

COROLLARY 14.3.12 *If T and T' are ternary rings generated by two four-points of an alternative plane π, then T is isomorphic to T'.*

This section is concluded with an observation that follows directly from Corollary 14.3.12. Alternative planes share with Desarguesian planes the powerful property that every two four-points are related by a collineation.

THEOREM 14.3.13 *If π is an alternative plane, then $C(\pi)$ is transitive on four-points.*

Proof. The proof follows from Theorem 9.3.1 and Corollary 14.3.12.

It is interesting to note that this theorem has never been proved by geometric methods. The consequences of transitivity on four-points will be studied in the next section.

EXERCISES

1. Prove Lemma 14.3.2.

2. Prove Lemma 14.3.6.

3. Prove Lemma 14.3.7.

*4. Prove geometrically that if $C(\pi)$ is transitive on triangles, then $C(\pi)$ is transitive on four-points.

14.4 PLANES WITH A COLLINEATION GROUP
      TRANSITIVE ON FOUR-POINTS

This section will examine those planes π satisfying the property that every pair of four-points can be related by a collineation. This property has wide applications in the study of Desarguesian and Pappian planes (see Chapters 5 and 6) and applies also to alternative planes, as has been shown by Theorem 14.3.13. The converse of this theorem (that is, that a plane satisfying the property that every two four-points are related by a collineation is an alternative plane) has not yet been established. It will be proved in this section that placing one of several additional restrictions on such a plane will ensure that it is an alternative plane. One of these restrictions is the condition of uniqueness, imposed on the collineation that relates two given four-points. Another restriction is the condition of finiteness, imposed on the plane π. A third restriction is the condition that π be coordinatized by a linear ternary ring. Other restrictions are stated in Lemma 14.4.1 and Theorems 14.4.2 and 14.4.3.

Notation. A plane with a collineation group that is transitive on four-points will be referred to as a *q.t. plane* ("quadruply transitive" plane).

LEMMA 14.4.1 *If π is a q.t. plane and is (p_0, L_0) transitive for some $p_0 \in L_0$, then π is an alternative plane.*

Proof. The proof is clear.

THEOREM 14.4.2 *Let π be a q.t. plane. If $C(\pi)$ contains an elation, then π is an alternative plane.*

Proof. Suppose that $f \in \text{El}(p, L)$ and that $f: q \rightarrow q' \neq q$. Let r, s be a pair of points on L different from p and let t be an arbitrary point on pq different from p, q, and q'. Since π is a q.t. plane, there exists a collineation $g: q, q', r, s \rightarrow q, t, r, s$. Thus $g \circ f \circ g^{-1} \in \text{El}(p, L)$ and $g \circ f \circ g^{-1}: q \rightarrow t$. It follows that π is (p, L) transitive and, by Lemma 14.4.1, π is an alternative plane.

THEOREM 14.4.3 *Let π be a q.t. plane. If $C(\pi)$ contains a harmonic homology, then π is an alternative plane.*

Proof. Let f be a harmonic homology in $\text{Hom}(p, L)$ and let $f: q \rightarrow q'$. Let r and s be points on L such that p, q, r, s is a four-point. Let $g: q, q', r, s \rightarrow p, q, r, s$ and observe that $h = g \circ f \circ g^{-1}$ is a harmonic homology in $\text{Hom}(q, L)$. The proof is completed by showing that $h \circ f$ is an elation and applying Theorem 14.4.2.

Clearly, $h \circ f$ is a central collineation with axis L. Let its center be the point t. Thus $h \circ f(t) = t$. Since $f^2 = h^2 = I$, it follows that $h \circ h \circ f(t) = h^2(f(t)) = f(t)$ and $h \circ h \circ f(t) = h(h \circ f(t)) = h(t)$. Therefore, $f(t) = h(t)$. Consequently, $h \circ f \circ f(t) = h(f^2(t)) = h(t) = f(t)$, so $(h \circ f)(f(t)) = f(t)$. If $t \notin L$, then $f(t) \notin L$, and so $h \circ f$ leaves two points not on L fixed. Thus $h \circ f = I$, so $h = f^{-1} = f$, a contradiction. Therefore, $t \in L$, and $h \circ f$ is an elation.

LEMMA 14.4.4 (Baer, 1946) *Let π be an arbitrary projective plane. If f is an involution on π, then either the fixed points of f form a proper subplane of π, or f is a central collineation.*

Proof. Let f be an involution on π. The first step is to show that every line in π contains a fixed point.

> *Case* 1. Suppose that L is not a fixed line. Then the point $p = L \cap f(L) \rightarrow f(p) = f(L) \cap f^2(L) = f(L) \cap L = p$, and so p is held fixed.

> *Case* 2. Suppose that L is a fixed line that contains no fixed points. It then follows that f fixes no other line M. (Why?) From Case 1 we may assume that every $M \neq L$ contains a fixed point. But this implies that there exist at least two fixed points p and q not on L and thus a fixed line $pq \neq L$. This contradiction implies the existence of a fixed point on L.

A similar argument shows that every point is incident with some fixed line.

The last step is to examine the arrangement of the fixed points. Clearly, there are three possible arrangements:

1. There exists a four-point, and therefore the fixed points form a subplane.

2. All the fixed points lie on a line L.

3. All the fixed points but one lie on a line L.

Proof that in (2) and (3) the whole line L is held pointwise fixed and thus that $f \in CC(L)$ is left to the student.

THEOREM 14.4.5 *If $C(\pi)$ is sharply transitive on four-points, then π is an alternative plane.*

Proof. Let p,q,r,s be a four-point and let f: $p,q,r,s \rightarrow q,p,r,s$. Then f^2: $p,q,r,s \rightarrow p,q,r,s$, and thus $f^2 = I$ since there is only one mapping that fixes p,q,r,s. Since f is an involution, by Lemma 14.4.4 either it holds a subplane π' pointwise fixed or it is a central collineation. If it holds π' fixed, it holds a four-point fixed and thus is the identity, a contradiction. Therefore, f is either an elation or a harmonic homology. In either instance, Theorems 14.4.2 and 14.4.3 imply that π is an alternative plane.

THEOREM 14.4.6   *The plane π satisfies the Fundamental Theorem of Projective Collineations* (FT-II) *if and only if π satisfies the Pappus Theorem.*

Proof. It has already been shown (Theorem 6.5.6) that the Pappus Theorem implies FT-II. Suppose that PC(π) is sharply transitive on four-points. Let p,q,r,s be a given four-point and let $f \in$ PC(π) such that f: $p,q,r,s \rightarrow q,p,r,s$. Because $f^2 \in$ PC(π), f^2 fixes p,q,r,s and PC(π) is sharply transitive on four-points; therefore, $f^2 = I$. Since f is an involution and cannot leave a subplane pointwise fixed (because it would then be the identity), it must be a central collineation. Therefore, π is an alternative plane.

Let A be the alternative ring that represents π. Then A is either a Cayley-Dickson algebra, a division ring, or a field. If A is a division ring, we know that PC(π) is not sharply transitive on four-points. If A is a Cayley-Dickson algebra, then there exists a quaternion ring Q in A. Since PC(π_Q) is not sharply transitive on four-points and since π_Q is isomorphic to a subplane of π_A, PC(π_A) is not sharply transitive on four-points. Therefore, A is a field and π_A is Pappian.

With the proof of this theorem, the equivalence of the Pappus Theorem and FT-II is finally established. The equivalence of the Pappus Theorem and FT-I was proved in Theorem 6.5.4. Theorem 6.5.4, along with Theorem 14.4.6, thus establishes the equivalence of FT-I and FT-II.

Next it will be shown that if π is a finite q.t. plane, then it is an alternative plane and thus a Pappian plane. This result was improved upon by Ostrom and Wagner (1959), who have shown that a doubly transitive finite plane is Pappian. The proof of their theorem is quite difficult, however, so only the proof of the more specialized result will be given here. It, too, is rather difficult, but it is a good example of the application of algebraic tools for geometric ends.

LEMMA 14.4.7 *If π is a finite q.t. plane, then every quadrangle generates a Pappian subplane $\pi(p)$ for some fixed prime p.*

Proof. Since π is a q.t. plane, all its subplanes are isomorphic to some subplane $\pi(n)$. That $\pi(n)$ is Pappian can be shown as follows:

Let p,q,r,s be a four-point of $\pi(n)$ and let f: $p,q,r,s \rightarrow q,p,r,s$. We know that f cannot leave $\pi(n)$ nor any of its subplanes fixed because $\pi(n)$ is the least subplane containing p,q,r,s. Therefore, f is a central collineation. It follows that $\pi(n)$ is a q.t. plane containing an elation or a harmonic homology, and so $\pi(n)$ is an alternative plane. But finite alternative planes are Pappian.

If n is not a prime, it must be a power of a prime p^k, and therefore $\pi(n)$ must contain a subplane $\pi(p)$. This contradicts the fact that every four-point generates $\pi(n)$. Thus n is a prime.

LEMMA 14.4.8 *A q.t. plane either is a Fano plane or satisfies Fano's Axiom.*

Proof. The proof is clear.

LEMMA 14.4.9 *Let π be a finite q.t. plane. Then π is of odd order if π satisfies Fano's Axiom.*

Proof. Suppose that π satisfies Fano's Axiom and that π is of order n. Let L be a line of π, let $s,t \in L$, and let $q,r \notin L$ such that $qr \cap L = t$. Finally, let $S = \{x \in L: x \neq s,t\}$. Clearly, S contains $n - 1$ points, and the proof is completed by showing that $n - 1$ is an even number.

Observe that $x \in S$ iff x,q,r,s is a four-point. Each four-point $\Sigma_x = x,q,r,s$ generates a subplane $\pi(\Sigma_x) = \pi_x$. Furthermore, by Lemma 14.4.7, $\pi_x \sim \pi_y$ for all $x,y \in S$. The order of π_x must be an odd prime p because if $p = 2$, then Fano's Axiom would not hold. Let $\mathscr{P}_x$ and $\mathscr{L}_x$ denote the points and lines of π_x. Then $\mathscr{P}_x \cap S = L_x \in \mathscr{L}_x$, so $\mathscr{P}_x \cap S$ has $p - 1$ points.

The next step is to show that L_x and L_y are disjoint or identical. If $z \in L_x \cap L_y$, then z,q,r,s is in both planes, and thus $\pi_z = \pi_y = \pi_x$.

Therefore, $L_x = L_y$. We have now partitioned S into sets L_x, each of even order. Therefore, S has an even number of elements. This concludes the proof that n is odd.

The next lemma has been made obsolete by Gleason's result (Gleason, 1956) that all finite Fano planes are Desarguesian. Since this difficult result is not proved here, the lemma must be proved independently.

LEMMA 14.4.10   *Let π be a finite q.t. plane. Then π is of even order if π is a Fano plane.*

Proof. Let L, q, r, and t be the same as in Lemma 14.4.9. Let $S = \{x \in L: x \neq t\}$, let $N = qr$, and let M be a third line concurrent with L and N at t (see Figure 87). Let $f = L \overset{r}{\wedge} M \overset{q}{\wedge} L$ and let $g = f \mid S$. Clearly, g is not the identity on S. That $g^2(x) = x$ can be shown as follows.

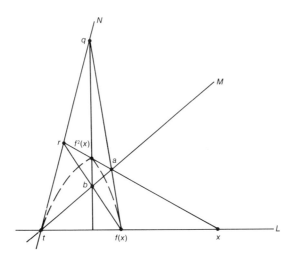

FIGURE 87

Let $x \in S$, $rx \cap M = a$, and $rf(x) \cap M = b$. Then q,r,a,b is a four-point and its diagonal points, $qr \cap ab = t$, $qa \cap rb = f(x)$, and $qb \cap ra = f^2(x)$, are collinear. Thus $f^2(x) \in L$, and therefore $x = L \cap ra = f^2(x) = g^2(x)$. We conclude that the points of S may be partitioned into pairs of points $(x, f(x))$: $x \in S$. Thus S has an even number of elements; call this number n. Since n is also the order of π, the proof is complete.

LEMMA 14.4.11 (Baer, 1946)   *Let π be an arbitrary projective plane of order n and let f be an involution whose fixed points form a subplane π'. Then π' is of order m where $m^2 = n$.*

Proof.   Recall Theorem 2.7.7 and note that either π' has order m, or π' has order k where $k^2 + k < n$. In the former situation, every line of π intersects points of π'; in the latter, this is not so. Lemma 14.4.4 shows that every line contains a fixed point. Since the only fixed points are those in π', it follows that π' fits the first description. Thus π' has order m where $m = n^2$.

The following theorem uses some elementary results in group theory that will not be developed here — specifically, results concerning the orders of subgroups of a finite group. The interested student can investigate these results in a chapter on Sylow subgroups and Sylow theorems in any text on group theory or advanced algebra.

THEOREM 14.4.12 (Wagner, 1958)   *If π is a finite q.t. plane, then π is Pappian.*

Proof.   Suppose that π is of order n.

Case 1. $n \neq m^2$ for any m. In this case every involution on π is a central collineation by Theorem 4.4.11 and Lemma 14.4.4. The collineation f where f: $p, q, r, s \rightarrow q, p, r, s$ is an involution on π. Thus, by Theorems 14.4.2 and 14.4.3, π is a finite alternative plane and is therefore Pappian.

Case 2. $n = m^2$.

Subcase A. π satisfies Fano's Axiom. Let q, r, s, t, L, S be the same as in Lemma 14.4.9 and let $G = \{f \in C(\pi)$: f leaves q, r, s, t fixed$\}$. Clearly, G is a group under composition. The fact that G is transitive on S may be seen by considering $x, y \in S$ and noting that since π is a q.t. plane, there exists $f \in C(\pi)$ such that f: $x, q, r, s \rightarrow y, q, r, s$. Since f: $t \rightarrow t$, $f \in G$. Let $G_x = \{f \in G$: f leaves q, r, s, x fixed for a given $x \in S\}$. It is easily seen that G_x is a subgroup of G for $x \in S$ and that the groups G_x are conjugate subgroups of G and thus have the same order, call it k. Since S has $n - 1$ elements, it follows (from an elementary result in group theory) that G is of order $(n - 1)k$.

Now let a, b, c be integers such that $2^a \| m + 1$; $2^b \| m - 1$; and $2^c \| k$ where $2^x \| y$ means that 2^x divides y but that 2^{x+1} does not divide y. By Lemma 14.4.9, we know that n, and therefore m, is odd. Thus $a, b \geq 1$. Now (by Sylow's first theorem) there exists a subgroup H of order 2^{a+b+c} since 2^{a+b+c} divides $(m + 1)(n - 1)k$ $= (n - 1)k$, the order of the group G. Let h be an element of order 2 in the center of H. In other words, h commutes with all members of H, $h^2 = e$ (the identity of G), and $h \neq e$. Since h is an involution, we can prove that π is a finite alternative plane and thus a Pappian plane if we can show that h is a central collineation. This fact can be established with a proof by contradiction.

Suppose that h leaves a subplane $\pi' = (\mathcal{P}', \mathcal{L}', \mathcal{I})$ fixed. Then π' is of order m where $m^2 = n$. Observe that h leaves s and t fixed because $h \in H$, so h leaves L fixed, and therefore $L \cap \mathcal{P}'$ is a line in π'. Also $s, t \in \mathcal{P}'$, so letting $S' = S \cap \mathcal{P}'$, S' has $m - 1$ points. Therefore, H is a permutation group on S' because $h(x) \in S'$ for all $x \in S'$ and $h \in H$. This fact may be seen by noting that if $g(x) = y \notin S'$ for some $g \in H$ and $x \in S'$, then $h(y) = h \circ g \circ g^{-1}(y)$ $= g \circ h \circ g^{-1}(y) = y$. However, this is a contradiction, because the only fixed points of h on L are in $S' \cup \{s, t\}$. Since H is a permutation group on S', H partitions S' into sets that are orbits of points on S' with respect to H. These partitions are denoted by S_i. Let $x_i \in S_i$ and let $H_i = \{f \in H: f \text{ fixes } x_i\}$. Since H_i is a subgroup of H, its order must be a power of 2, say, 2^{d_i}. Also H_i is a subgroup of G_{x_i}, so $d_i \leq c$. Denote by j_i the number of points in S_i; then $2^{d_i} \cdot j_i$ $= 2^{a+b+c}$, and therefore 2^{a+b} divides j_i. We have shown that the number of elements in every orbit S_i is divisible by 2^{a+b}. It follows that 2^{a+b} divides $m - 1$, the number of elements in S'. This contradicts the fact that $2^b \| m - 1$. Thus h is a central collineation and π is Pappian.

Subcase B. π does not satisfy Fano's Axiom. By Lemma 14.4.8, π is a Fano plane. The reasoning here is much the same as in Subcase A. Let q, r, t, L, S be the same as in Lemma 14.4.10. Thus $S = \{x \in L: x \neq t\}$. Let $G = \{f \in C(\pi): f \text{ leaves } q, r \text{ and } L\}$

fixed}. It is then easily shown that G is doubly transitive on the points of S. Also let $G_{x,y} = \{f \in G: f \text{ leaves fixed } x,y \in S\}$. Since all the groups $G_{x,y}$ where $x,y \in S$ are conjugate, they have the same number of elements, say, k. Thus G is of order $n(n-1)k$. Finally, let a and b be such that $2^a \parallel m$ and $2^b \parallel k$. Lemma 14.4.8 implies that n and therefore m is even, so $a \geq 1$. Let H be a Sylow subgroup of order 2^{2a+b}. Such a subgroup exists because $2^{2a+b} = 2^{2a} \cdot 2^b = m^2 \cdot k = nk$, and nk divides $n(n-1)k$, the order of G. As in Subcase A, let $h \in H$ be an element of order 2 in the center of H. The proof is concluded by showing that h is a central collineation.

Suppose that h leaves a proper subplane π' pointwise fixed. Since h holds L fixed, $L \cap \mathcal{P}'$ is a line of π'. Letting $S' = S \cap \mathcal{P}'$, we see that S' contains m points since π' is of order m and since $t \in \mathcal{P}'$.

As in Subcase A, H is a permutation group on S'; therefore, H partitions S into disjoint sets that are denoted by S_i. Let $x_i \in S_i$ and let $H_i = \{f \in H: f \text{ fixes } x_i\}$. Now H_i is of order 2^{d_i} for some j_i, and since H_i is a subgroup of H, $d_i \leq b$. Denote by j_i the number of points of S_i. Then $2^{d_i} \cdot j_i = 2^{2a+b}$, and thus 2^{2a} divides j_i. Therefore, 2^{2a} divides m, the number of elements in S'. But $2^a \parallel m$ and $b \geq 1$, a contradiction. Thus h is an involutory central collineation and so π is Pappian. This completes the proof.

Finally, it will be shown that if a q.t. plane is coordinatized by a linear ternary ring, then it is an alternative plane.

THEOREM 14.4.13 *If π is a q.t. plane and $T(u,v,o,e)$ is a linear ternary ring for some four-point u,v,o,e, then π is an alternative plane.*

Proof. Let abc and $a'b'c'$ be two triangles centrally perspective from p such that $a \neq a'$, $b \neq b'$, $c \neq c'$, $p \neq a,b,c,a',b',c'$, $ab \neq a'b'$, $ac \neq a'c'$, and $bc \neq b'c'$. Points $ab \cap a'b'$, $ac \cap a'c'$, and p lie on a line, call it L. Let g be a collineation such that $g: d,p,a,f \to u,v,o,e$ where $d = bc \cap b'c'$; f is any point such that $f \in cc'$; and d,p,a,f is a four-point. Thus $g(a)g(b)g(c)$ and $g(a')g(b')g(c')$ are centrally perspective triangles from v that are of the form of triangles bpw and oqr of Construction 10.1.1. By Corollary 10.1.3, $g(a)g(b)g(c)$ and $g(a')g(b')g(c')$

are axially perspective from uv, and since g is a collineation, abc and $a'b'c'$ are axially perspective from L. It follows easily that π satisfies the Little Desargues Theorem and thus is an alternative plane.

It should be noted that if one ternary ring of a q.t. plane is linear, then all ternary rings are linear because all must be isomorphic by Theorem 9.3.1.

There are several properties of a projective plane π that measure the "uniformity" or "homogeneity" of the plane: one is the degree of transitivity of $C(\pi)$; another is the relationship of the coordinatizing ternary rings of π; a third is the relationship of the subplanes of π. The following properties exhibit various degrees of "uniformity" of π:

1. $C(\pi)$ is transitive on four-points.

2. For every two four-points u,v,o,e and u',v',o',e' in π, the ternary rings $T(u,v,o,e)$ and $T(u',v',o',e')$ are isomorphic.

3. For every two quadrangles Σ and Σ' in π, the subplanes $\pi(\Sigma)$ and $\pi(\Sigma')$ are isomorphic.

4. $C(\pi)$ is transitive on triangles.

5. For every two four-points u,v,o,e and u',v',o',e' in π, the ternary rings $T(u,v,o,e)$ and $T(u',v',o',e')$ are isotopic.

6. $C(\pi)$ is doubly transitive.

7. $C(\pi)$ is transitive.

If a plane satisfies the first property, then that plane satisfies all seven properties; thus, according to this list, a q.t. plane is the most homogeneous of all projective planes. Properties 1 and 2 are equivalent by Theorem 9.3.1. Property 3 is weaker than properties 1 and 2 since the free extension plane $\Sigma_o{}^+$, where Σ_o is a four-point, satisfies property 3 (every four-point generates the whole plane) but is not a q.t. plane. For finite planes, properties 1, 2, and 3 are probably equivalent, although this fact has not yet been proved; it is known that only if every four-point generates $\pi(2)$ is a finite plane satisfying property 3 a q.t. plane (Gleason, 1956). Property 4 implies property 5 and property 6, but whether the respective converses are true is not yet known. The surprising fact is that the only known planes that satisfy properties 1, 2, 4, 5, and 6 are alternative planes. Thus, the Little Desargues

Theorem might be equivalent to each of these properties. Example 4.6.12 and exercise 5.2.5(d) show that property 7 is weaker than property 1; however, the two properties might be equivalent for finite planes. It is known that for finite planes, properties 1, 2, 4, and 6 are equivalent (Ostrom and Wagner, 1959).

BIBLIOGRAPHY

BOOKS

Albert, A. A., and Sandler, R.
  1968. *An introduction to finite projective planes*. New York:
    Holt, Rinehart and Winston.

Artin, E.
  1957. *Geometric algebra*. New York and London: Interscience.

Blattner, J. W.
  1968. *Projective plane geometry*. San Francisco: Holden-Day.

Bumcrot, R.
  1969. *Modern projective geometry*. New York:
    Holt, Rinehart and Winston.

Dembowski, P.
  1968. *Finite geometries*. New York: Springer-Verlag.

Hall, M.
  1954a. *Projective planes and related topics*. Pasadena: Calif. Inst. Techn.
  1959. *Theory of groups*. New York: Macmillan.
  1967. *Combinatorial theory*. Waltham, Mass.: Blaisdell.

Hartshorne, R.
  1967. *Foundations of projective geometry*. New York: Benjamin.

Pedoe, C.
   1963. *An introduction to projective geometry.* London: Pergamon Press.
Pickert, G.
   1955. *Projektive ebenen.* Berlin: Springer.
Room, T. G., and Kirkpatrick, P. B.
   1971. *Miniquaternion geometry.* New York: Cambridge University Press.
Seidenberg, A.
   1962. *Lectures in projective geometry.* Princeton, N.J.: Van Nostrand.
Veblen, O., and Young, J. W.
   1910. *Projective geometry.* Boston: Ginn.

JOURNALS

Albert, A. A.
   1960. Finite division algebras and finite planes. *Proc. Symp. Appl. Math.*
      10: 53–70.
Andre, J.
   1955. Projektive Ebenen über Fastkörpern. *Math. Zeitschr.* 62: 137–160.
Baer, R.
   1942. Homogeneity of projective planes. *Amer. J. Math.* 64: 137–152.
   1946. Projectivities with fixed points on every line of the plane. *Bull. Amer.
      Math. Soc.* 52: 273–286.
Barlotti, A.
   1957. Le possibili configurazioni del sistema delle coppie punto-retta (A,a)
      per cui un piano grafico risulta (A,a)-transitivo. *Boll. Un. Mat. Ital.*
      12: 212–226.
Bose, R. C.
   1942. An affine analogue of Singer's theorem. *J. Ind. Math. Soc.* 6: 1–15.
———, Shrikhande, S. S., and Parker, E. T.
   1960. Further results on the construction of mutually orthogonal Latin
      squares and falsity of Euler's conjecture. *Canad. J. Math.* 12: 189–
      203.
Bruck, R. H.
   1960. Quadratic extensions of cyclic planes. *Proc. Symp. Appl. Math.*
      10: 15–44.
———, and Kleinfeld, E.
   1951. The structure of alternative division rings. *Proc. Amer. Math. Soc.*
      2: 878–890.

————, and Ryser, H. J.
   1949. The nonexistence of certain finite projective planes. *Canad. J. Math.*
      1: 88–93.

Burn, R. P.
   1968. Bol quasifields and Pappus theorem. *Math. Z.* 106: 351–365.

Dickson, L. E.
   1905. On finite algebras. *Nachr. kgl. Ges. Wiss. Göttingen*, pp. 358–393.

Evans, T. A., and Mann, H. B.
   1951. On simple difference sets. *Sankhya*. 11: 357–364.

Gleason, A. M.
   1956. Finite Fano planes. *Amer. J. Math.* 78: 797–807.

Hall, M.
   1943. Projective planes. *Trans. Amer. Math. Soc.* 54: 229–277.
   1947. Cyclic projective planes. *Duke Math. J.* 14: 1079–1090.
   1949. Correction to "Projective planes." *Trans. Amer. Math. Soc.* 65:
      473–474.
   1953. Uniqueness of the projective plane with 57 points. *Proc. Amer.
      Math. Soc.* 4: 912–916.
   1954b. Correction to "Uniqueness of the projective plane with 57 points."
      Proc. Amer. Math. Soc. 5: 994–997.

————, Swift, J. D., and Killgrove, R.
   1959. On projective planes of order nine. *Math. Comp.* 13: 233–246.

Hoffman, A. J.
   1952. Cyclic affine planes. *Canad. J. Math.* 4: 295–301.

Hughes, D. R.
   1957. A class of non-Desarguesian projective planes. *Canad. J. Math.* 9:
      378–388.
   1959. Collineation groups of non-Desarguesian planes, I. The Hall Veblen-
      Wedderburn systems. *Amer. J. Math.* 81: 921–938.

Killgrove, R. B.
   1965. Completions of quadrangles in projective planes II. *Canad. J. Math.*
      17: 155–165.

Kleinfeld, E.
   1953. Alternative division rings of characteristic 2. *Proc. Nat. Acad. Sci.
      U.S.A.* 37: 818–820.

Klingenberg, W.
   1955. Beweis des Desarguesschen Satzes aus der Reidemeisterfigur und
      verwandte Sätze. *Abh. Hamburg* 19: 158–175.

Knuth, D. E.
   1965. Finite semifields and projective planes. *J. Algebra* 2: 182–217.

Lenz, H.
    1954. Kleiner Desarguesscher Satz and Dualität in projektiven Ebenen.
        Jahresber. Deutsche Math. Ver. 57: 20–31.

Mendelsohn, N. S.
    1956. Non-Desarguesian projective plane geometries which satisfy the
        harmonic point axiom. *Canad. J. Math.* 8: 532–562.

Moufang, R.
    1933. Alternativkörper und der Satz vom vollständigen Vierseit. *Abh.*
        Hamburg 9: 207–222.
    1935. Zur Struktur von Alternativkörpern. *Math. Ann.* 110: 416–430.

Moulton, F. R.
    1902. A simple non-Desarguesian plane geometry. *Trans. Amer. Math.*
        Soc. 3: 192–195.

Neumann, H.
    1955. On some finite non-Desarguesian planes. *Arch. Math.* 6: 36–40.

Ostrom, T. G.
    1956. Double transitivity in finite projective planes. *Canad. J. Math.* 8:
        563–567.
    1957. Transitivities in projective planes. *Canad. J. Math.* 9: 389–399.

———, and Wagner, A.
    1959. On projective and affine planes with transitive collineation groups.
        Math. Z. 71: 186–199.

Panella, G.
    1965. Una classe di sistemi cartesiani. *Atti Accad. Naz. Lincei Rendic.*
        38: 480–485.

Passman, D. S.
    1967. *Lecture notes on permutation groups.* New Haven, Conn.:
        Yale University Press.

Pickert, G.
    1956. Eine nichtdesarguessche Ebene mit einem Körper als Koordinaten-
        bereich. *Publ. Math. Debrecen* 4: 157–160.
    1959. Der Satz von Pappos mit Festelementen. *Arch. Math.* 10: 56–61.

Rosati, L. A.
    1958. I gruppi di collineazioni dei piani di Hughes. *Boll. Un. Math. Ital.*
        13: 505–513.
    1960. Su una generalizzazione dei piani di Hughes. *Atti Accad. Naz. Lincei*
        Rendic. 29: 303–308.

San Soucie, R. L.
    1955. Right alternative division rings of characteristic 2. *Proc. Amer.*
        Math. Soc. 6: 291–296.

Sandler, R.
1963. The collineation groups of free planes. *Trans. Amer. Math. Soc.* 107: 129–139.

Schafer, R. D.
1943. Alternative algebras over an arbitrary field. *Bull. Amer. Math. Soc.* 49: 549–555.

Singer, J.
1938. A theorem in finite projective geometry and some applications to number theory. *Trans. Amer. Math. Soc.* 43: 377–385.

Skornyakov, L. A.
1950. Alternative fields. *Ukrain. Math. J.* 2: 70–85.
1951a. Right-alternative fields. *Isv. Akad. Nauk* SSSR 15: 177–184.
1951b. Projective planes. *Uspehi Mat. Nauk* 6: 112–154. Amer. Math. Soc. Transl. No. 99. 1953.

Spencer, J. C. D.
1960. On the Lenz-Barlotti classification of projective planes. *Quart. J. Math.* 11: 241–257.

Stevenson, F. W.
1971. The collineation group of the Veblen-Wedderburn plane of order nine. *Canad. J. Math.* 22: 967–973.

Tarry, G.
1901. Le problème des 36 officiers. *C. R. Assoc. Franc. Avanc. Sci. Nat.* 2: 170–203.

Veblen, O., and Wedderburn, J. H. M.
1907. Non-Desarguesian and non-Pascalian geometries. *Trans. Amer. Math. Soc.* 8: 379–388.

Wagner, A.
1956. On finite non-Desarguesian planes generated by 4 points. *Arch. Math.* 7: 23–27.
1958. On projective planes transitive on quadrangles. *J. London Math. Soc.* 33: 25–33.

Wedderburn, J. H. M.
1905. A theorem on finite algebras. *Trans. Amer. Math. Soc.* 6: 349–352.

Zassenhaus, H.
1935. Über endliche Fastkörper. *Abh. Hamburg.* 11: 187–220.

Zemmer, J. L.
1964. Nearfields, planar and non-planar. *Math. Student* 32: 145–150.

INDEX OF
SYMBOLS AND NOTATION

INDEX